The Magic of Origami Polyhedra

Books by John Montroll
www.johnmontroll.com
Instagram: @montrollorigami

Origami Symphonies

Origami Symphony No. 1: The Elephant's Trumpet Call
Origami Symphony No. 2: Trio of Sharks & Playful Prehistoric Mammals
Origami Symphony No. 3: Duet of Majestic Dragons & Dinosaurs
Origami Symphony No. 4: Capturing Vibrant Coral Reef Fish
Origami Symphony No. 5: Woodwinds, Horns, and a Moose
Origami Symphony No. 6: Striped Snakes Changing Scales
Origami Symphony No. 7: Musical Monkeys
Origami Symphony No. 8: An Octet of Cats
Origami Symphony No. 9: Ode to Australia
Origami Symphony No. 10: Lucky & Dangerous Sides of Origami

Animal Origami

Arctic Animals in Origami
Origami Aquarium
Dogs in Origami
Perfect Pets Origami
Dragons and Other Fantastic Creatures in Origami
Bugs in Origami
Horses in Origami: Second Edition
Origami Birds: Second Edition
Origami Gone Wild
Dinosaur Origami
Origami Dinosaurs for Beginners
Prehistoric Origami: Dinosaurs and other Creatures: Third Edition
Mythological Creatures and the Chinese Zodiac Origami
Origami Sea Life: Third Edition
Bringing Origami to Life: Second Edition
Origami Sculptures: Fourth Edition
African Animals in Origami: Third Edition
North American Animals in Origami: Third Edition
Origami for the Enthusiast: Second Edition
Animal Origami for the Enthusiast: Second Edition

Geometric Origami

The Magic of Origami Polyhedra
Origami Stars: Second Edition
Galaxy of Origami Stars: Second Edition
Origami and Math: Simple to Complex: Second Edition
Origami & Geometry
3D Origami Platonic Solids & More: Second Edition
3D Origami Diamonds
3D Origami Antidiamonds
3D Origami Pyramids
A Plethora of Polyhedra in Origami: Third Edition
Classic Polyhedra Origami
A Constellation of Origami Polyhedra
Origami Polyhedra Design

General Origami

Origami Fold-by-Fold
DC Super Heroes Origami
Origami Worldwide
Teach Yourself Origami: Third Edition
Christmas Origami: Second Edition
Storytime Origami
Origami Inside-Out: Third Edition

Dollar Bill Origami

Dollar Origami Treasures: Second Edition
Dollar Bill Animals in Origami: Second Revised Edition
Dollar Bill Origami
Easy Dollar Bill Origami

Simple Origami

Fun and Simple Origami: 101 Easy-to-Fold Projects: Second Edition
Origami Twelve Days of Christmas: And Santa, Too!
Super Simple Origami
Easy Dollar Bill Origami
Easy Origami
Easy Origami 2
Easy Origami 3
Easy Origami Coloring Book
Easy Origami Animals
Easy Origami Polar Animals
Easy Origami Ocean Animals
Easy Origami Woodland Animals
Easy Origami Jungle Animals
Meditative Origami

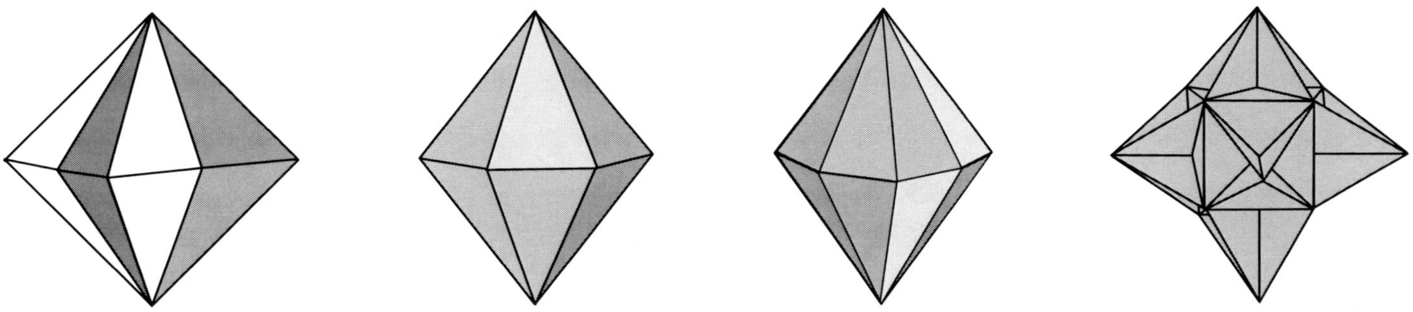

The Magic of Origami Polyhedra

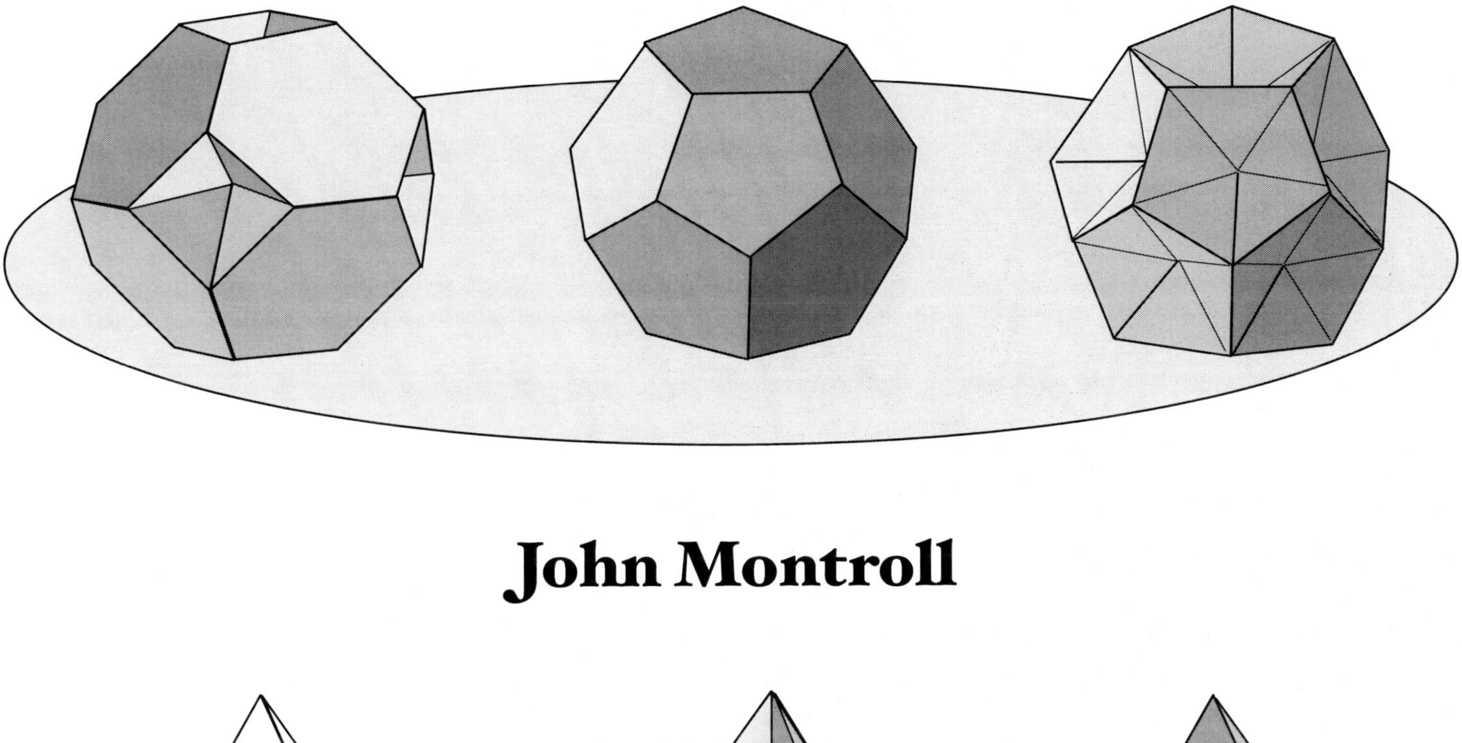

John Montroll

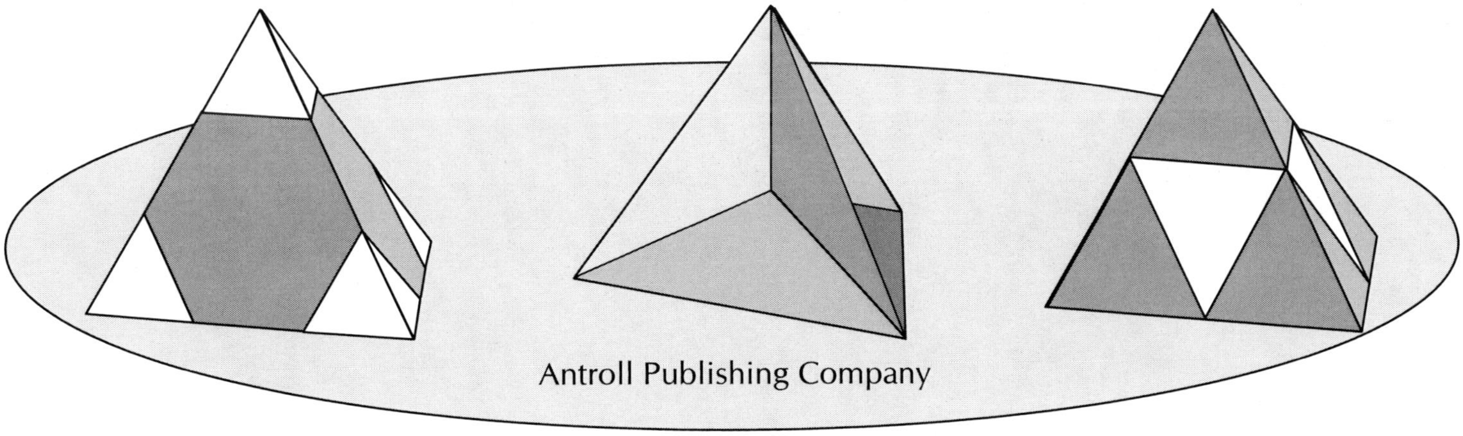

Antroll Publishing Company

To Cindy

The Magic of Origami Polyhedra

ISBN-10: 1-877656-68-2
ISBN-13: 978-1-877656-68-2

Antroll Publishing Company

Introduction

Polyhedra are beautiful shapes. Divided into eight chapters, 36 polyhedra models are shown. From the Platonic Solids to color-change models, dimpled and sunken shapes, dipyramids and more, the models range from simple to very complex. By folding each polyhedron from a single square sheet adds magic to origami.

The folding methods follow different forms of symmetry to keep these as easy to fold as possible, given their inherent complexity. In designing these models, I take great care to have these accomplished in as few steps as possible. A Cube is shown in 14 steps, then a Cube of Triangles (a cube with color-change patterns) is also shown in 14 steps. Complex as the Icosahedron is, I have designed a version that can be folding in only 24 steps. The mind-boggling Sunken Dodecahedron is diagrammed in only 35 steps.

The diagrams are drawn in the internationally approved Randlett-Yoshizawa style. You can use any kind of square paper for these models, but the best results will be achieved with standard origami paper, which is colored on one side and white on the other (in the diagrams in this book, the shading represents the colored side). Large sheets, such as nine inches squared, are easier to use than small ones.

Origami supplies can be found in arts and craft shops, or at Dover Publications online: www.doverpublications.com. You can also visit OrigamiUSA at www.origamiusa.org for origami supplies and other related information including an extensive list of local, national, and international origami groups.

Please follow me on Instagram @montrollorigami to see posts of my origami.

I am a pioneer in designing polyhedra where each is from a single square. I hope you enjoy the variety of models presented here.

John Montroll
www.johnmontroll.com

Contents

★ Simple
★★ Intermediate
★★★ Complex
★★★★ Very Complex

Colorful Octahedra

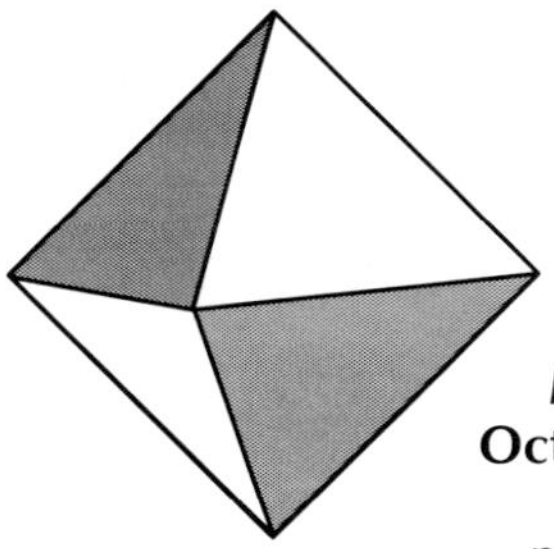

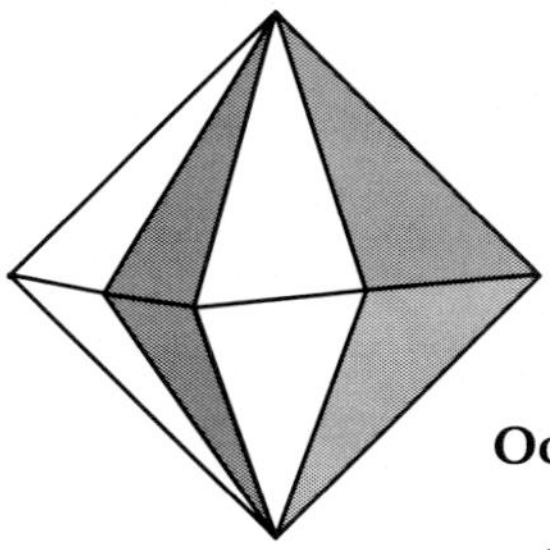

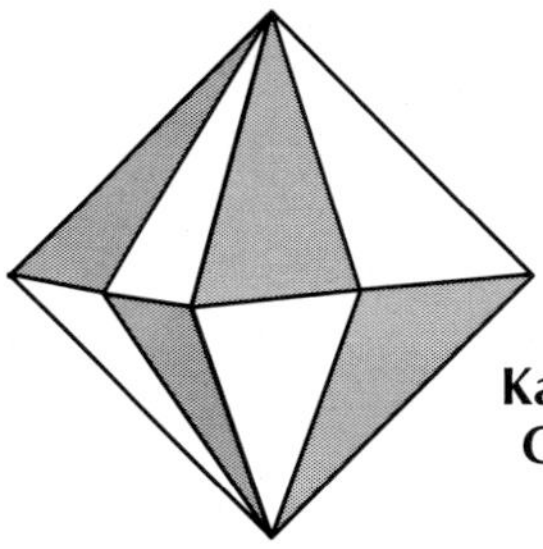

Deltahedra

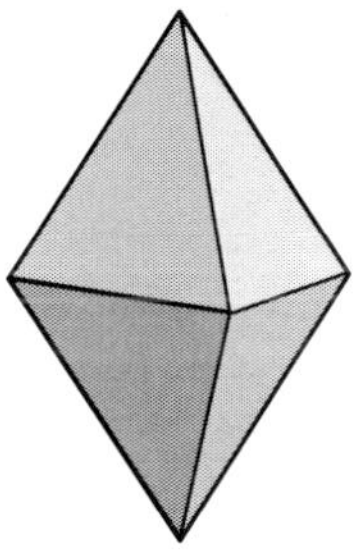

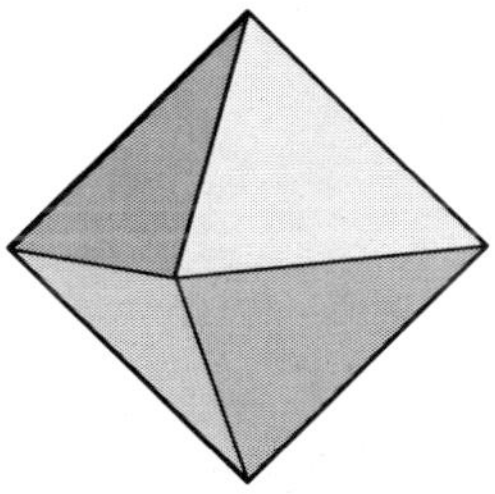

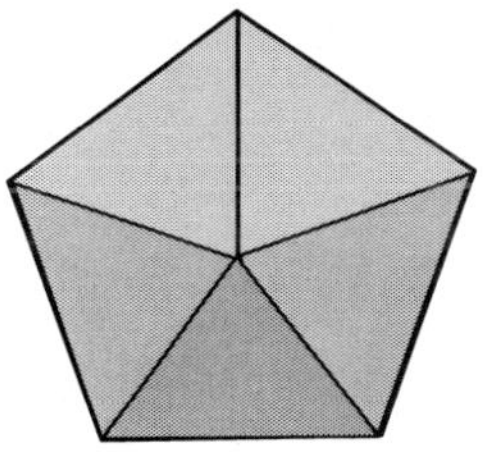

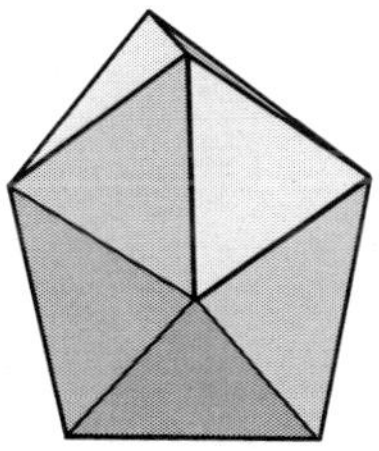

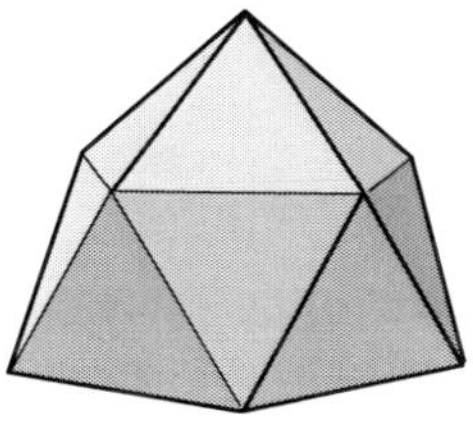

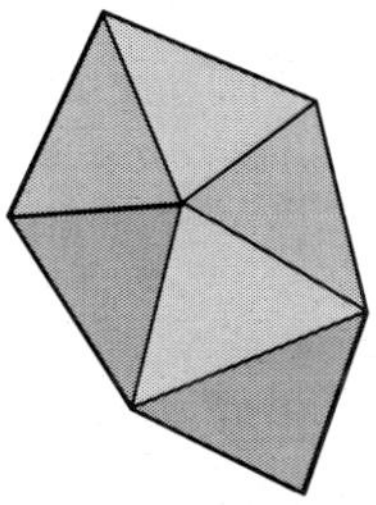

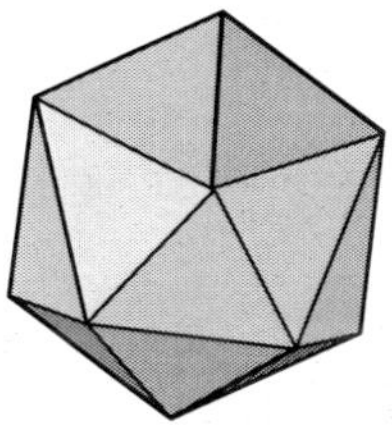

Dimpled Polyhedra

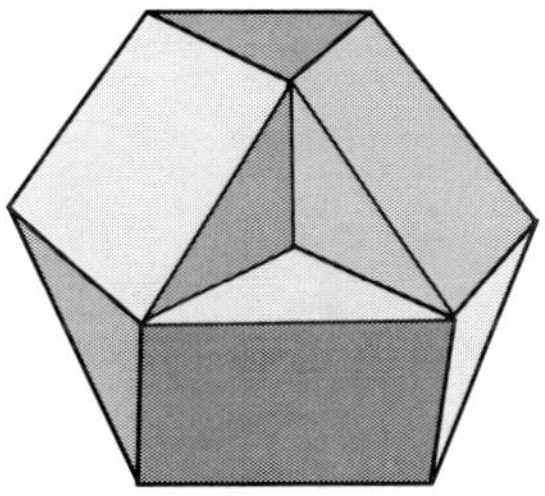

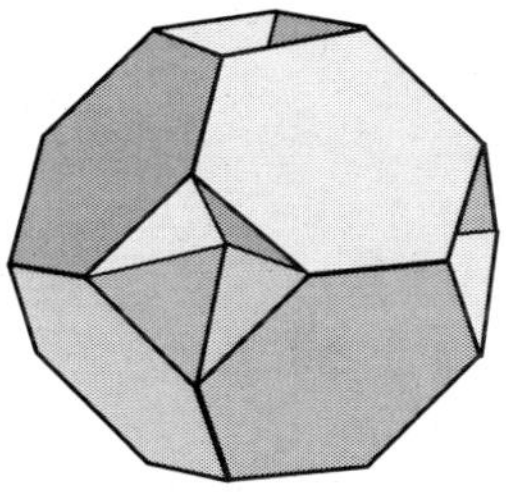

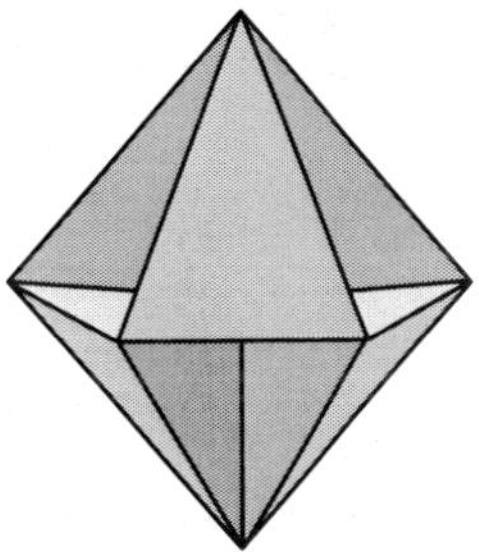

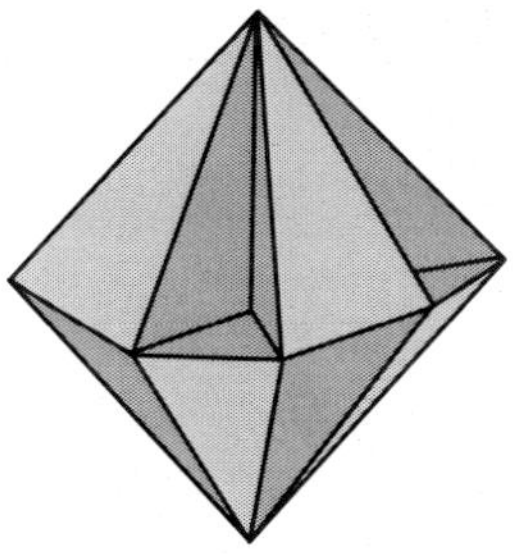

more ⟶

Sunken Polyhedra

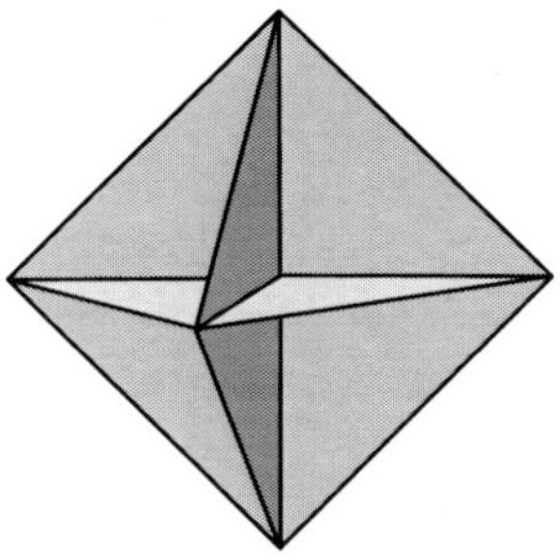

Sunken Octahedron
★★
page 91

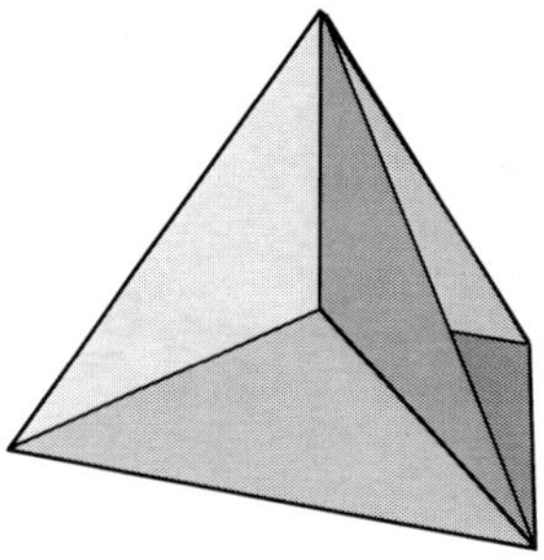

Sunken Tetrahedron
★★★
page 93

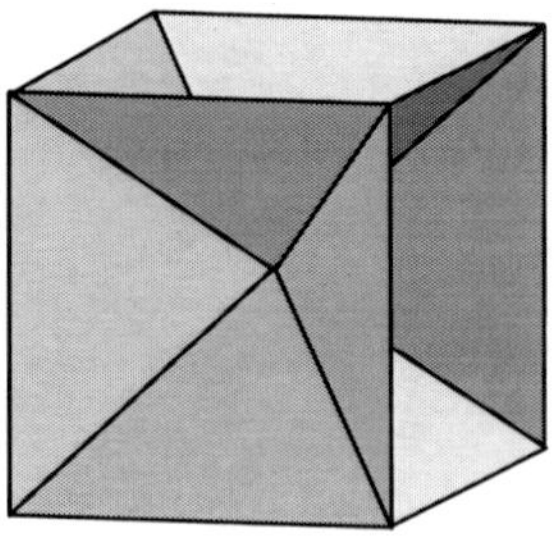

Sunken Cube
★★★
page 96

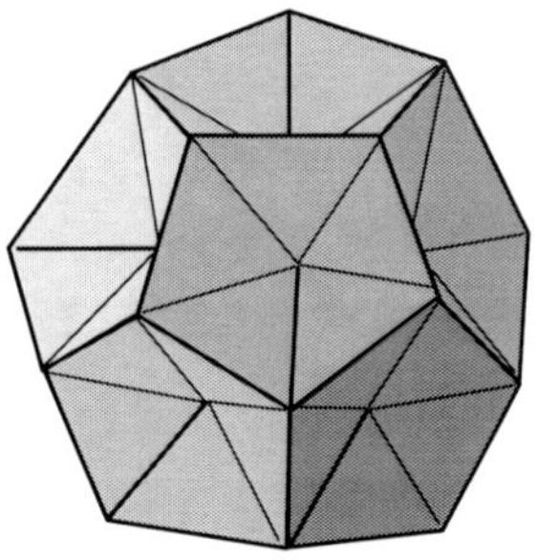

Sunken Dodecahedron
★★★
page 99

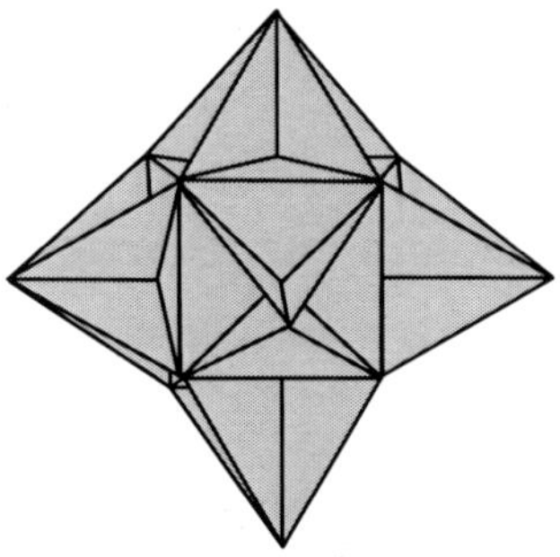

Sunken Stellated Cube
★★★
page 103

Pentagons in Polyhedra

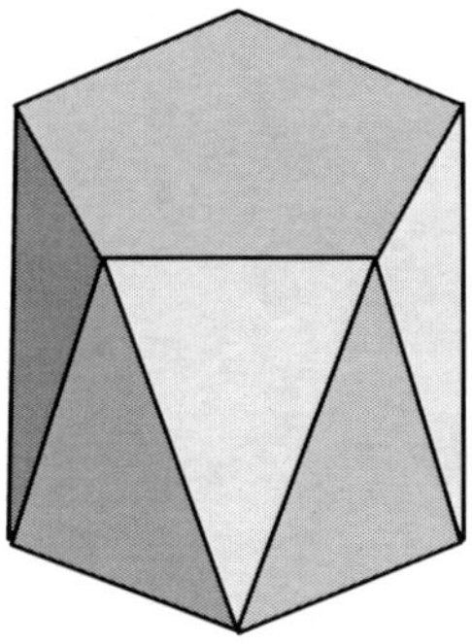

Golden Pentagonal Antiprism
★★★
page 107

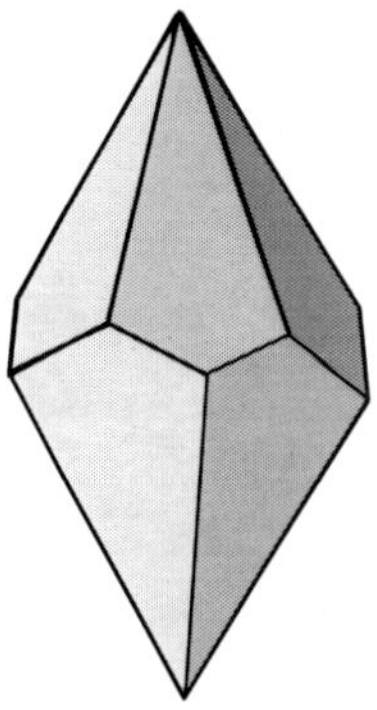

Pentagonal Trapezohedron
★★★
page 111

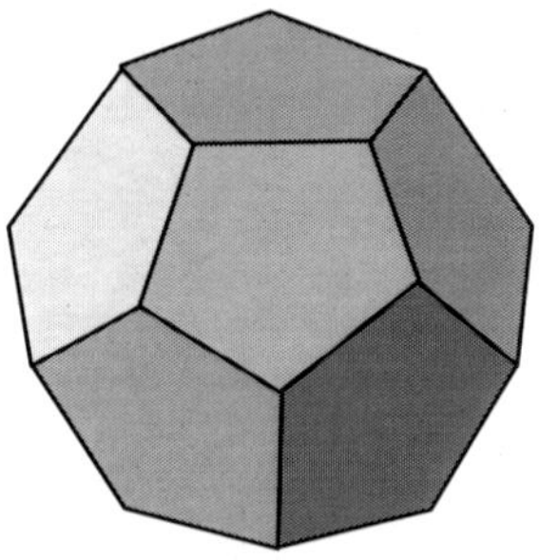

Dodecahedron
★★★★
page 115

Symbols

Lines

Valley fold, fold in front.

Mountain fold, fold behind.

Crease line.

X-ray or guide line.

Arrows

Fold in this direction.

Fold behind.

Unfold.

Fold and unfold.

Turn over.

Sink or three dimensional folding.

Place your finger between these layers.

Basic Folds

Squash Fold.

In a squash fold, some paper is opened and then made flat. The shaded arrow shows where to place your finger.

1 2 3

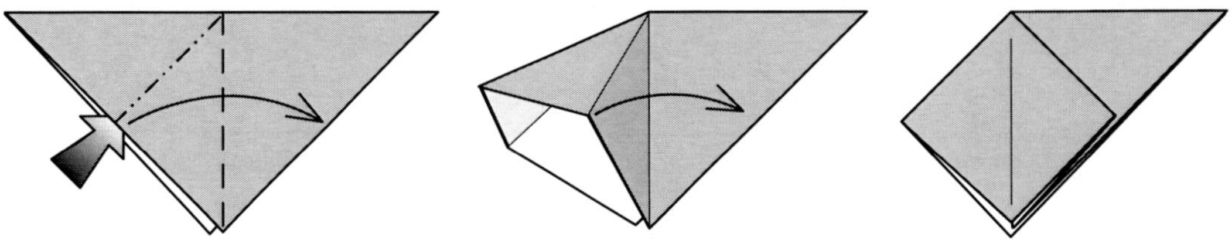

Squash-fold. A 3D step.

Inside Reverse Fold.

In an inside reverse fold, some paper is folded between layers. Here are two examples.

1 2

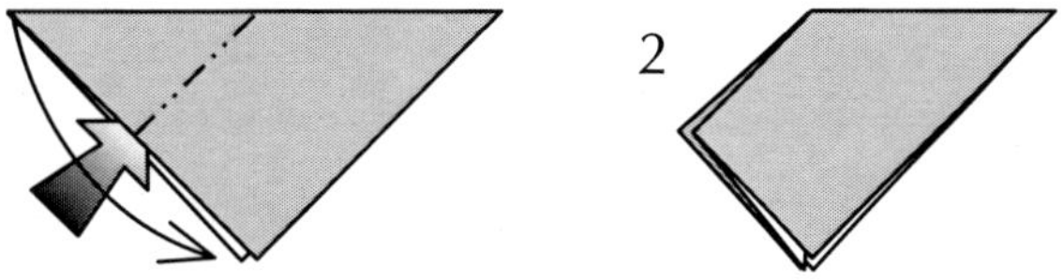

Reverse-fold.

1 2

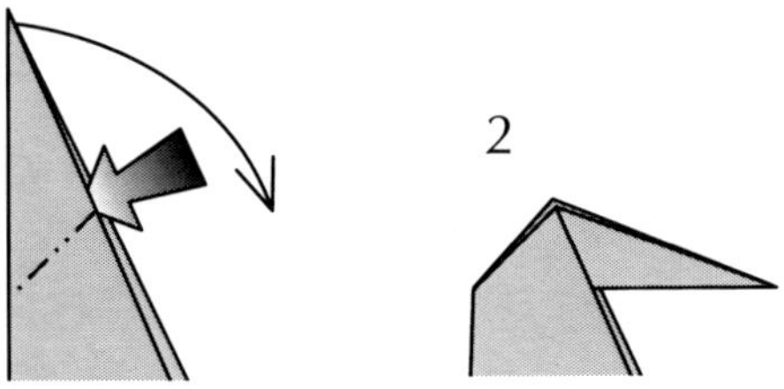

Reverse-fold.

Preliminary Fold.

The Preliminary Fold is the starting point for many models. The maneuver in step 3 occurs in many other models.

1 2 3

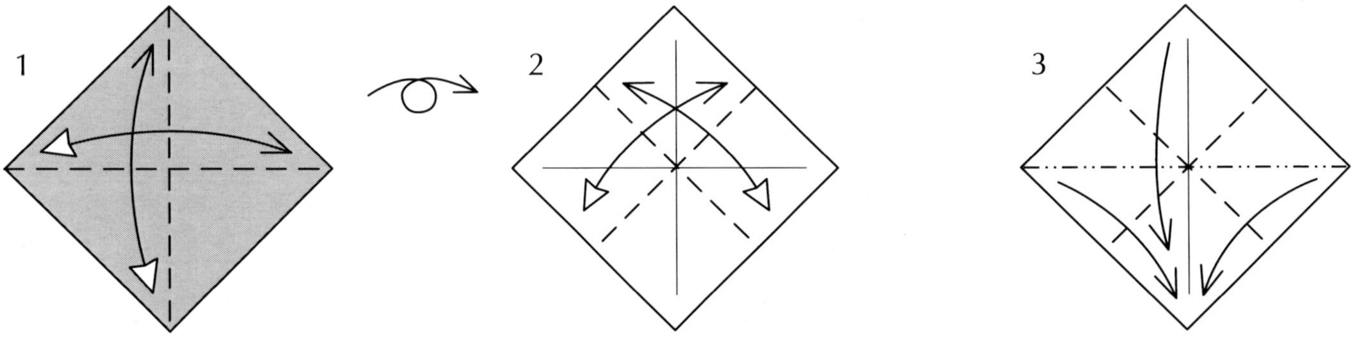

Fold and unfold.

Fold and unfold.

Collapse the square by bringing the four corners together.

4 5

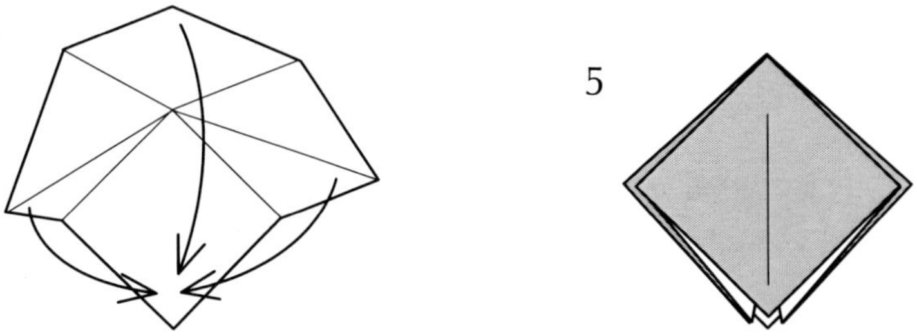

This is 3D.

Preliminary Fold

Symmetry

Each model in this collection is folded from a square sheet of paper. To design these complex three-dimensional shapes, I want to satisfy several conditions that include:
1. The model is not too difficult to fold.
2. The paper is used efficiently to produce a shape as large as possible.
3. The model holds together well.
4. Each face has no stray lines, or a minimal number.
5. The folding procedure is as simple as possible, with fewest number of steps.

Symmetry simplifies the folding procedure. Symmetry can be seen by looking at the crease pattern of a model. The crease pattern shows the unfolded square sheet where the shaded regions represent the faces of the model, and the orientation of these regions shows the form of symmetry.

Five forms of symmetry are used in this work, and here are some examples.

Even Symmetry: Two sides of the square are mirror images. The cube uses even symmetry.

Cube

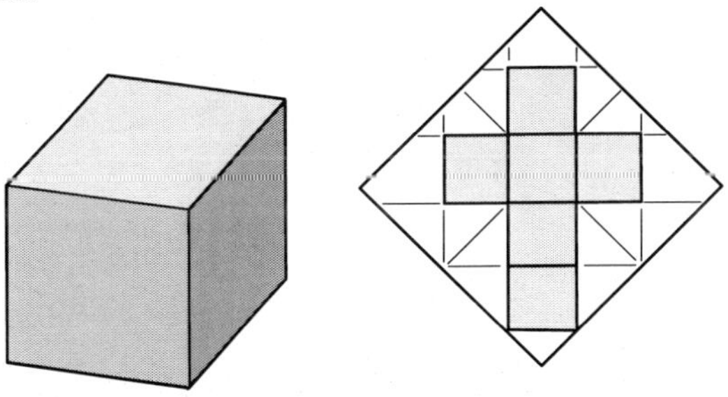

Odd Symmetry: The crease pattern is the same when rotated 180°. The icosahedron uses odd symmetry.

Icosahedron

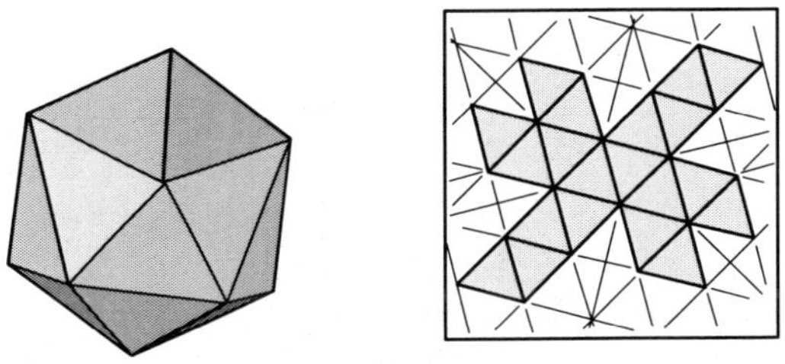

Square Symmetry: The crease pattern is the same when rotated 90°. The octahedron uses square symmetry.

Octahedron

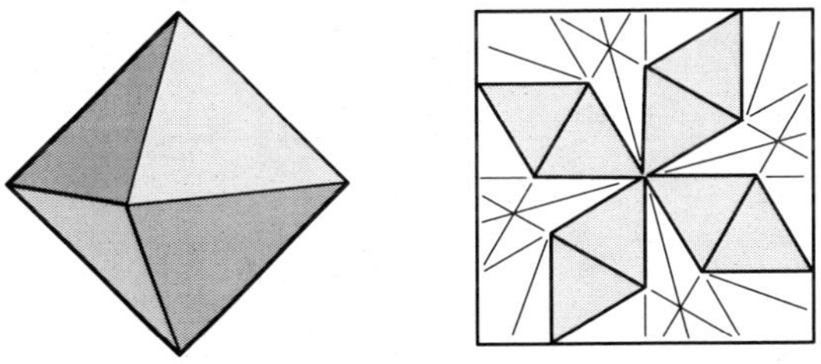

3/4 Square Symmetry: The crease pattern is the same when rotated 90°, but one of four quadrants is not used. The sunken tetrahedron uses 3/4 square symmetry.

Sunken Tetrahedron

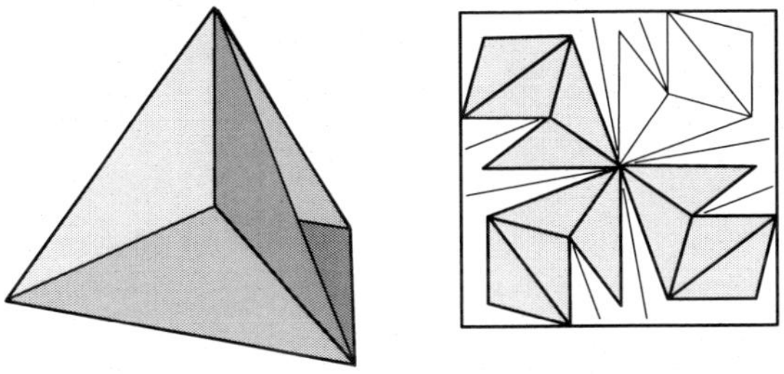

No Symmetry: The crease pattern shows no symmetry. The tetrahedron has no symmetry.

Tetrahedron

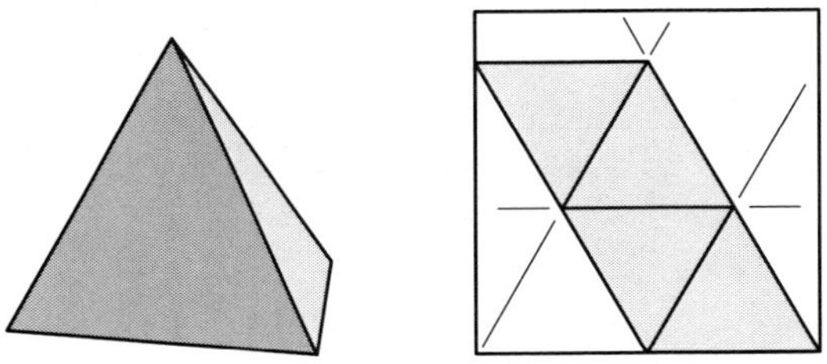

Exploring Tetrahedra

Composed of four equilateral sides, the tetrahedron is one of the simplest of polyhedra. Color patterns highlight the magical side of origami.

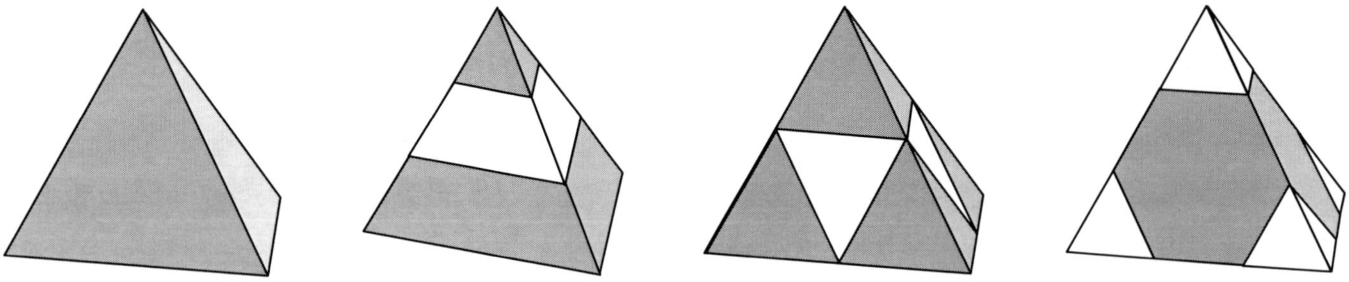

Tetrahedron

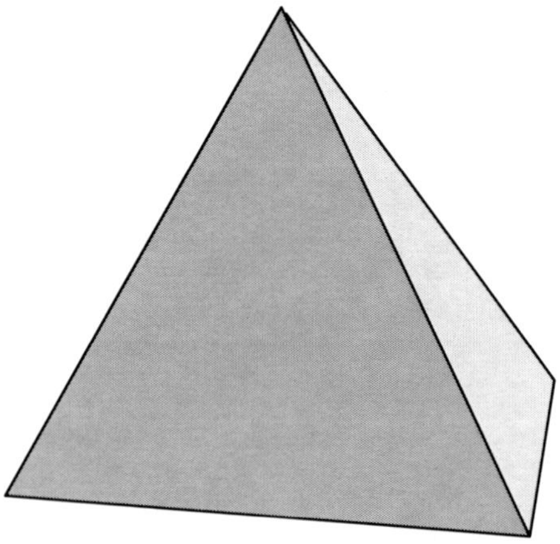

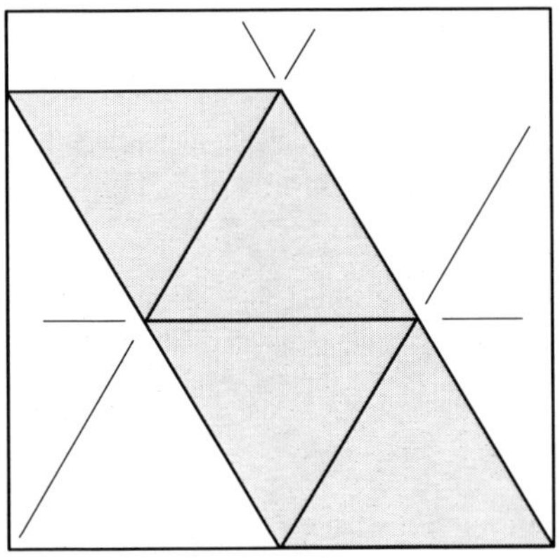

As the simplest of the five Platonic solids, Plato believed the tetrahedron represented fire because of its sharpness and simplicity.

The crease pattern shows a band of four triangles. Without the tab at the top, the model would have odd symmetry. As is, there is no symmetry. This is the perfect place to begin this journey.

1

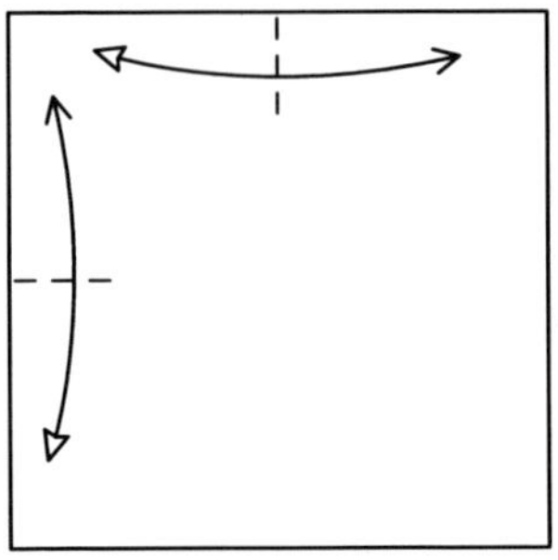

Fold and unfold.

2

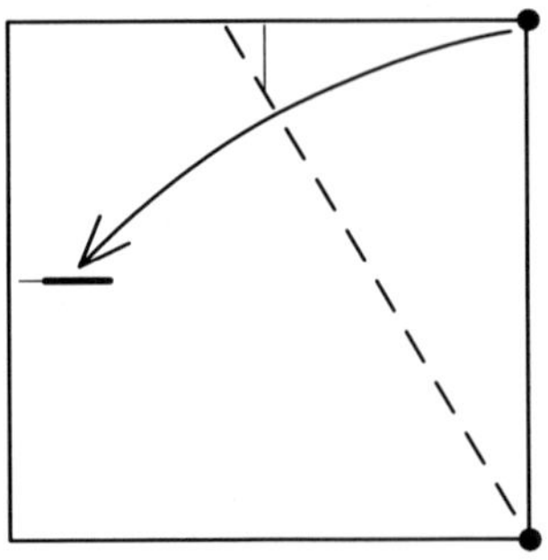

Bring the top corner to the line.

3

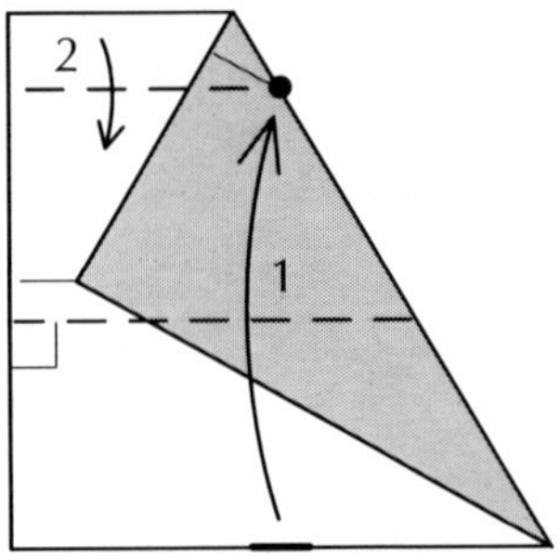

1. Bring the bottom edge to the dot.
2. Fold down.

4

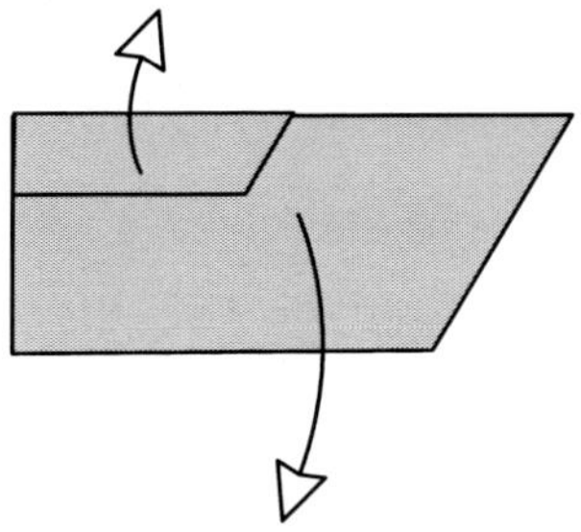

Unfold.

5

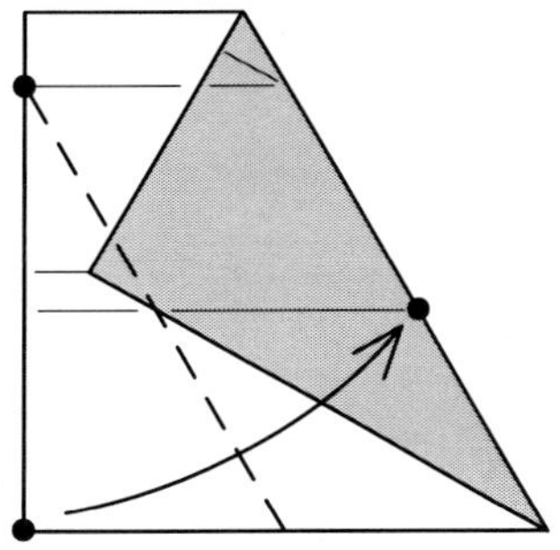

Bring the bottom dot to the right dot.

6

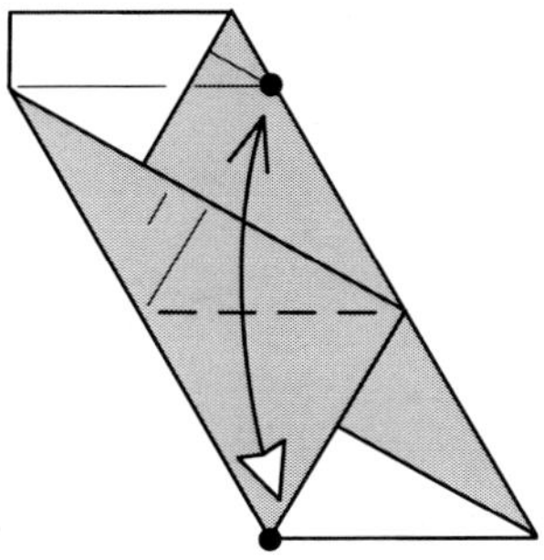

Fold and unfold.

7

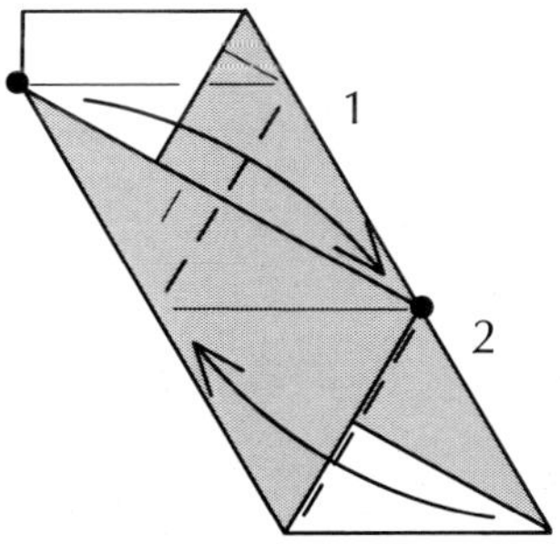

1. The dots will meet.
2. Fold along the edge.

8

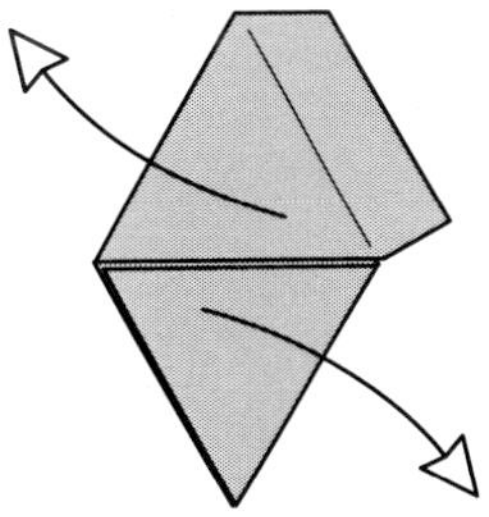

Unfold.

9

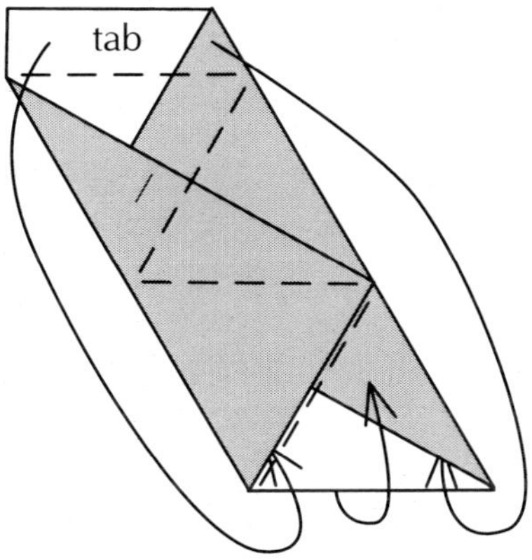

The left and right sides of the tab tuck into two pockets. Fold along the creases and tuck the tab inside the pockets.

10

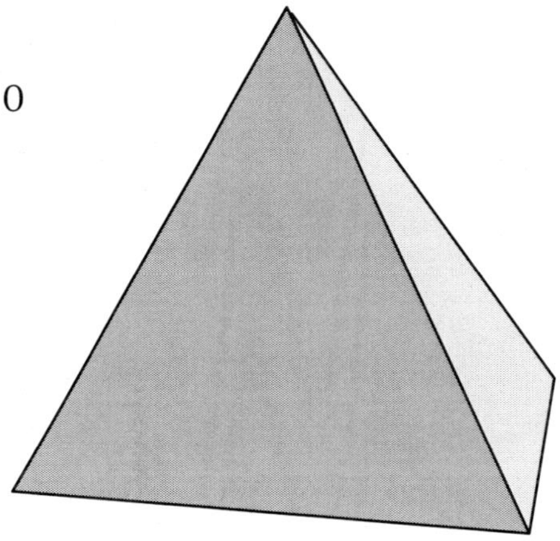

Tetrahedron

Striped Tetrahedron

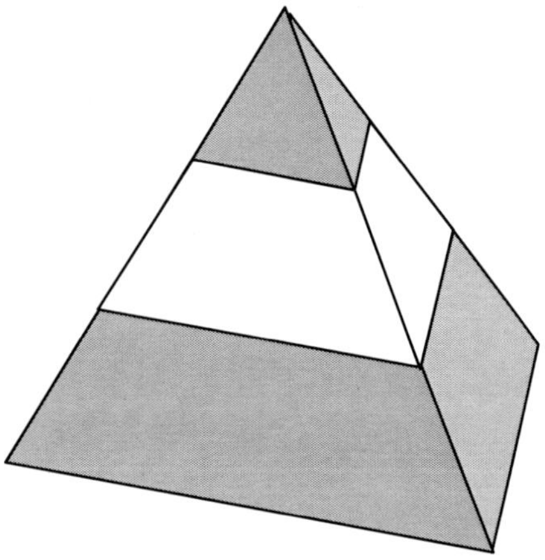

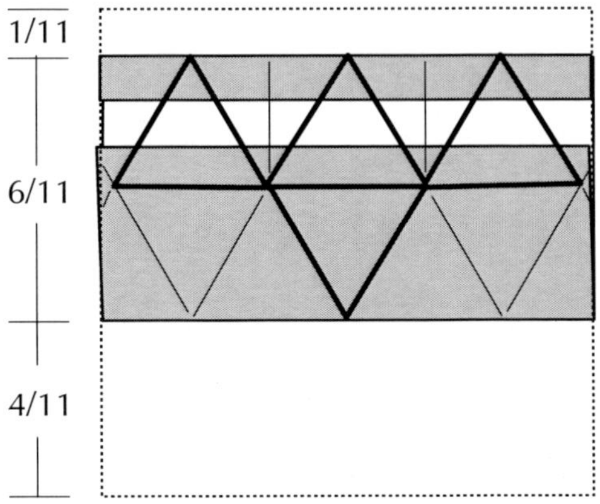

A few folds set up the three stripes. Even symmetry is used. To achieve the stripes, as shown in the diagram above, the square paper is divided into 11ths.

1

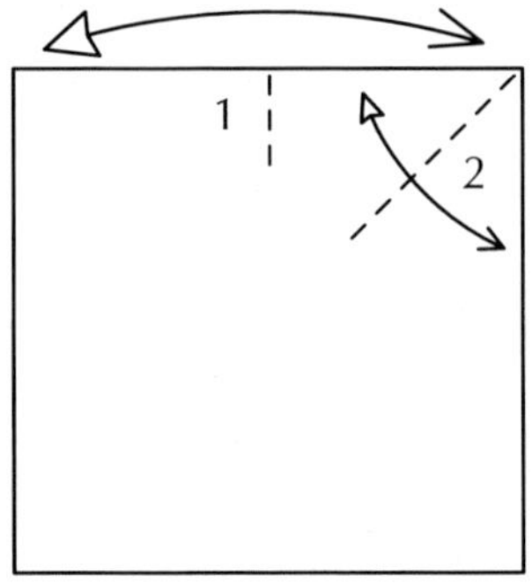

Fold and unfold at 1 and 2.

2

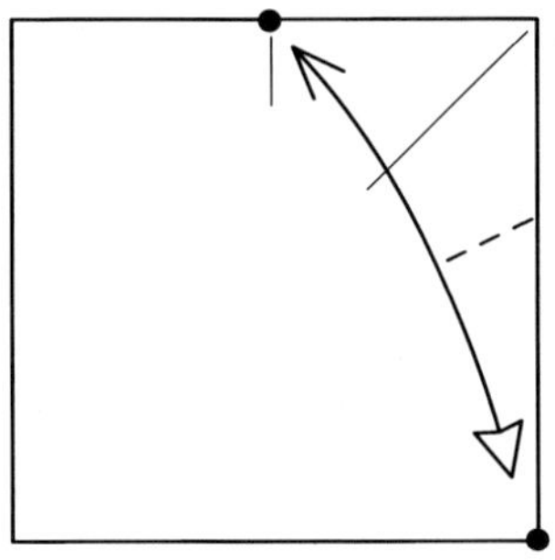

Fold and unfold on the right.

3

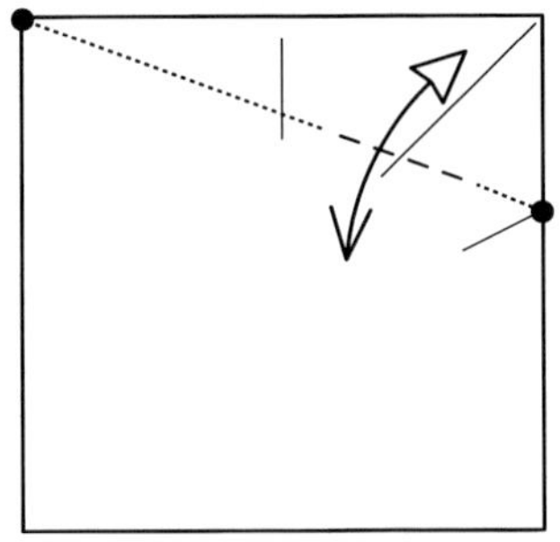

Fold and unfold along the intersection.

4

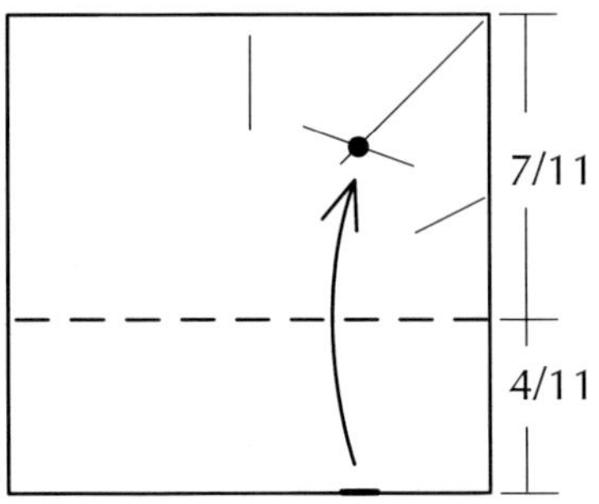

Divisions of 1/11 are found.

5

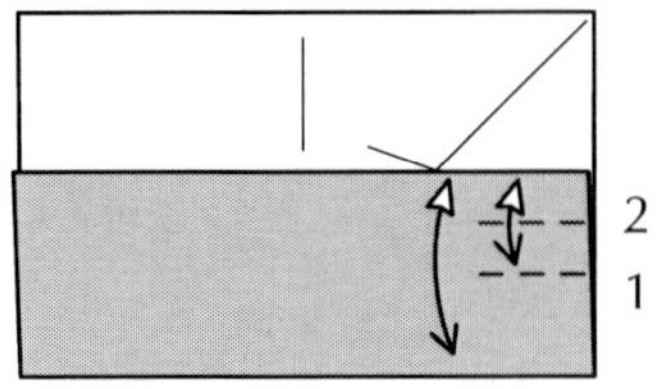

Fold and unfold on the right.

6

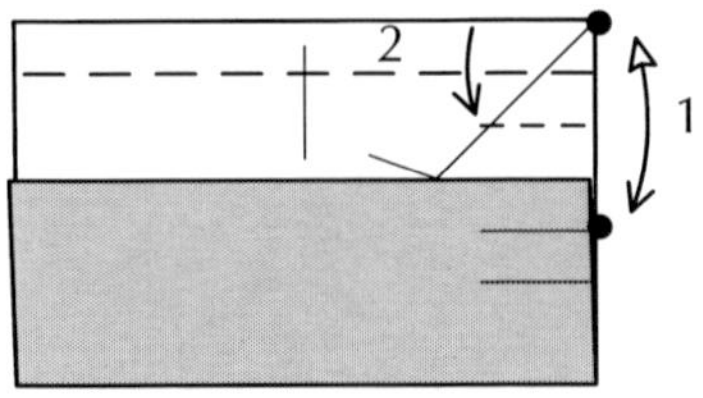

1. Fold and unfold.
2. Fold to the crease.

7

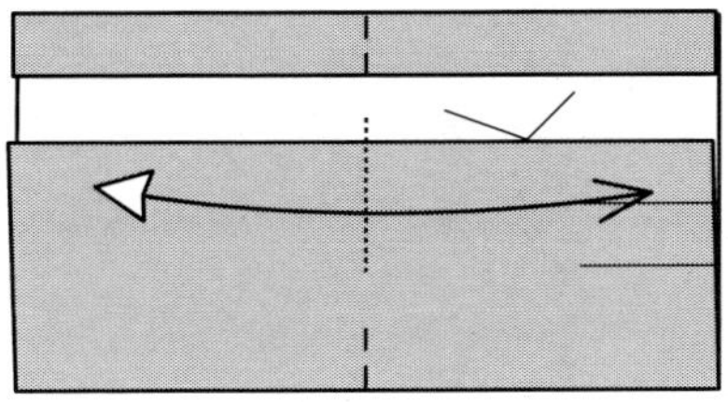

Fold and unfold at the top and bottom.

8

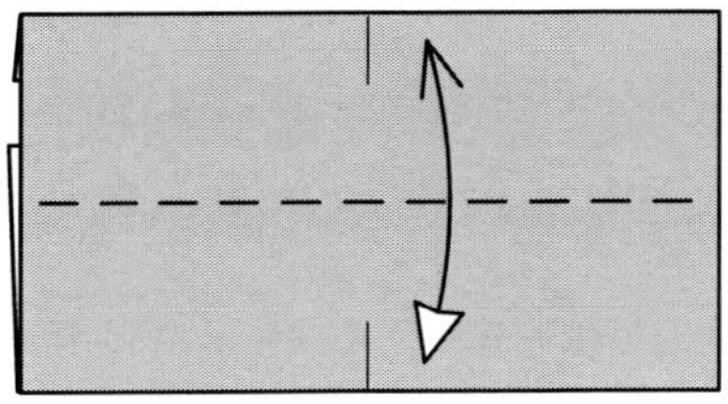

Fold in half and unfold.

9

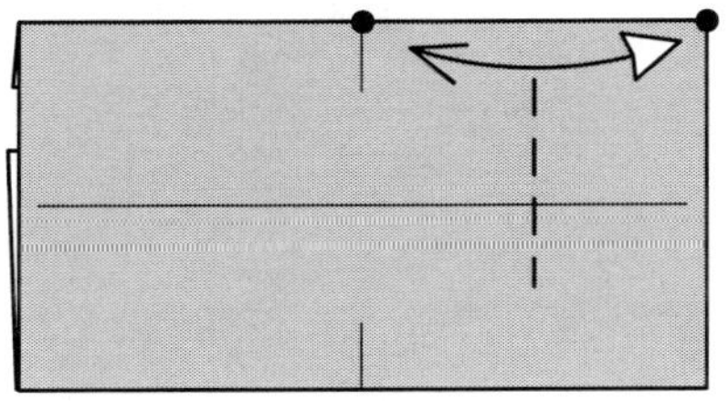

Fold and unfold.

10

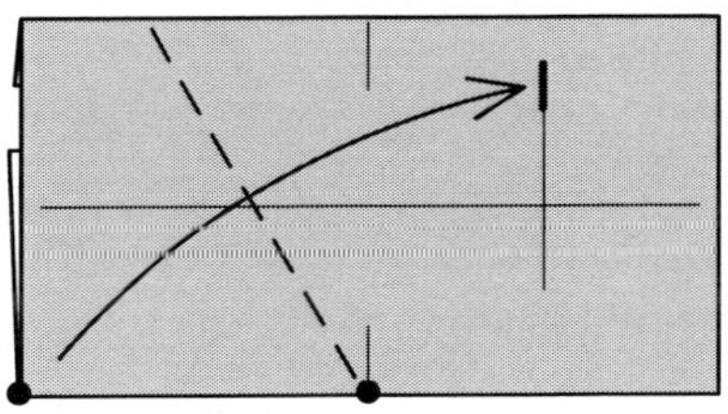

Bring the lower left corner to the line.

11

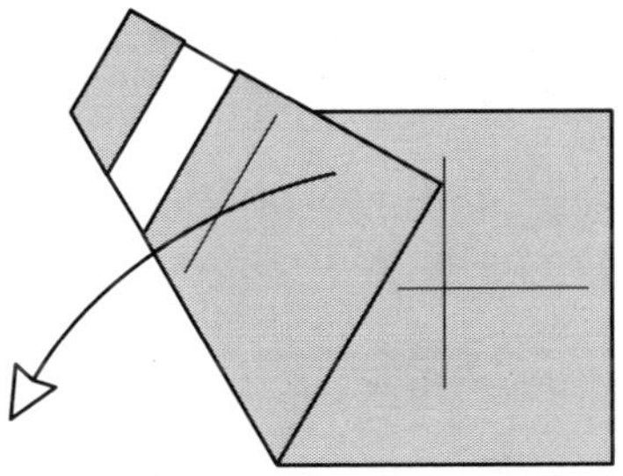

Unfold.

12

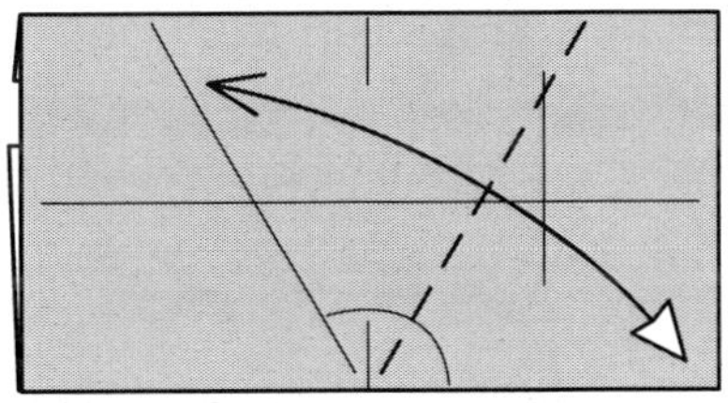

Fold and unfold.

13

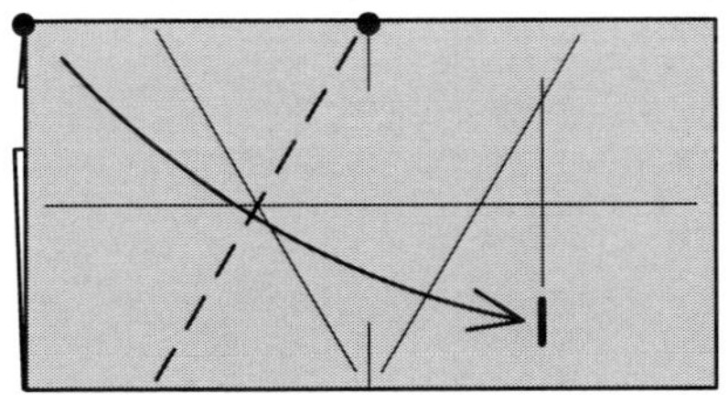

Repeat steps 10–12 in the opposite direction.

14

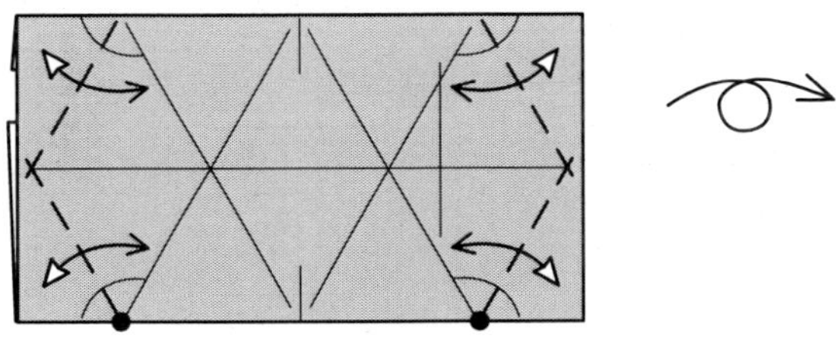

Fold and unfold.

15

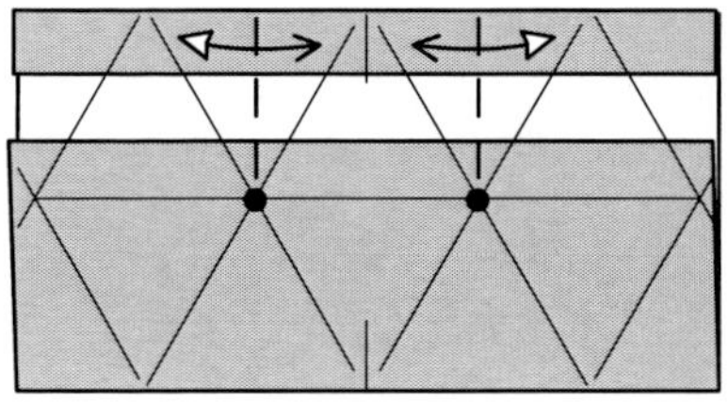

Fold and unfold.

16

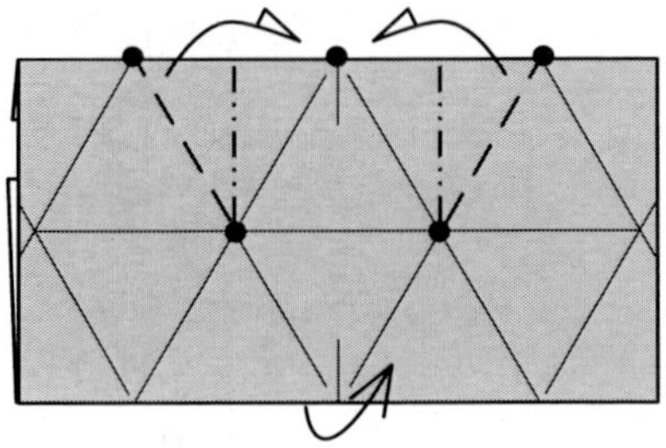

Push in at the lower dots.
The upper dots will meet.

17

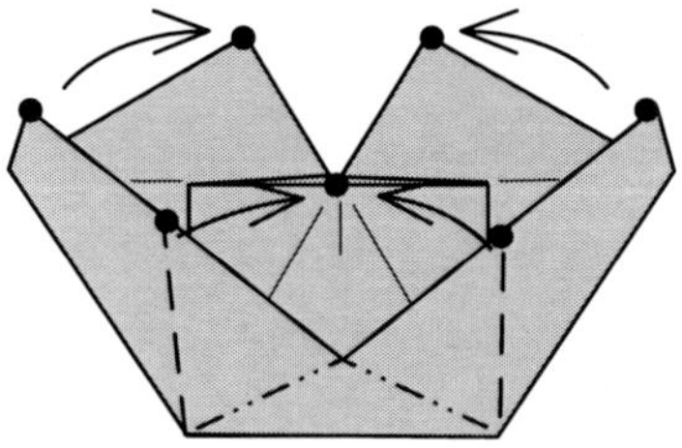

This is 3D. The groups
of dots will meet.

18

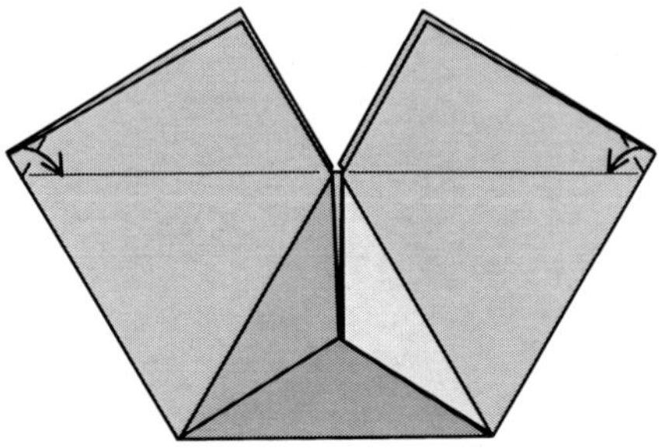

Fold along the creases.

19

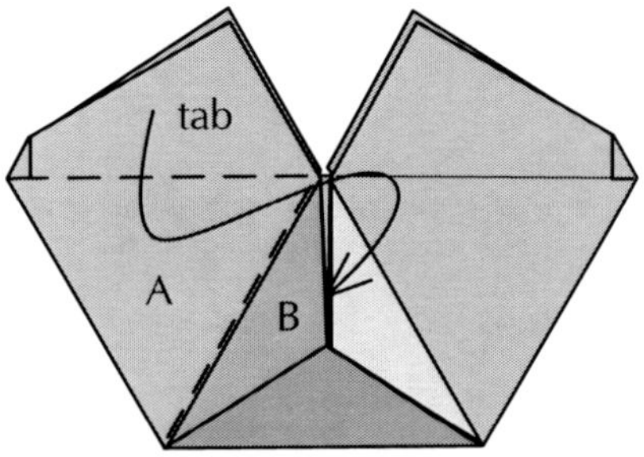

Fold all the layers together
to tuck the tab inside.
Region A will cover B.

20

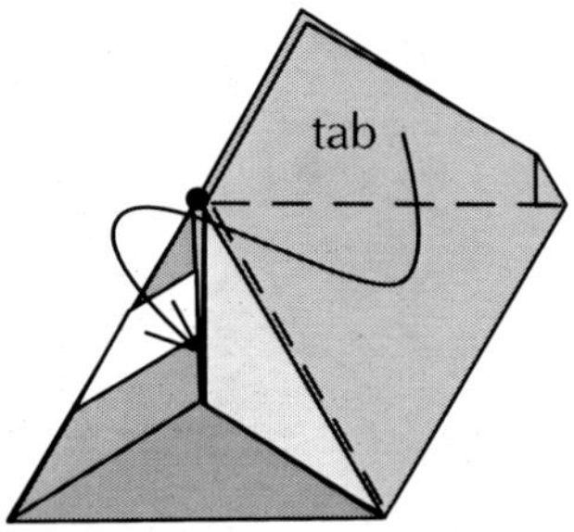

Tuck inside. Rotate
the dot to the top.

21

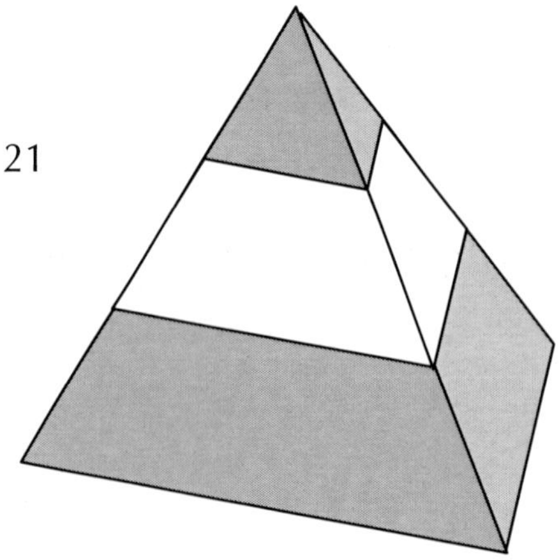

Striped Tetrahedron

Tetrahedron of Triangles

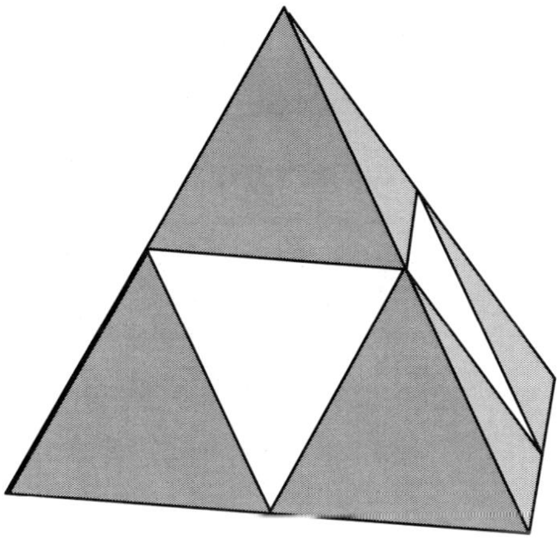

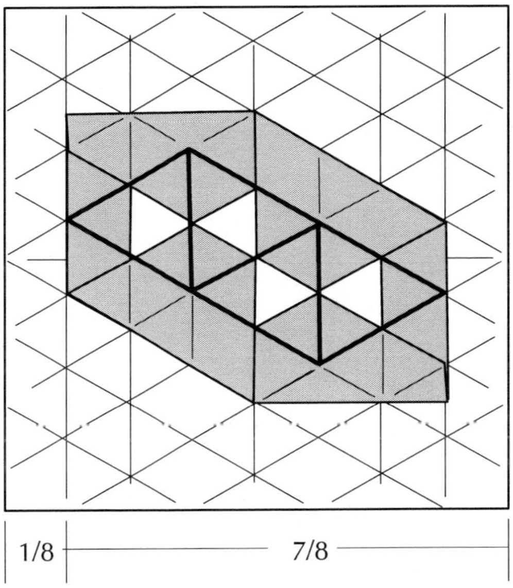

Each triangular face is composed of four triangles, with a color change in the center of each triangle. Odd symmetry is used. The paper is divided into 8ths.

1

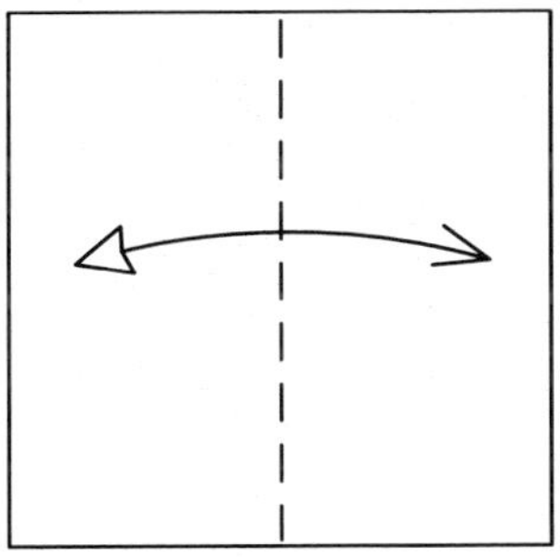

Fold and unfold.

2

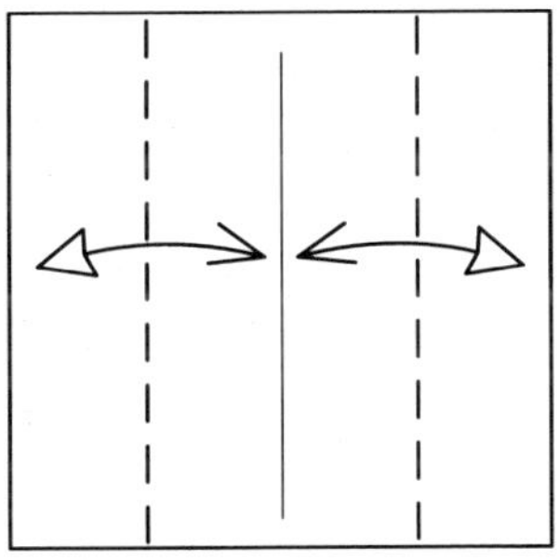

Fold and unfold.

3

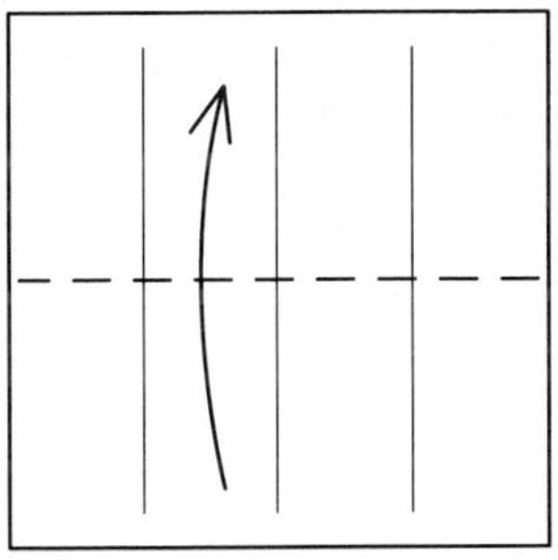

4

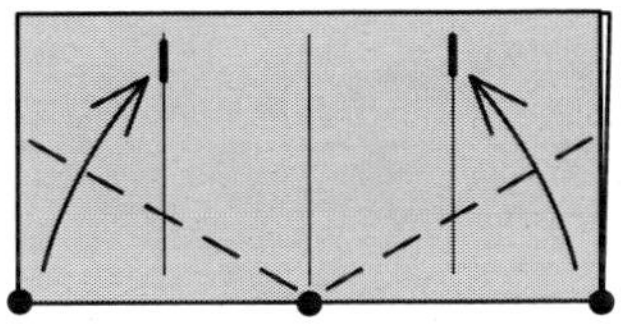

Bring the corners to the lines.

5

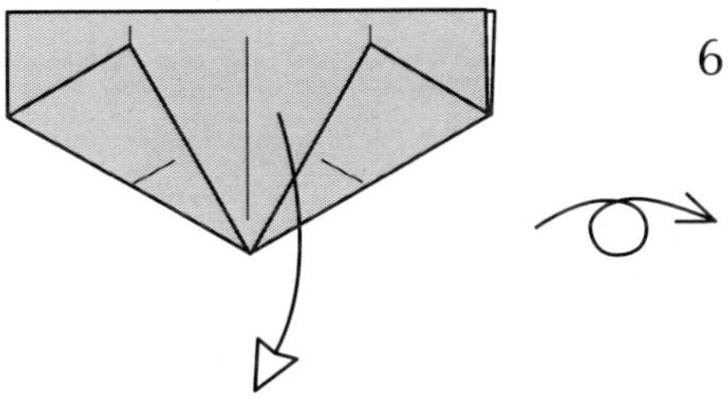

Unfold.

6

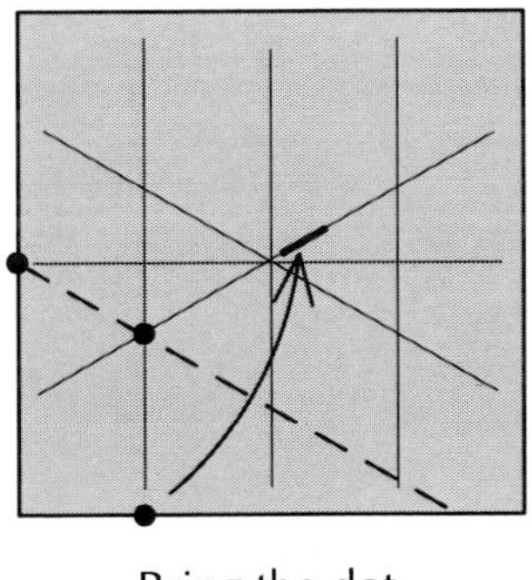

Bring the dot to the line.

7

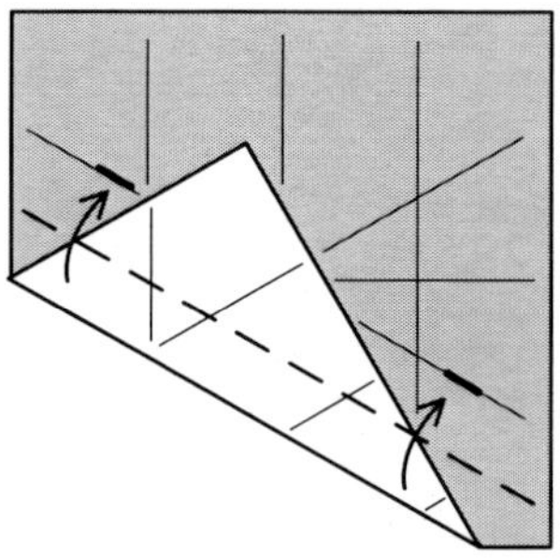

Fold to the bold line.

8

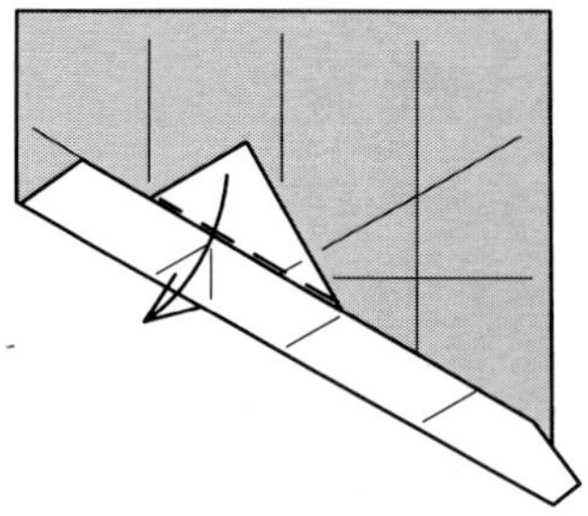

9

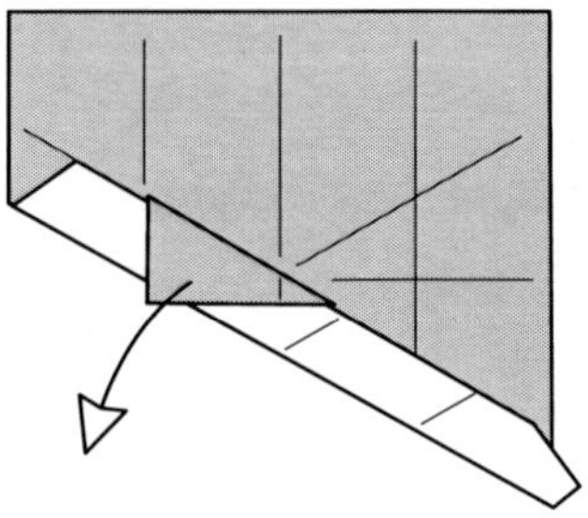

Unfold.

10

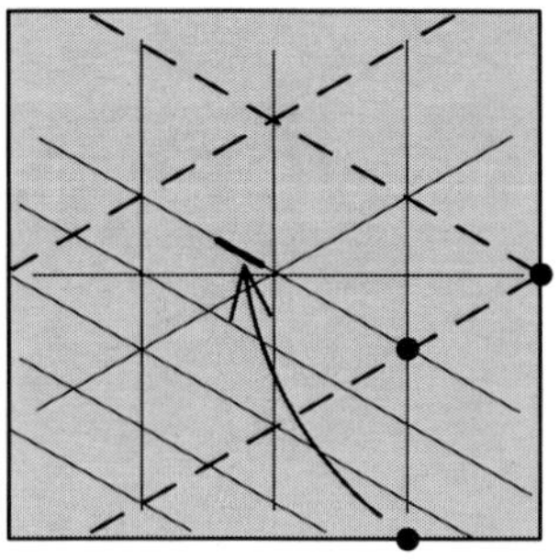

Repeat steps 6–9 three times, on the right and above.

11

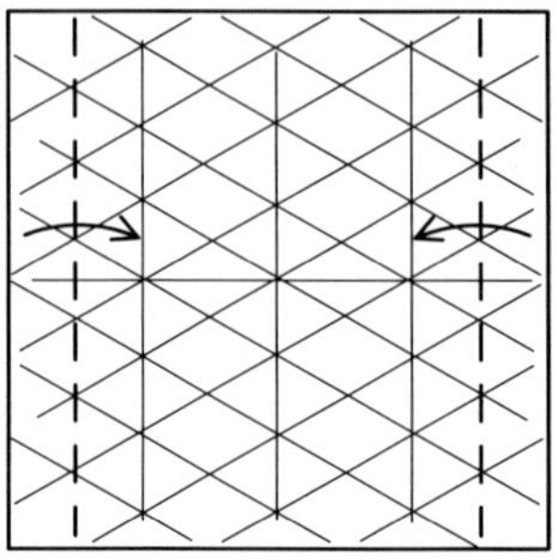

Fold to the creases.

12

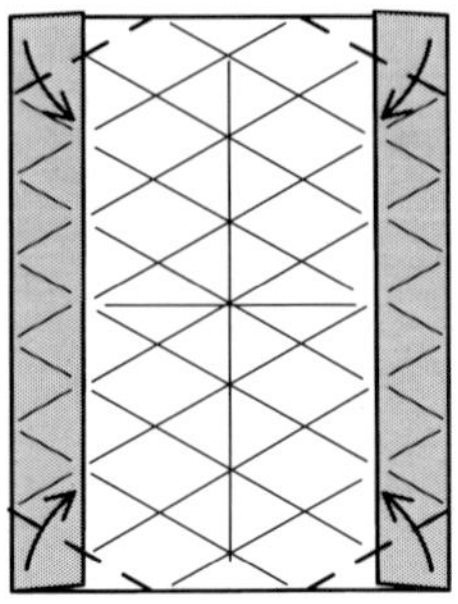

Fold along the creases.

13

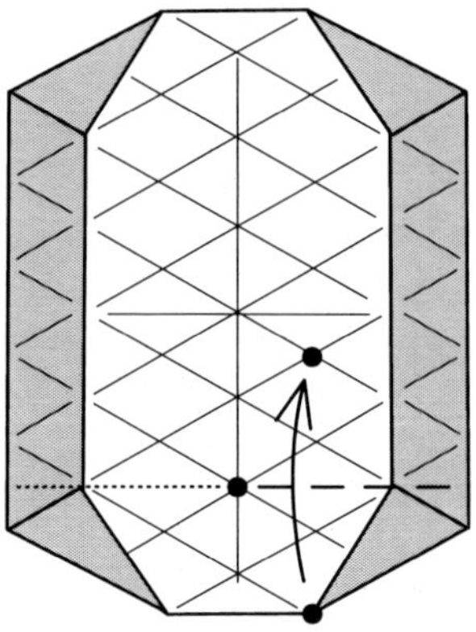

14

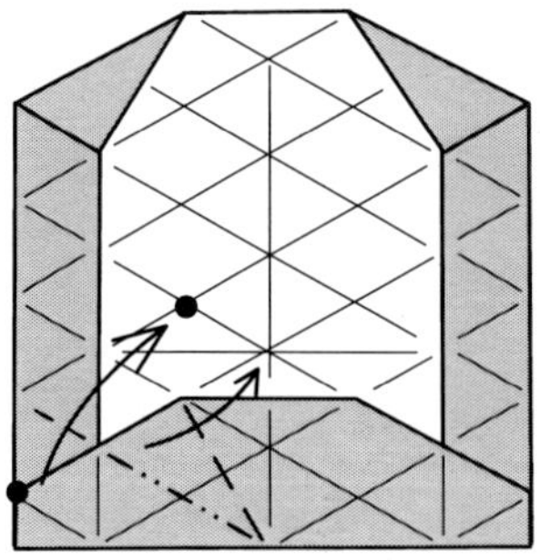

Squash-fold.

15

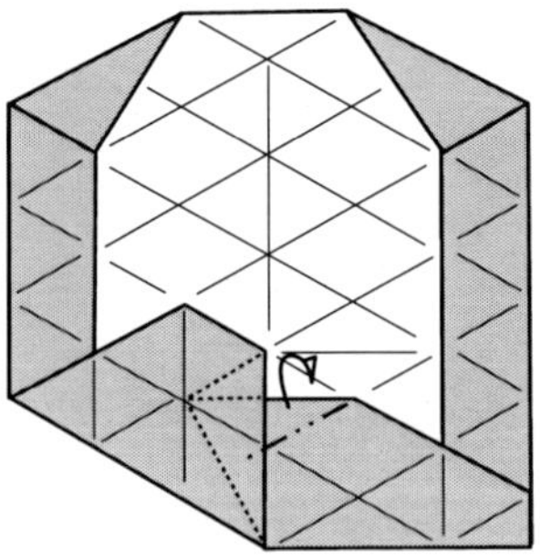

Squash-fold.

16

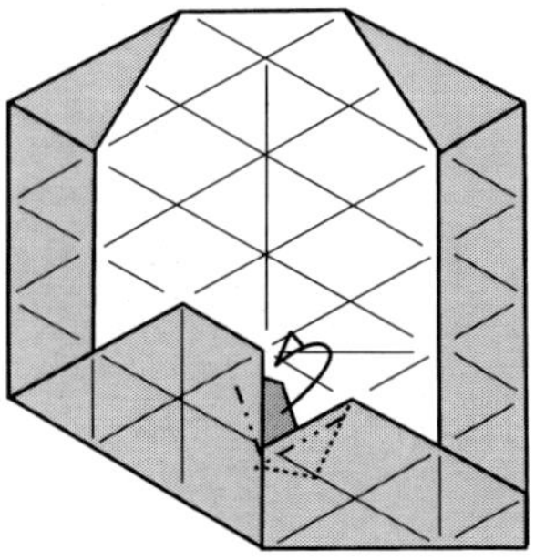

Make a small, hidden squash fold, so the darker paper is hidden. Rotate 180°.

17

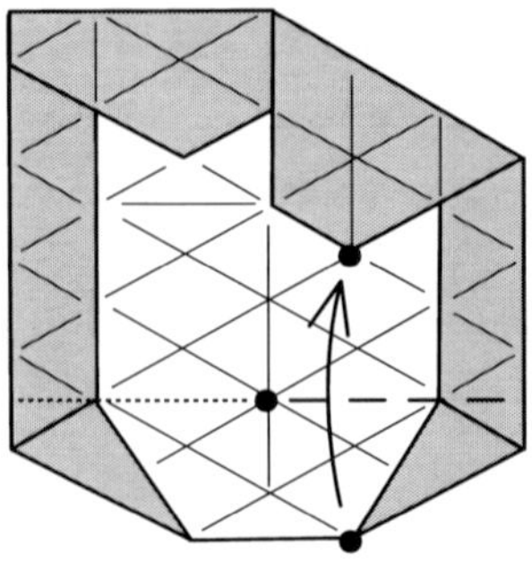

Repeat steps 13–16. Rotate.

18

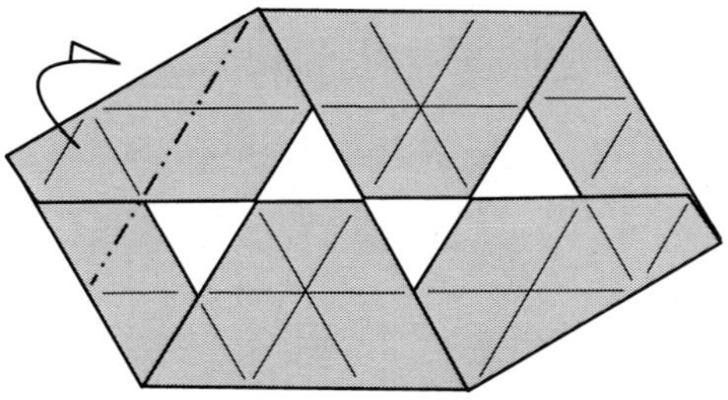

Fold along the crease.

19

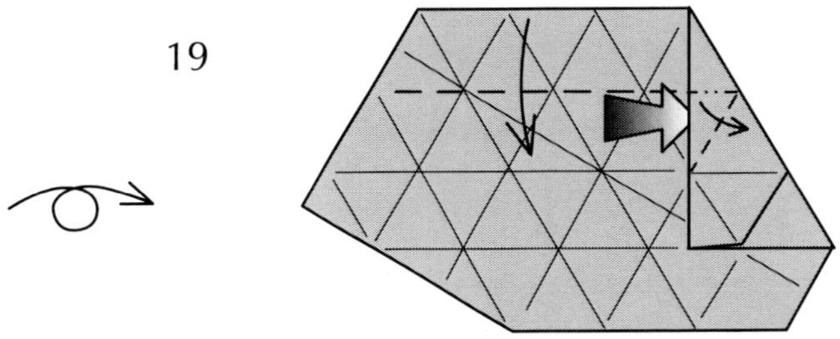

Squash-fold.

20

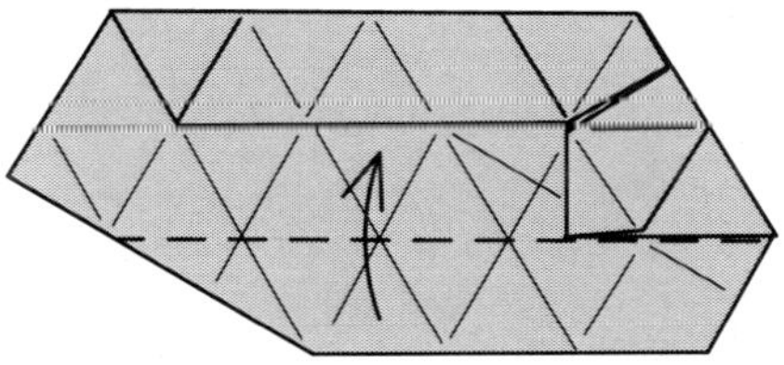

Fold along the crease.

21

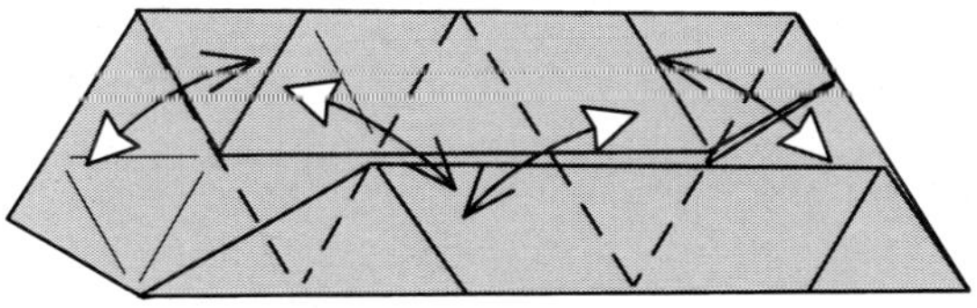

Fold and unfold
along the creases.

22

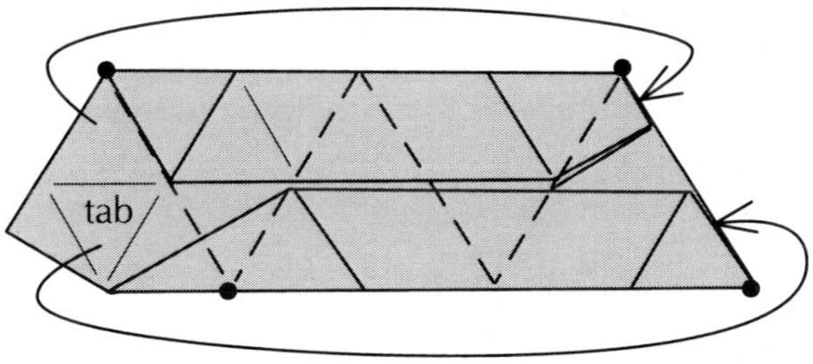

Tuck the tab into the pockets.
The pairs of dots will meet.

23

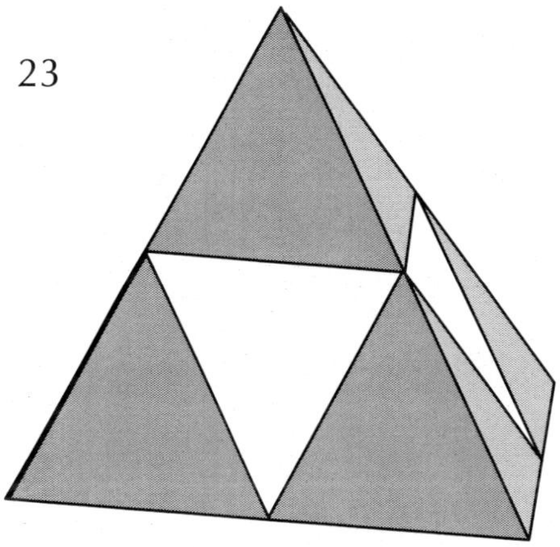

Tetrahedron of Triangles

Capped Tetrahedron

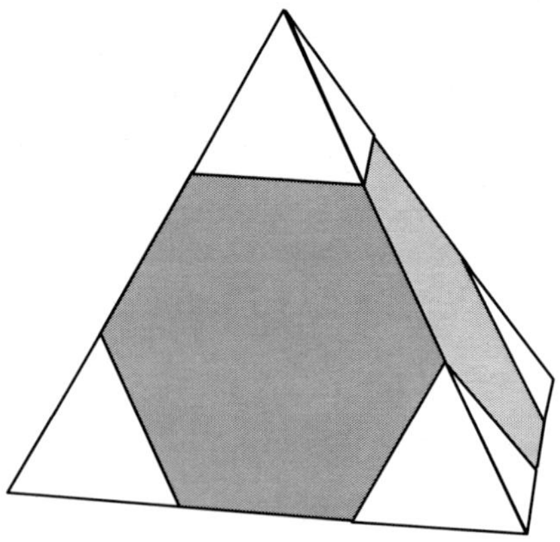

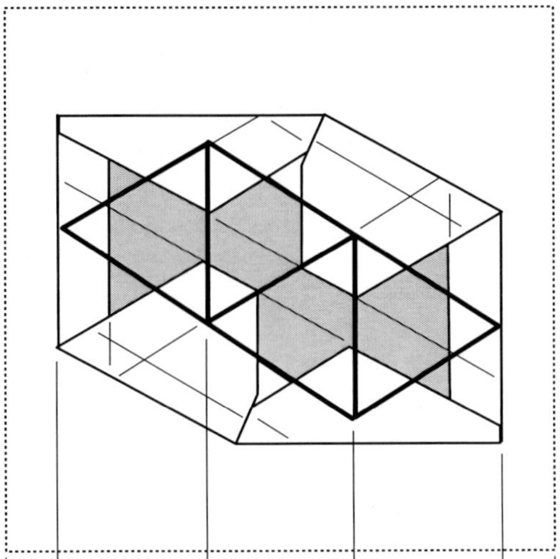

This tetrahedron has colored hexagons with white caps on each of the four faces.

The figure above shows step 28 embedded in the original square, showing how the colors are placed on the crease pattern. Odd symmetry is used. The paper is divided into 11ths.

1

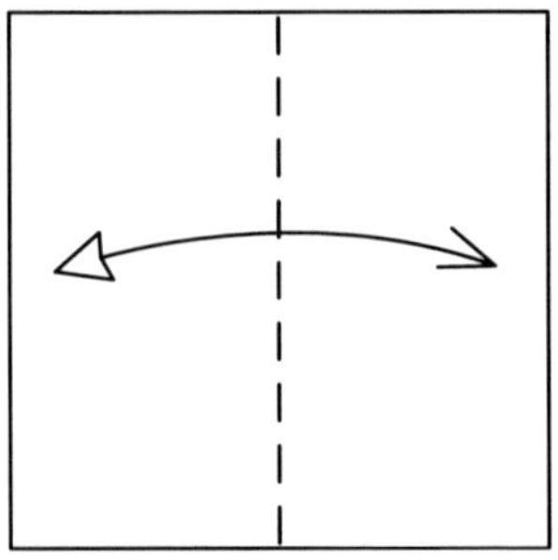

Fold and unfold.

2

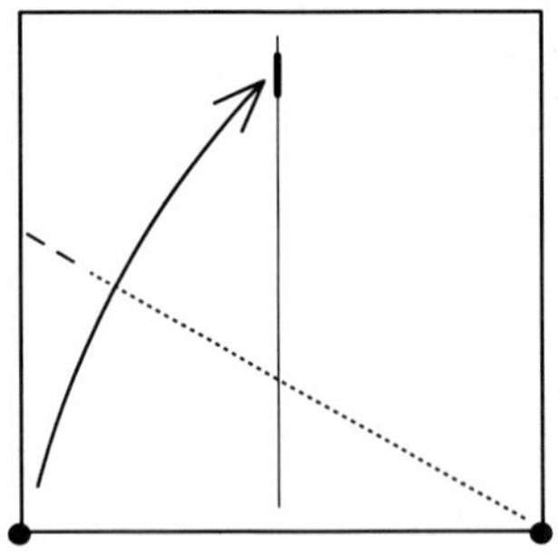

Bring the left dot to the line. Crease on the left.

3

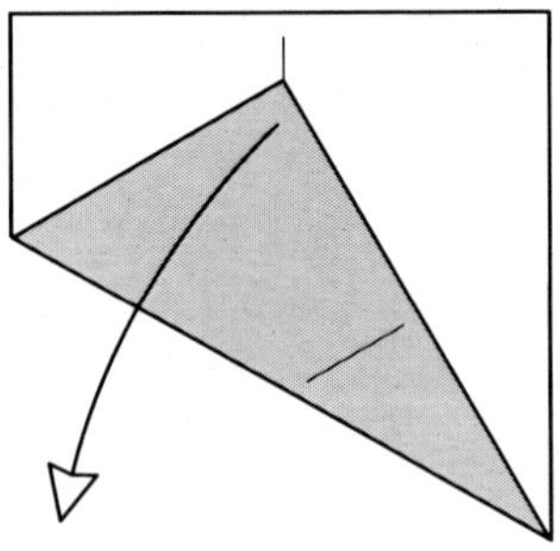

Unfold and rotate 180°.

4

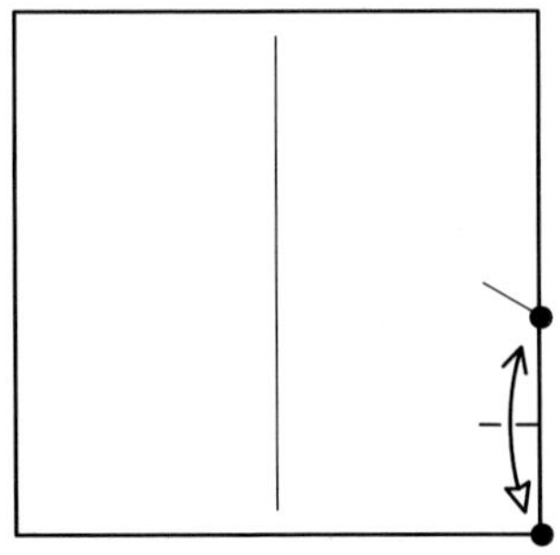

5

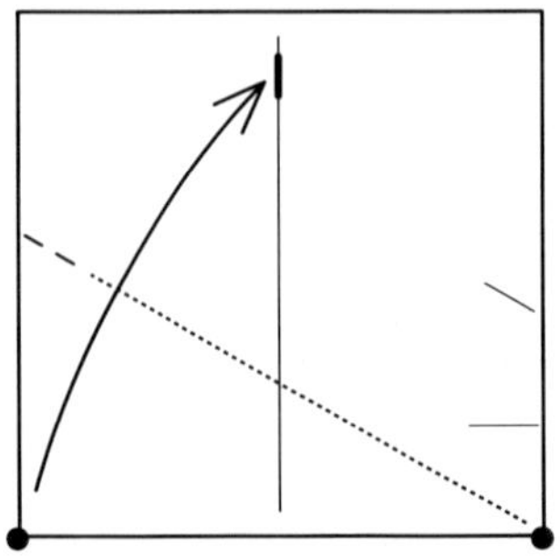

Repeat steps 2–4.

6

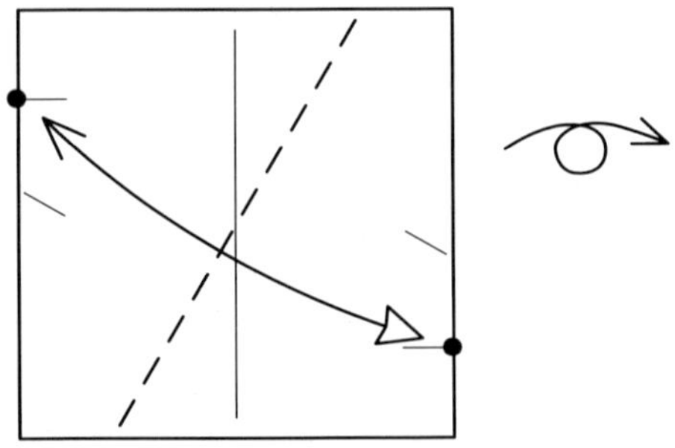

Fold and unfold.

7

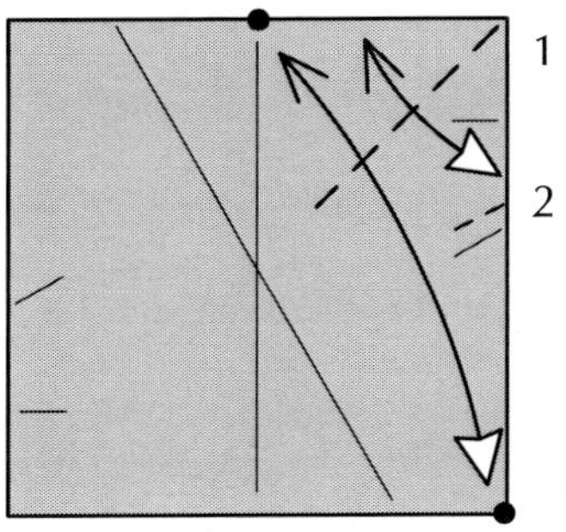

1. Fold and unfold on the diagonal.
2. Fold and unfold on the right.

8

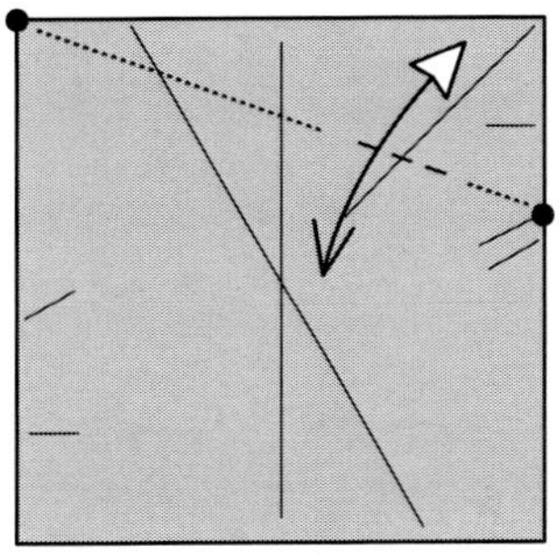

Fold and unfold at the intersection.

9

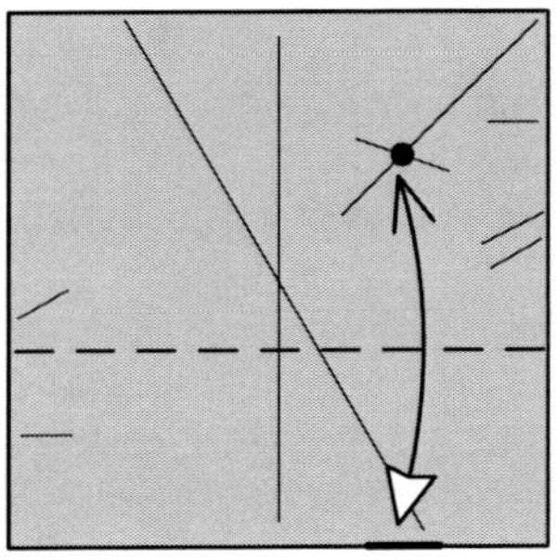

Fold and unfold.

10

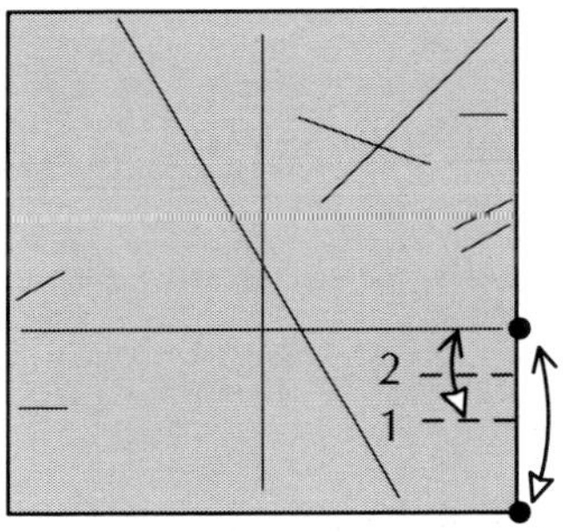

Fold and unfold twice.

11

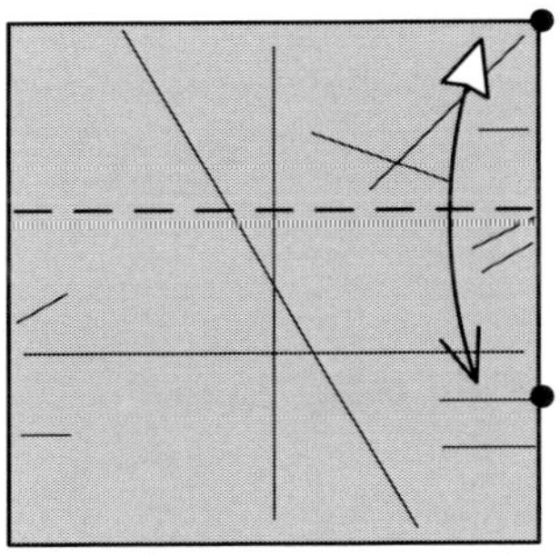

Fold and unfold.

12

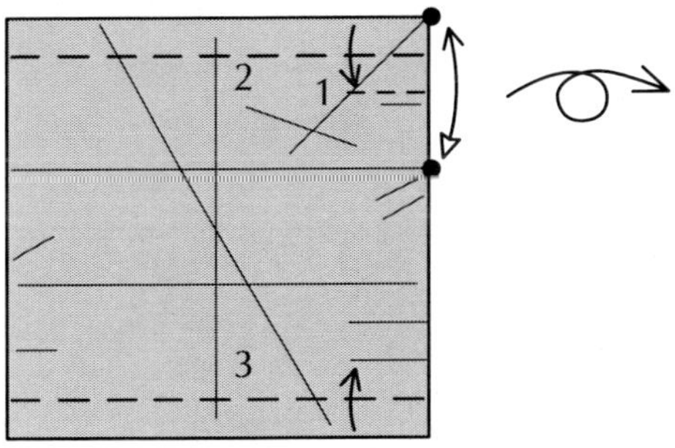

1. Fold and unfold.
2, 3. Fold to the marks.
Rotate 90°.

13

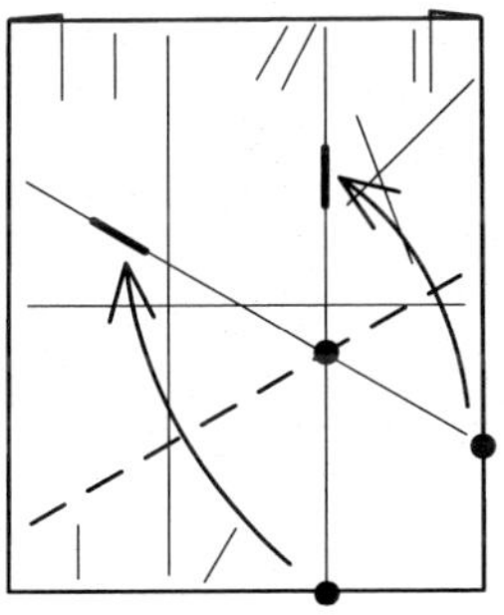

14

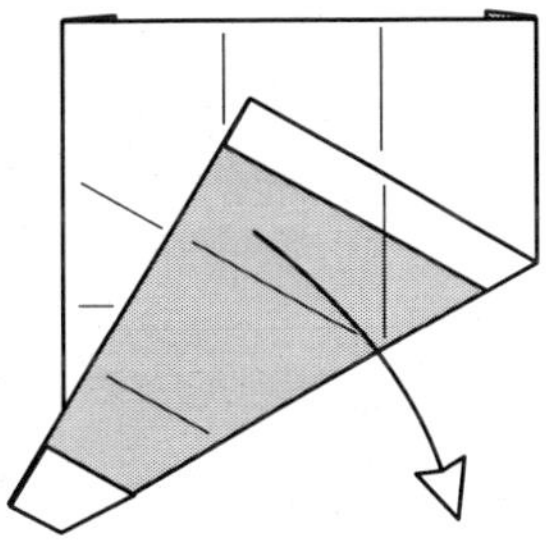

Unfold.

15

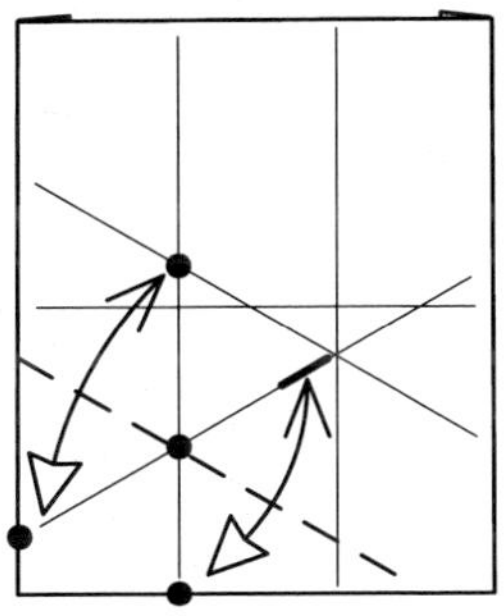

Fold and unfold.
Rotate 180°.

16

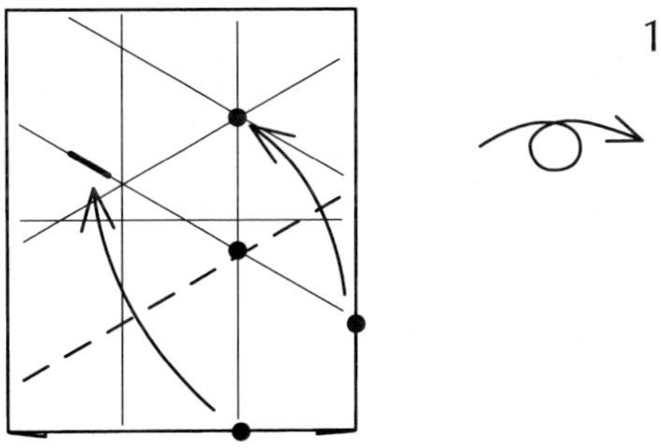

Repeat steps 13–15.

17

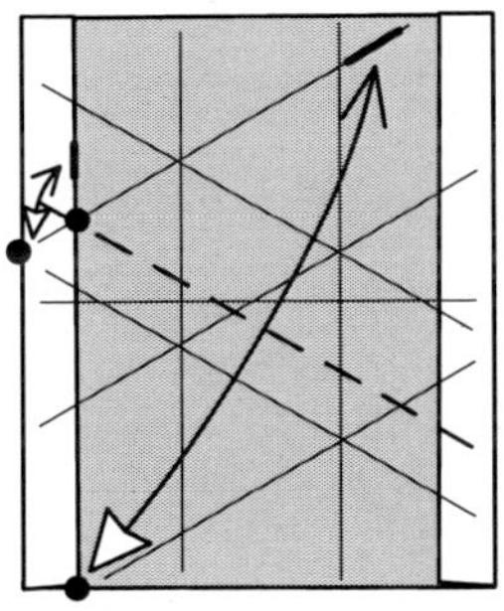

Fold and unfold.
Rotate 180°.

18

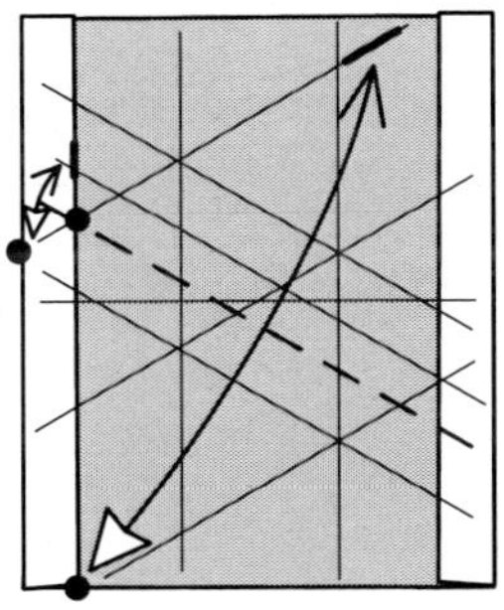

Repeat step 17.

19

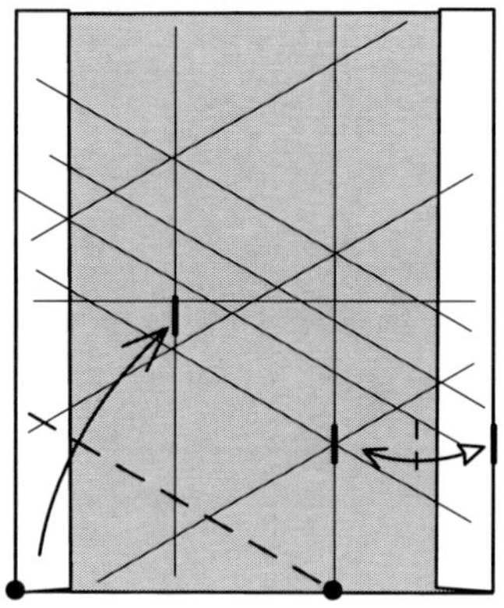

1. The dot will meet the line.
2. Fold and unfold.

20

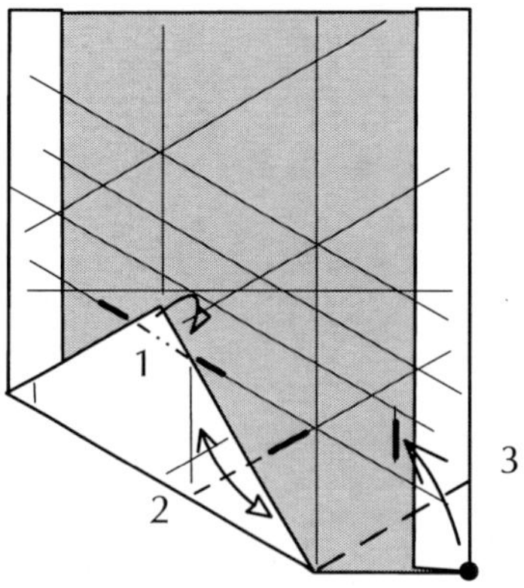

1. Fold behind along the bold line.
2. Fold and unfold along the bold line.
3. The dot will meet the line.

21

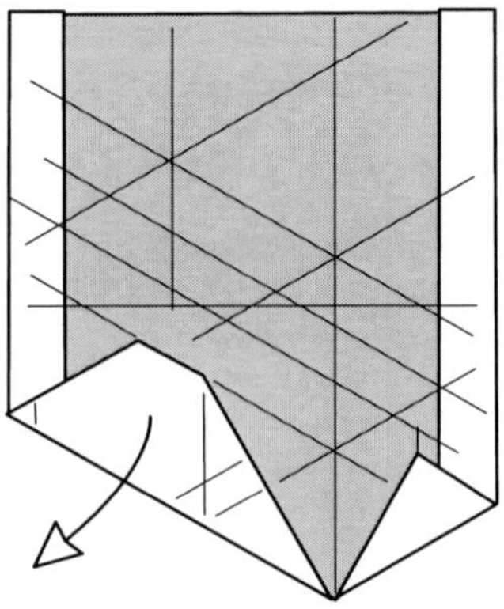

Unfold and rotate 180°.

22

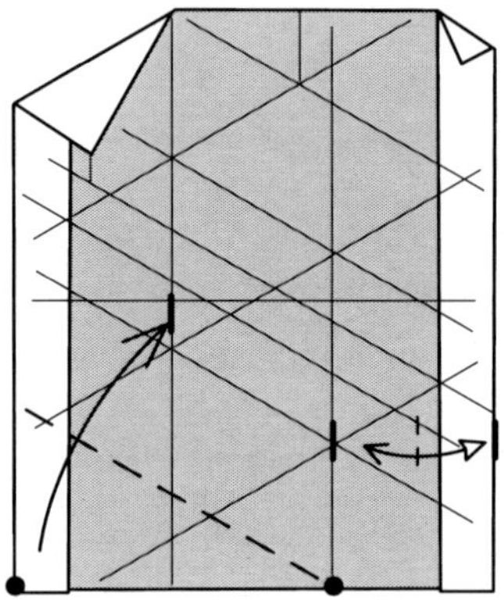

Repeat steps 20–21.

23

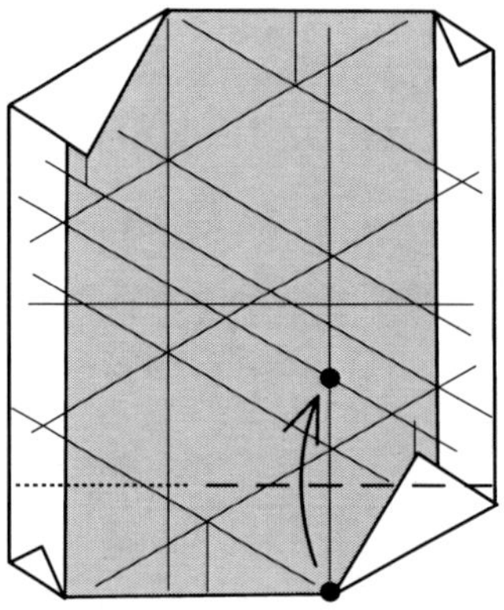

24

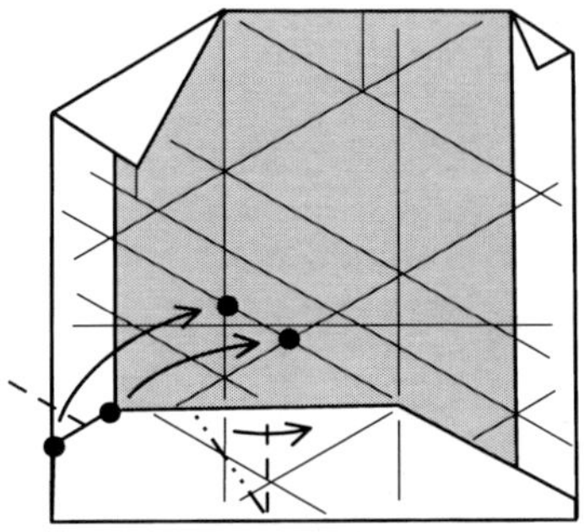

Squash-fold so the pair of dots meet.

25

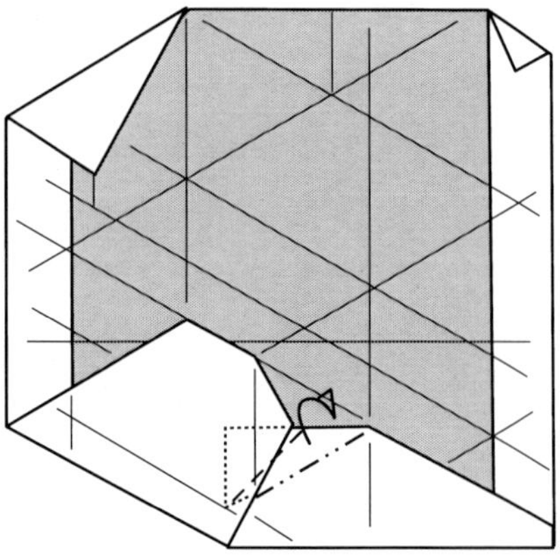

This is similar to a squash fold.

26

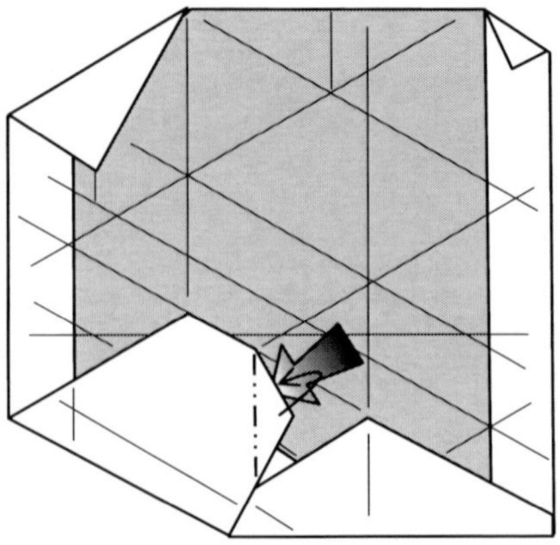

Reverse-fold and rotate 180°.

27

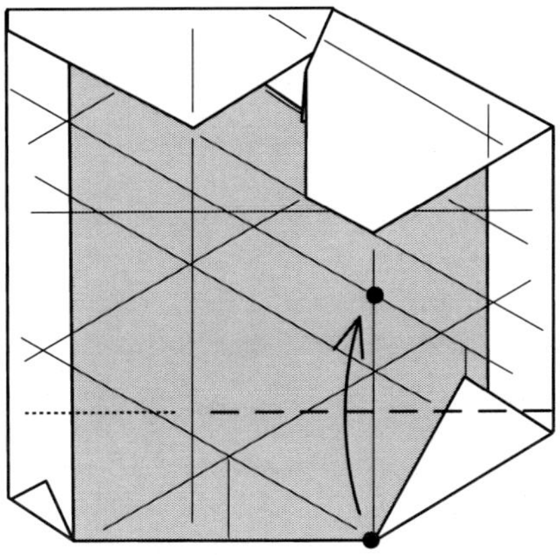

Repeat steps 23–26.

28

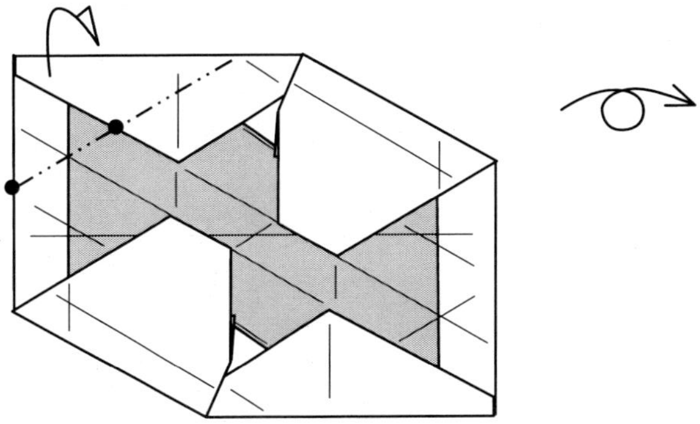

Fold along the crease between the dots. Rotate.

29

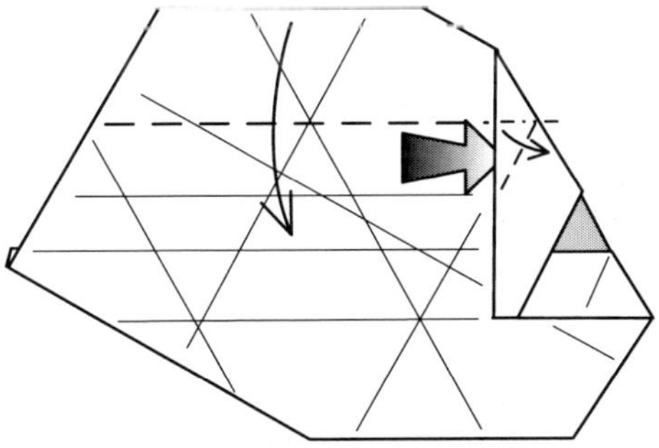

Squash-fold.

30

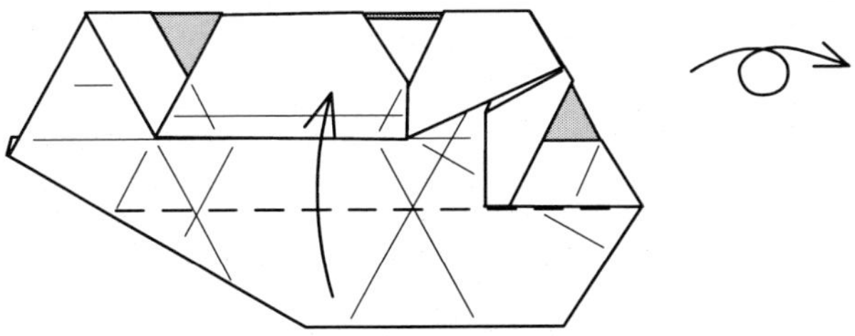

Fold along the crease.

31

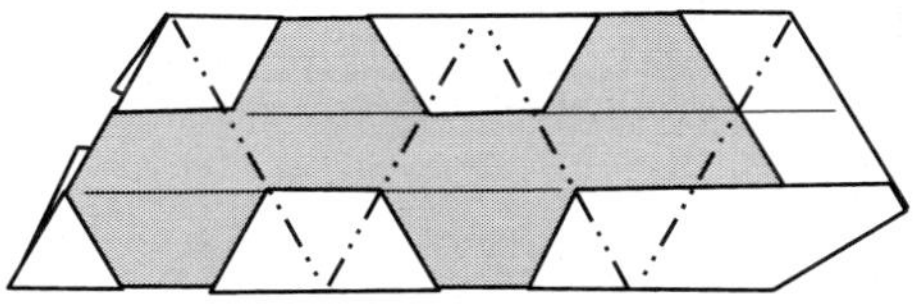

Fold and unfold along the creases.

32

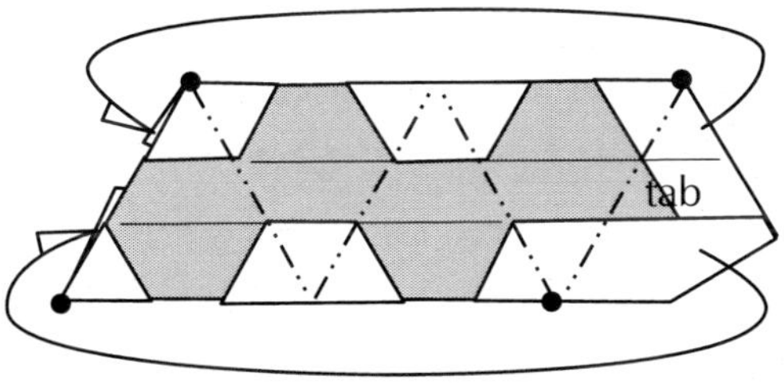

Tuck the tab into the pockets. The pairs of dots will meet.

33

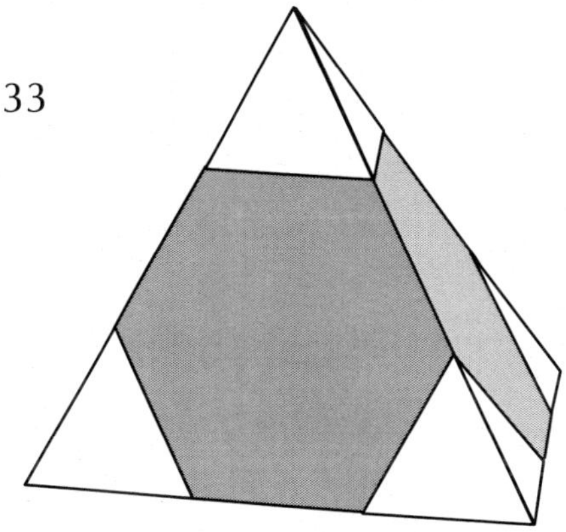

Capped Tetrahedron

Exploring Cubes

Composed of six squares, the cube should be easy to design from a square sheet of paper, yet it is not so simple. It is fun to add color patterns on the faces of the cube.

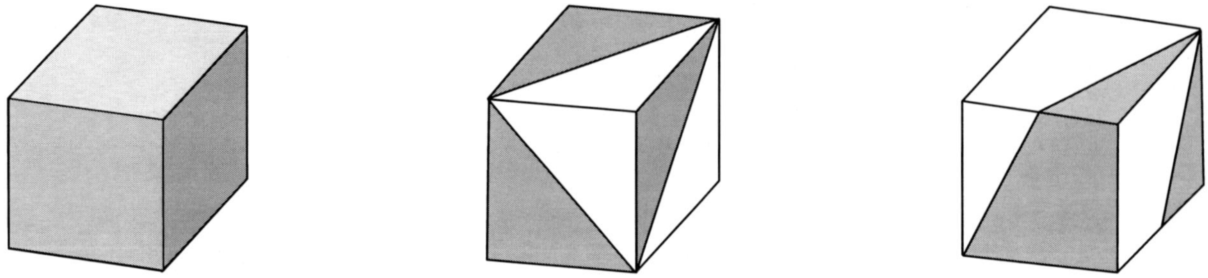

Cube

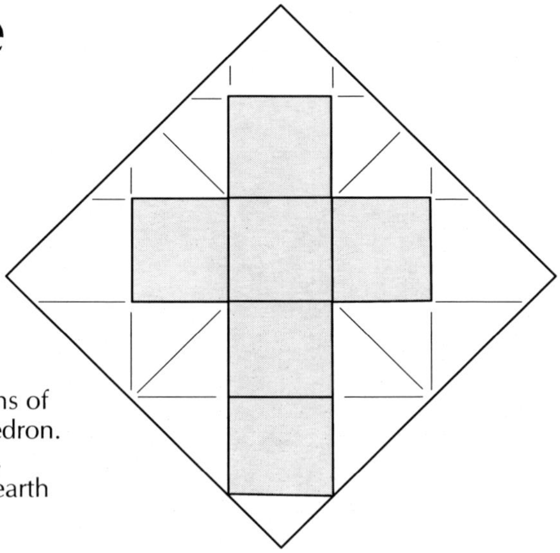

In keeping with the naming conventions of polyhedra, this could be called a hexahedron. Plato believed this regular polyhedron, composed of six squares, symbolized earth because of its stability.

Six squares forming a cross become the six sides of the cube. Even symmetry is used.

1

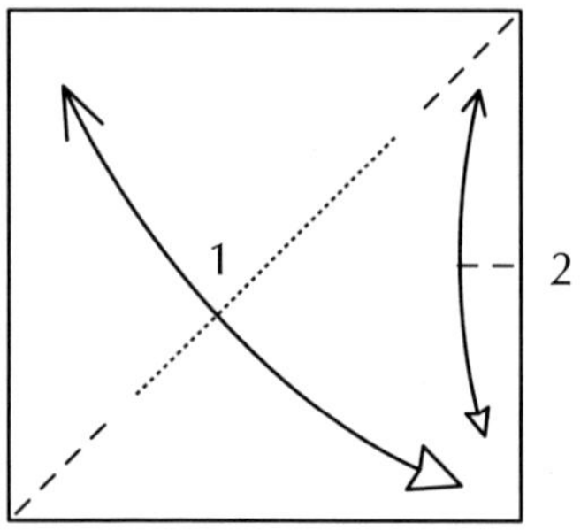

1. Fold and unfold on the corners.
2. Fold and unfold on the right.

2

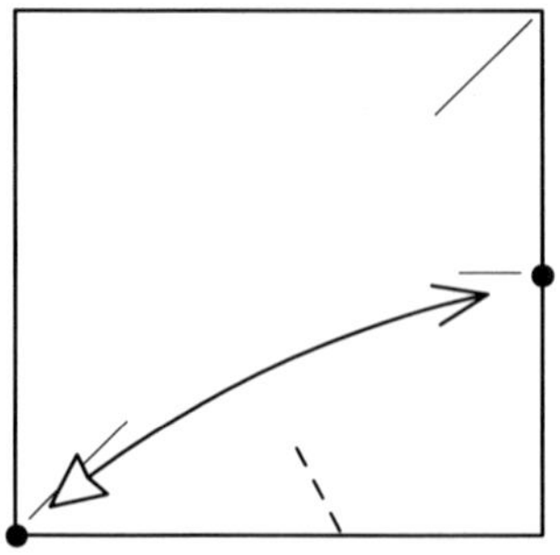

Fold and unfold on the bottom.

3

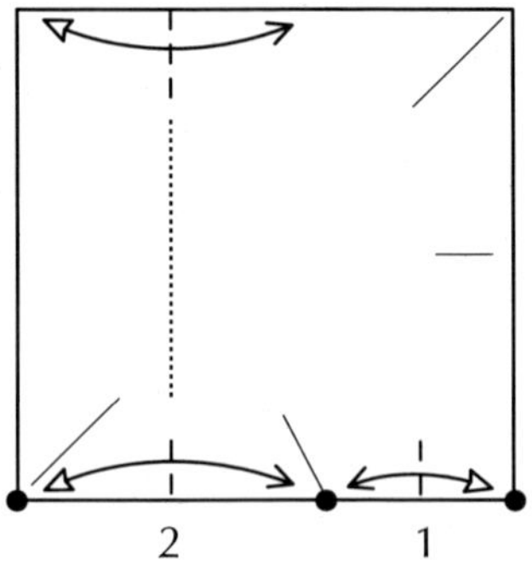

1. Fold and unfold on the bottom.
2. Fold and unfold on the top and bottom.

Rotate 45°.

4

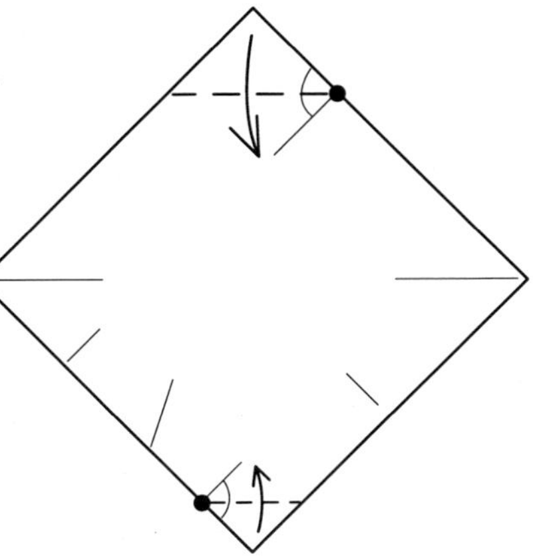

5

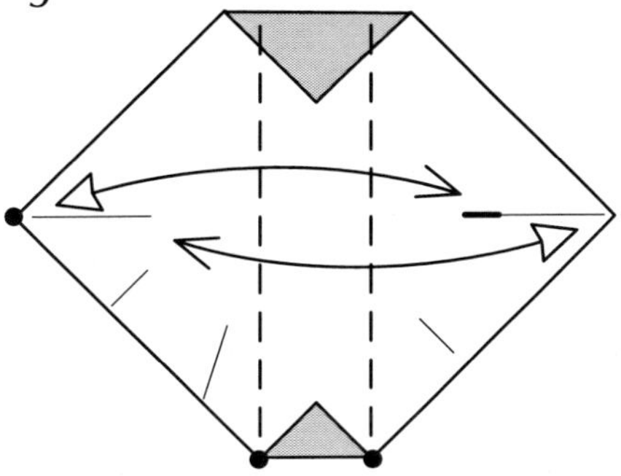

Fold and unfold.

6

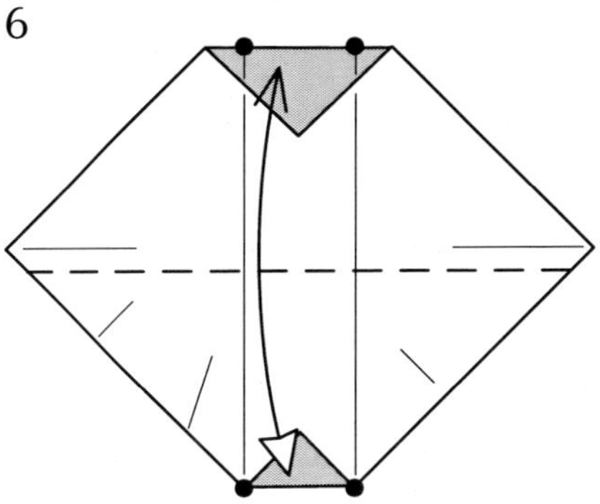

Fold and unfold.

7

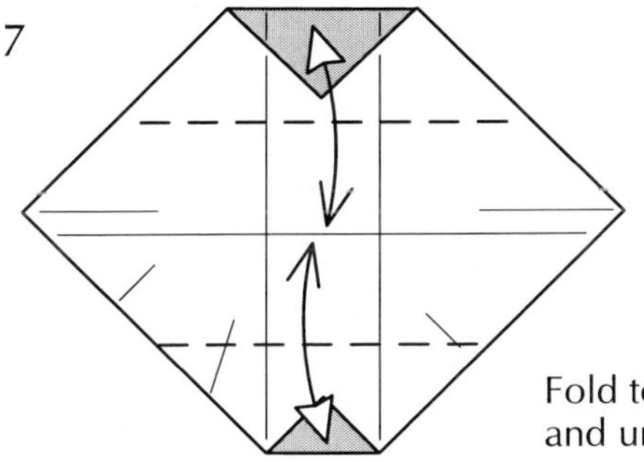

Fold to the center
and unfold.

8

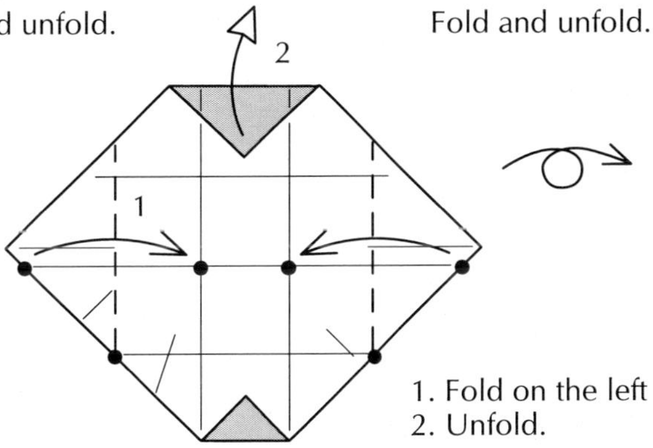

1. Fold on the left and right.
2. Unfold.

9

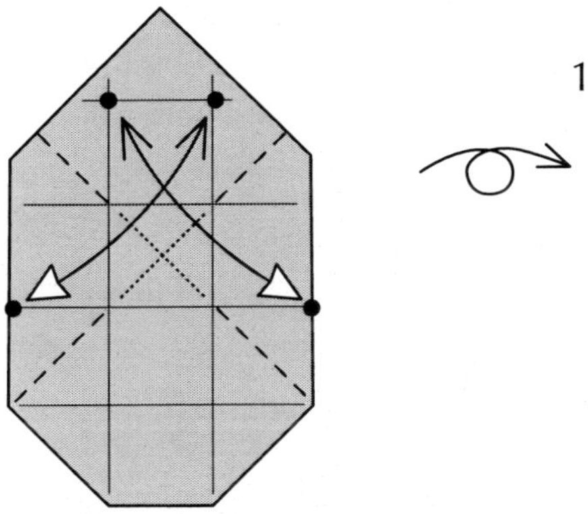

Fold and unfold but not in
the center. Rotate 180°.

10

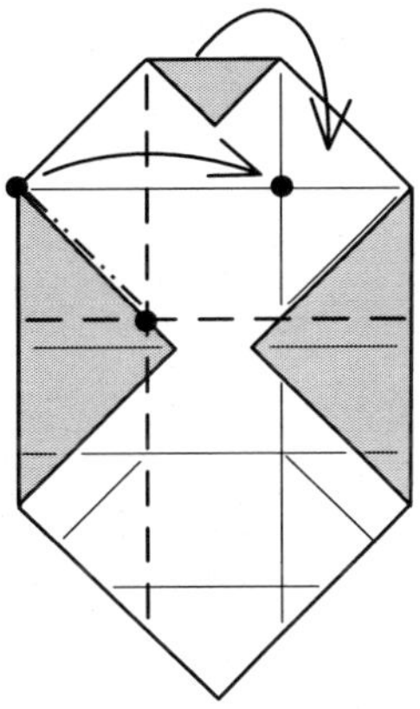

Push in at the
lower dot.

11

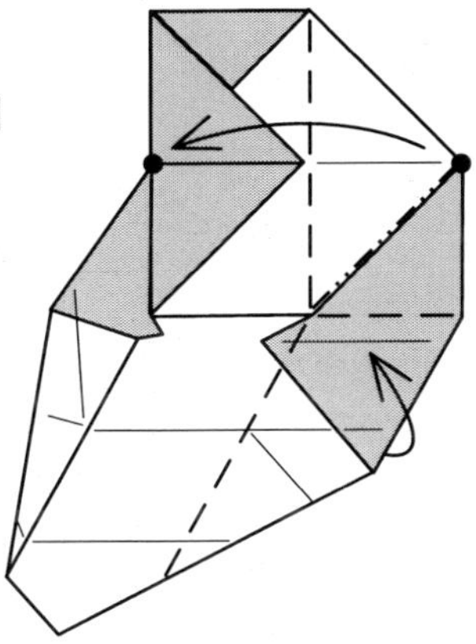

This is 3D.

12

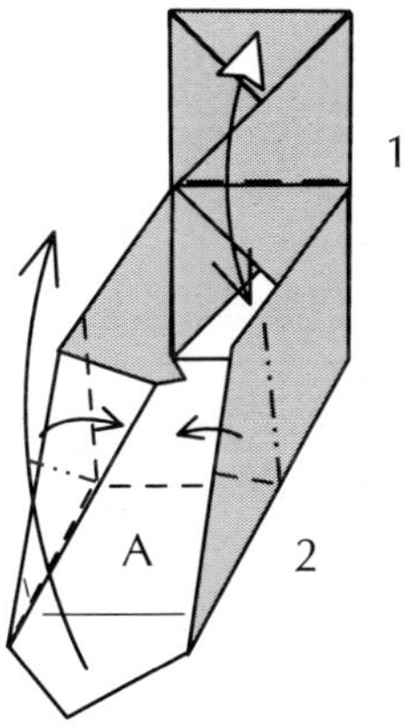

1. Fold and unfold.
2. Lift up at A and flatten
the layers behind A.

13

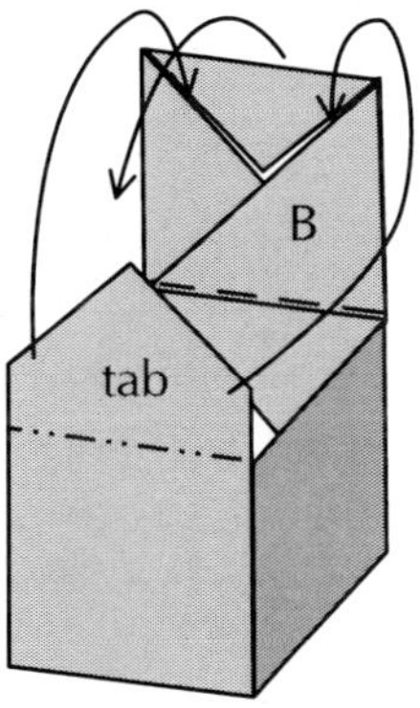

Tuck the tab into the pockets
of B. The top square of the cube
is B. Flatten the layers of B.

14

Cube

Cube of Triangles

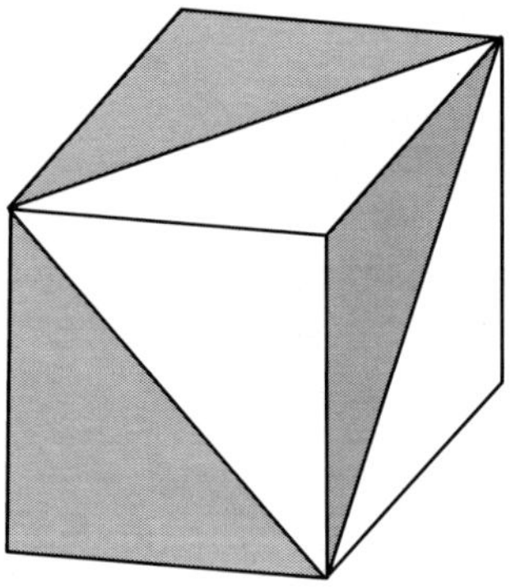

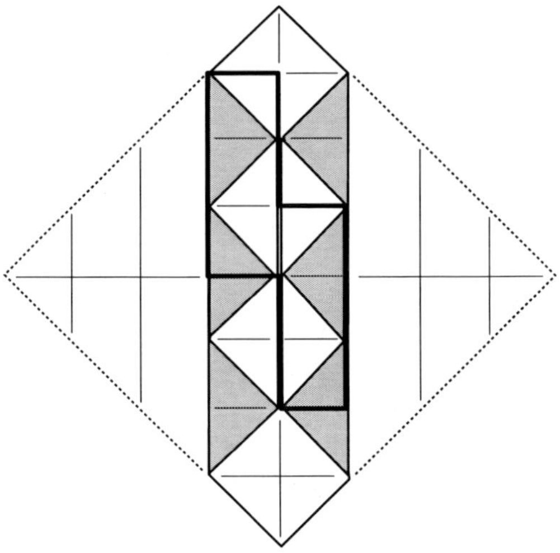

Each of the six faces of this cube show two triangles of different colors. Two bands of three squares are used in the layout. This is close to odd symmetry. Both this colorful cube and the solid cube are shown in 14 steps, adding magic to this cube.

1

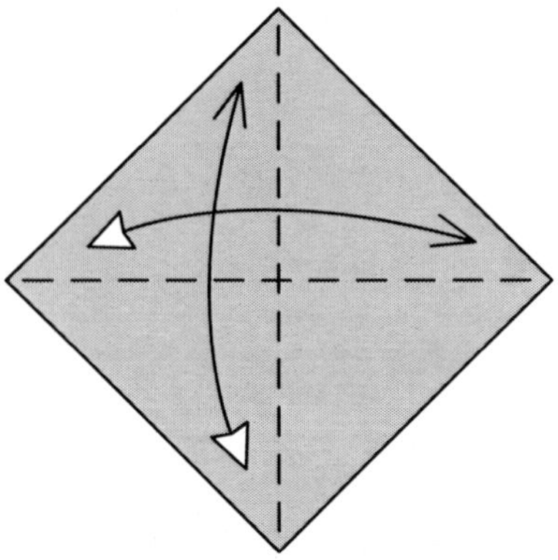

Fold and unfold.

2

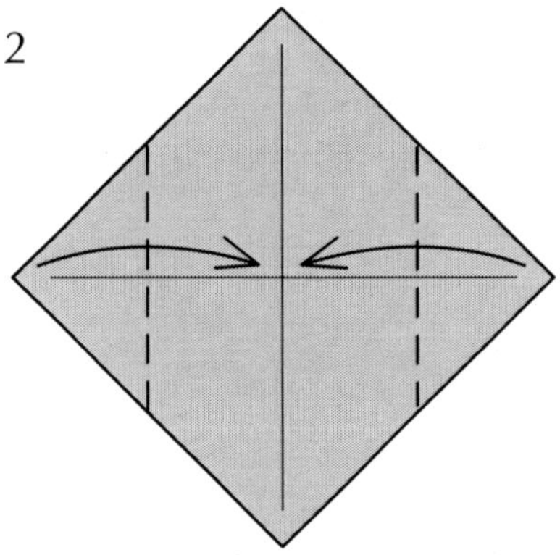

Fold to the center.

3

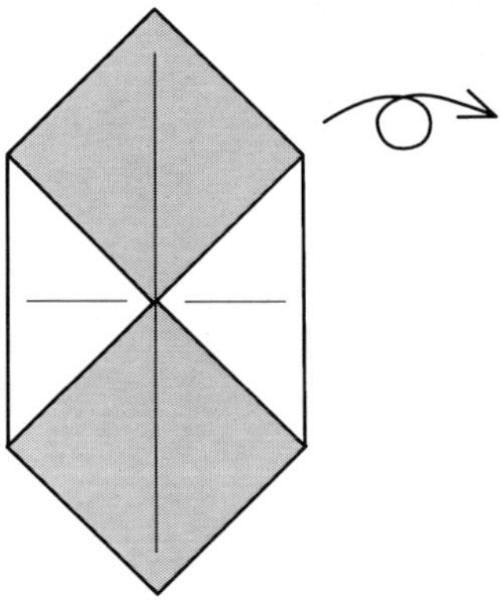

4

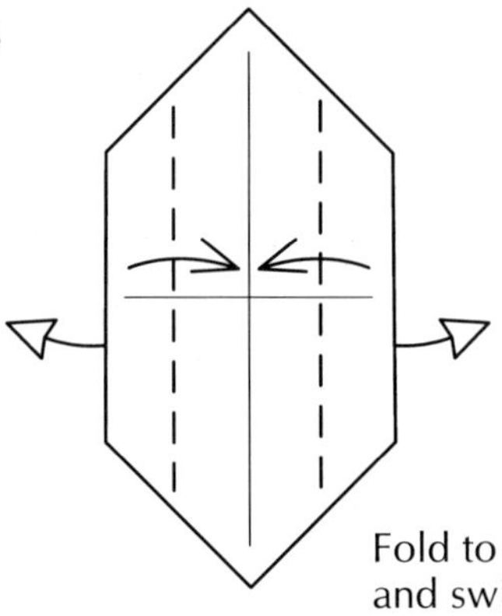

Fold to the center and swing out from behind.

5

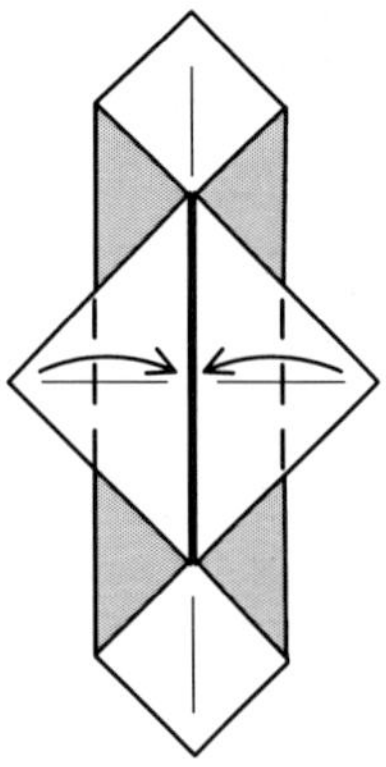

Fold to the center.

6

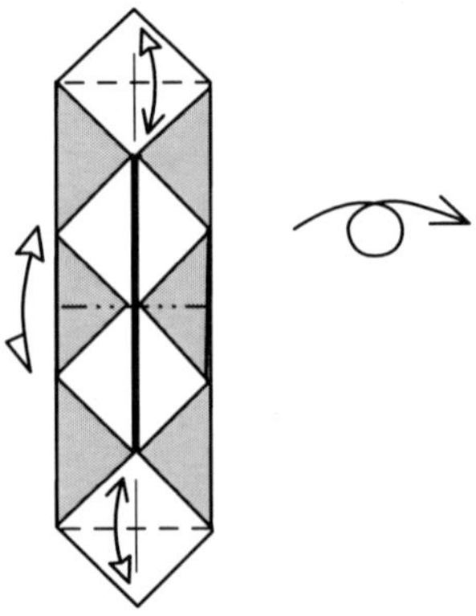

Fold and unfold.

7

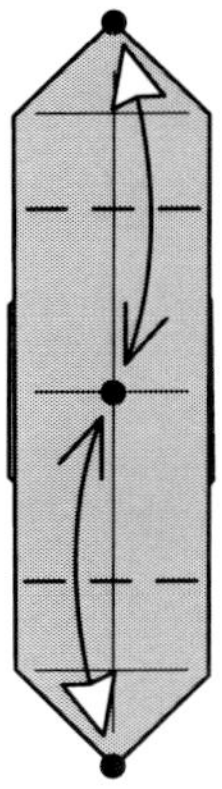

Fold and unfold.

8

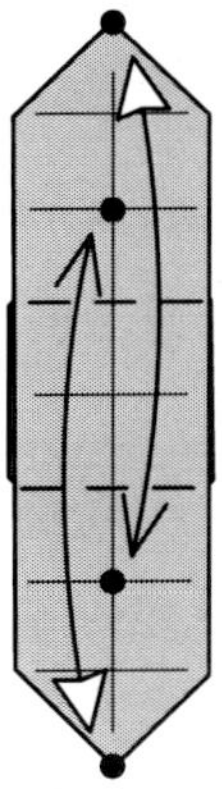

Fold and unfold.

9

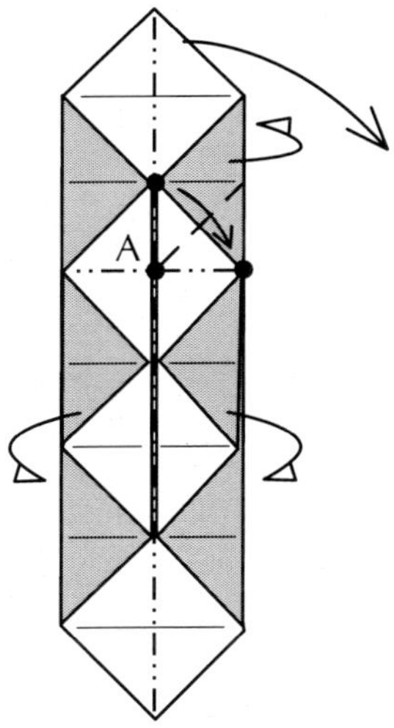

Puff out at dot A. The model will become 3D.

10

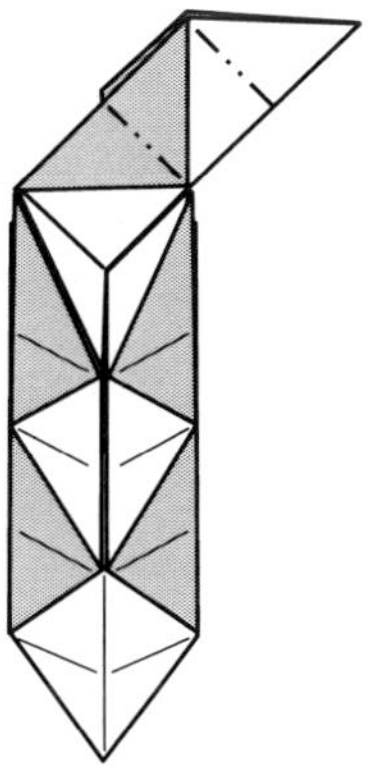

Fold and unfold along the creases. Rotate the top to the bottom.

11

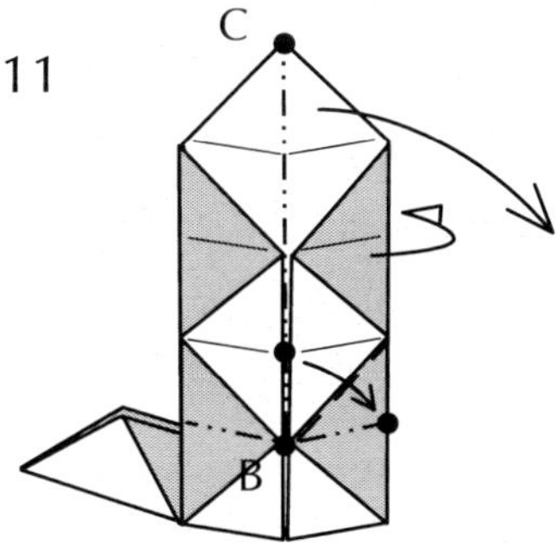

Puff out at dot B. Rotate so dot C goes to the top.

12

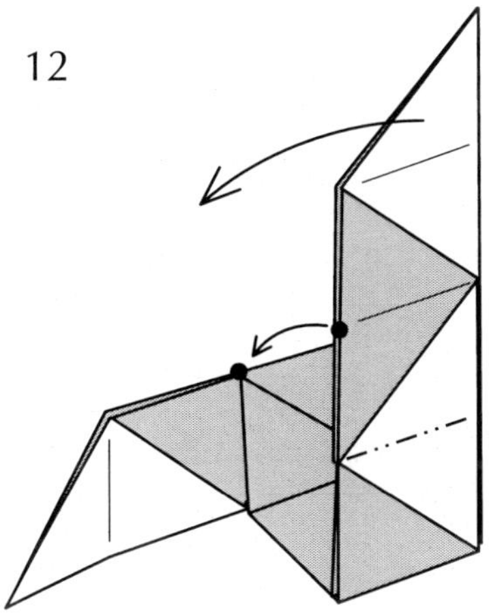

13

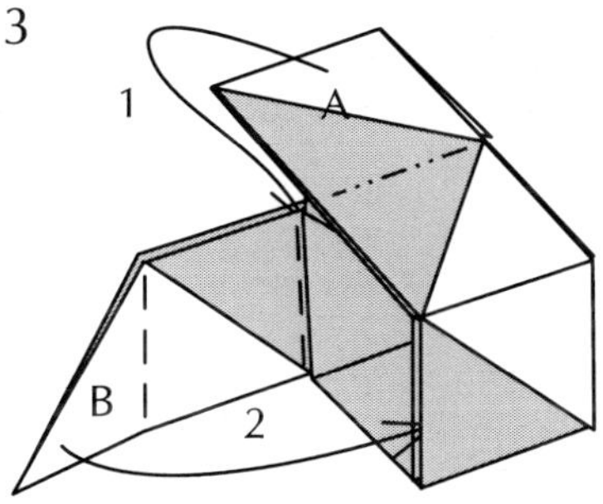

1. Tuck A into the pocket.
2. Tuck B into the pocket.

14

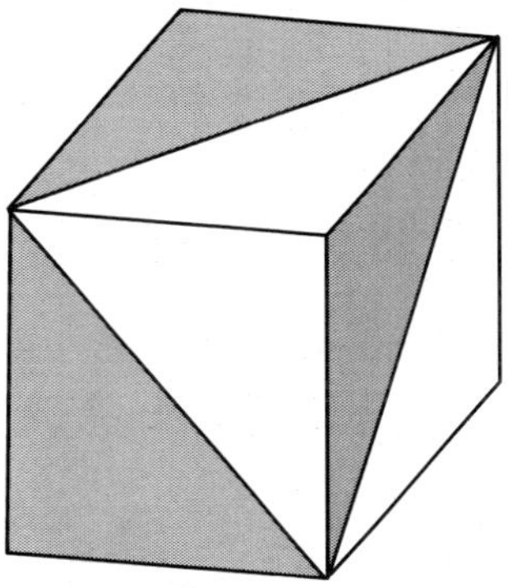

Cube of Triangles

Marble Cube

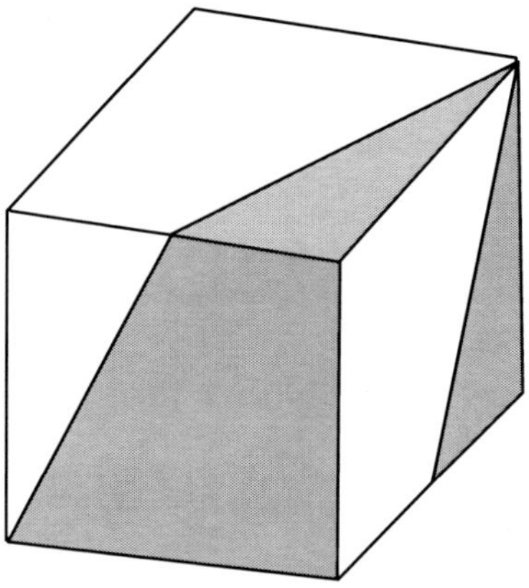

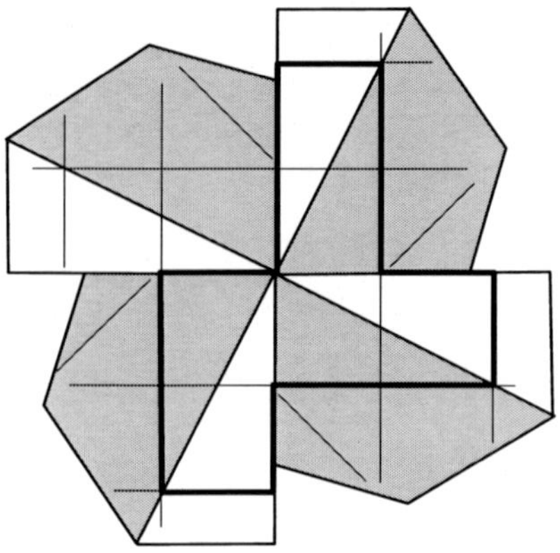

A pinwheel is formed to show both sides of the paper, with just the correct angles so a diagonal covers two squares, starting from the center of the paper. The model uses 3/4 square symmetry. The figure above, from step 10, shows the layout.

The Golden Mean, represented by the Greek letter phi (Φ) is the number 1.618034. It appears in Greek art and nature. It is an interesting number because its reciprocal (1/Φ) is the same as Φ - 1, or .618034. For this marble cube (from a 1 x 1 square), to make the color change line up in the right place, the number Φ - 1 is the key landmark. This landmark is found in step 3.

1

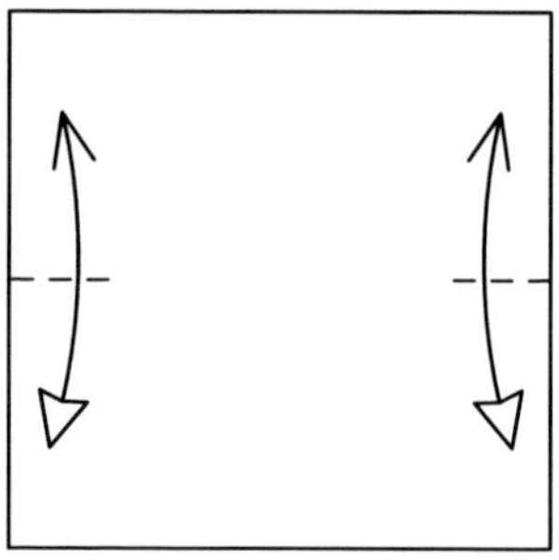

Fold and unfold on the left and right.

2

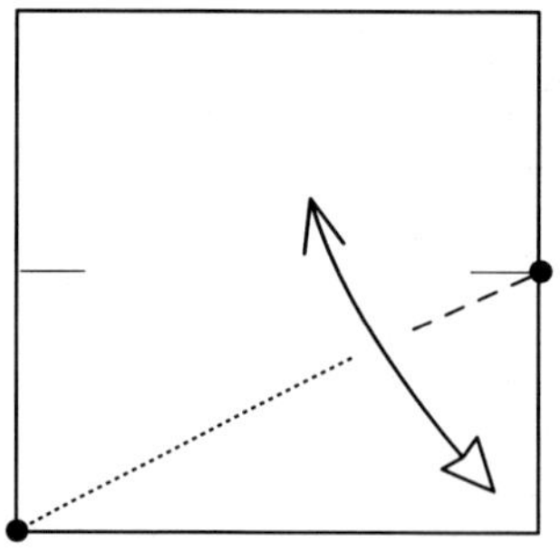

Fold and unfold on the right.

3

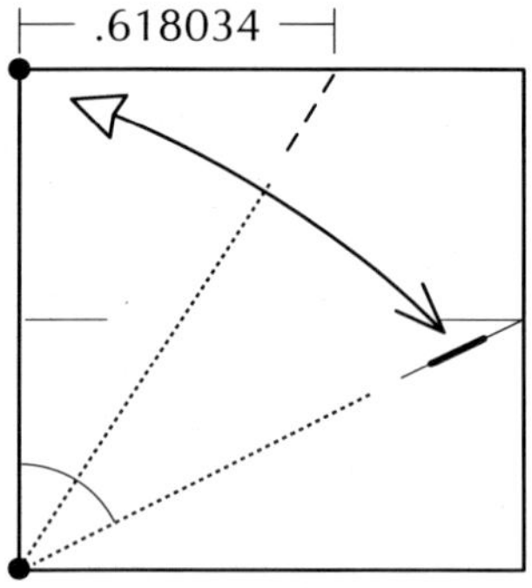

Bisect the angle. Fold and unfold on the top. Rotate 180°.

4

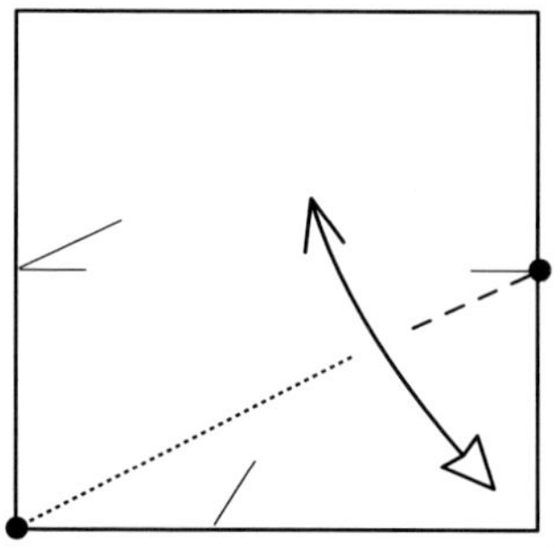

Repeat steps 2–3.

5

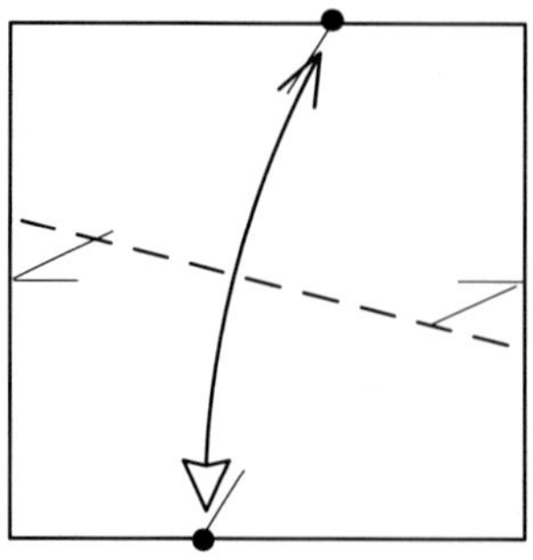

Fold and unfold. Rotate 90°.

6

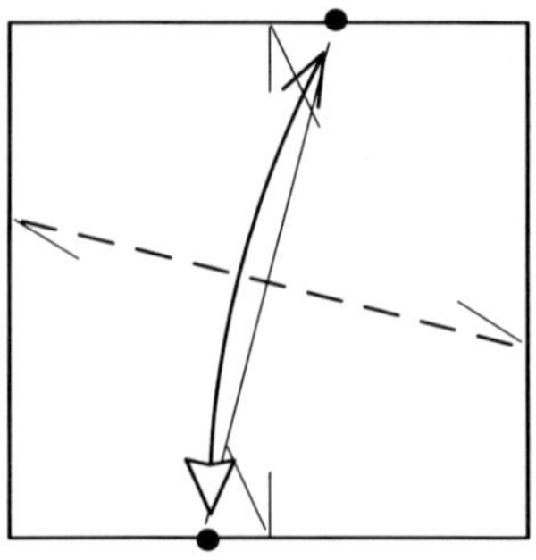

Fold and unfold.

7

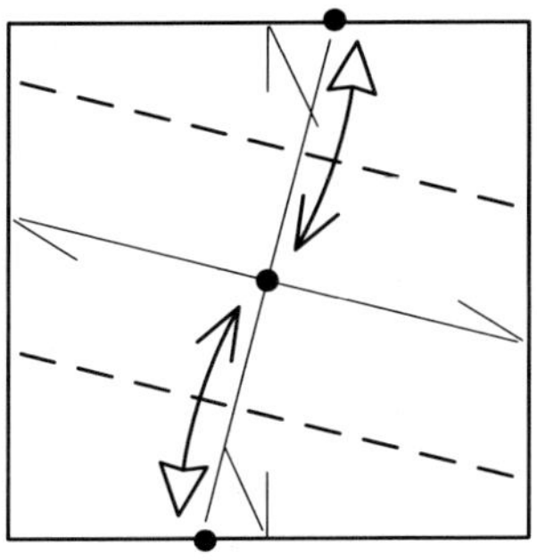

Fold and unfold.
Rotate 90°.

8

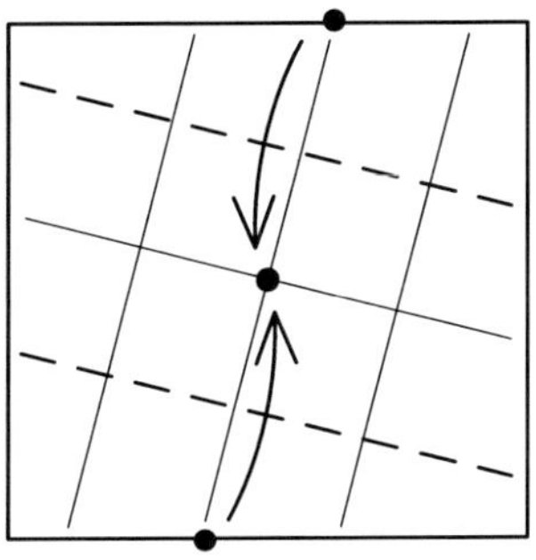

9

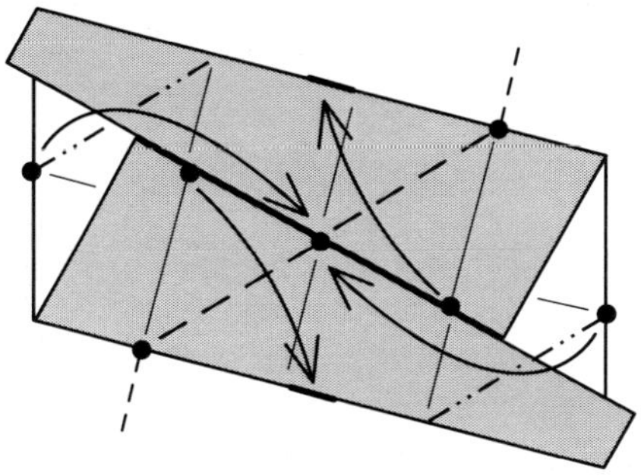

This is a combination of squash folds. The dots on the left and right will meet in the center.

10

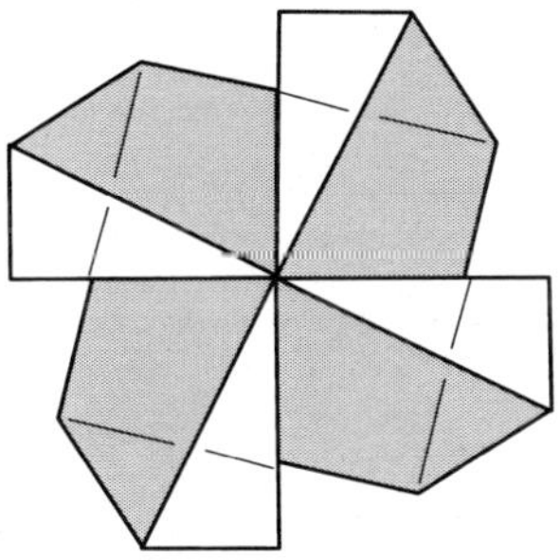

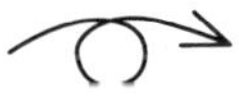

11

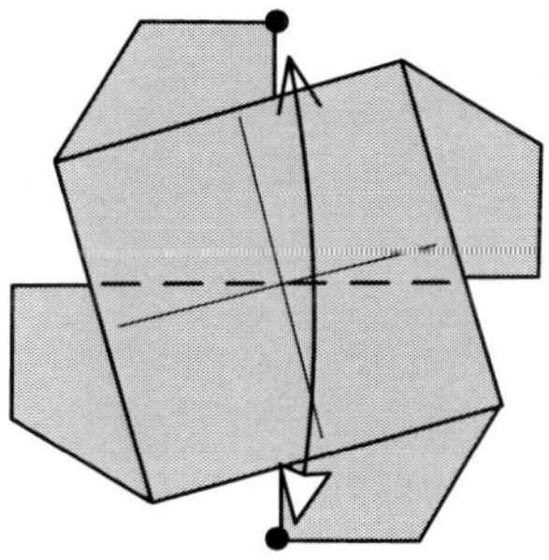

Fold and unfold.
Rotate 90°.

12

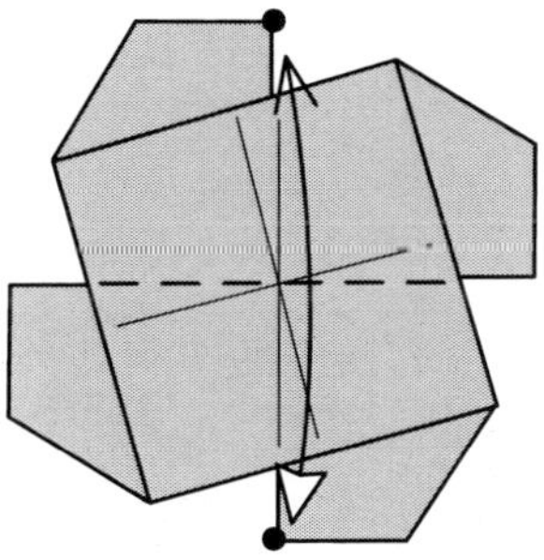

Fold and unfold.

13

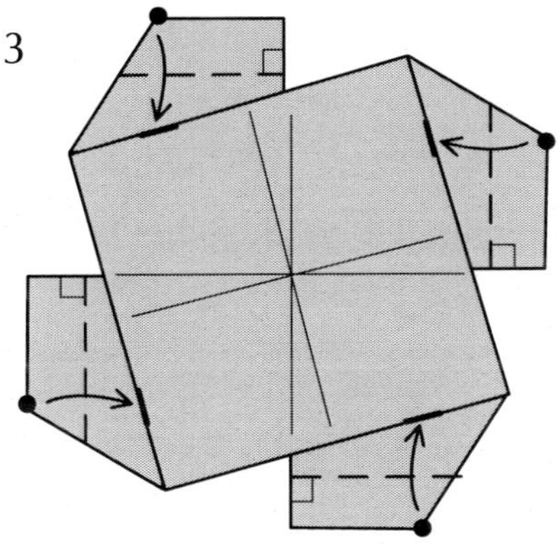

Bring the dots to the lines.

14

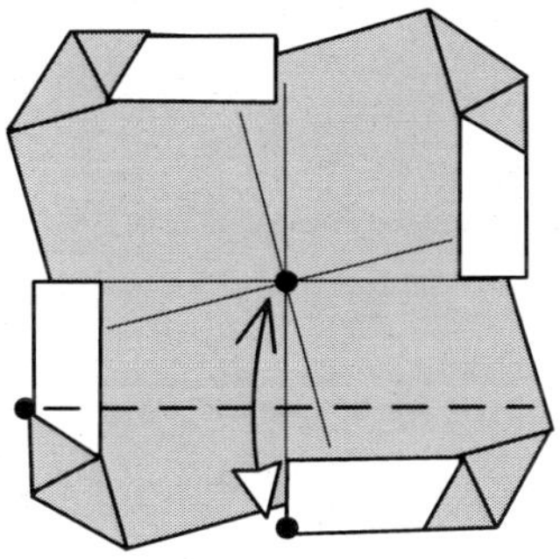

Fold and unfold.
Rotate 90°.

15

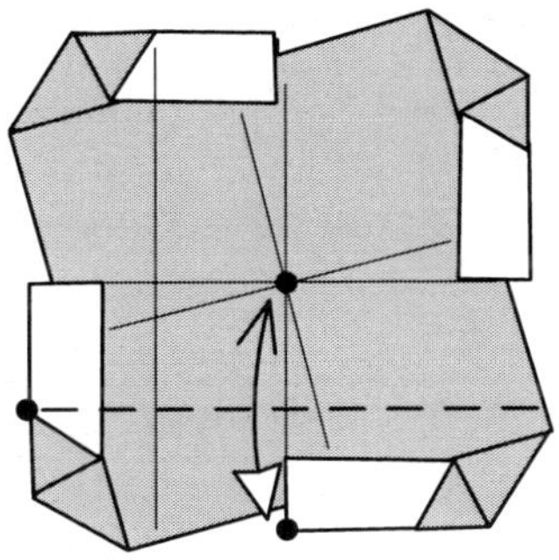

Repeat step 14 three times.

16

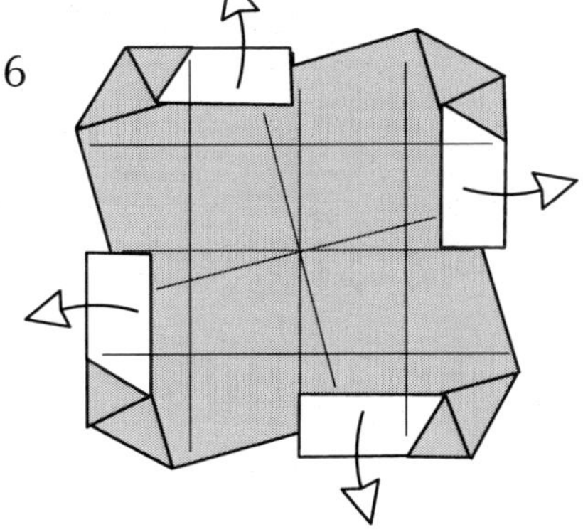

Unfold.

17

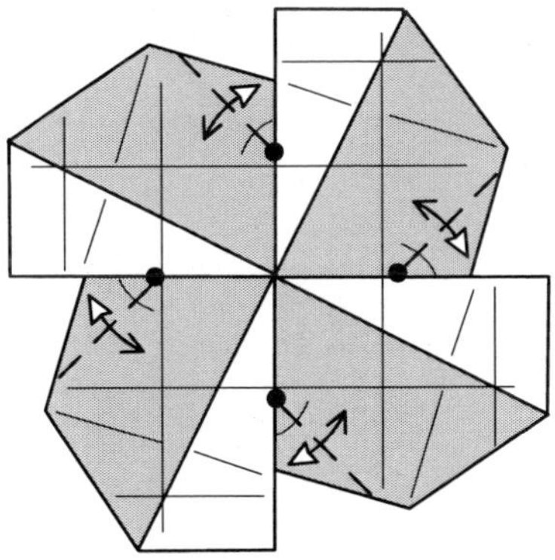

Fold and unfold.

18

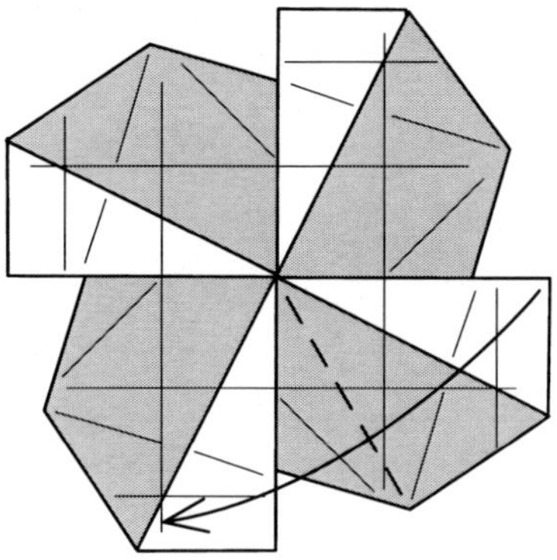

19

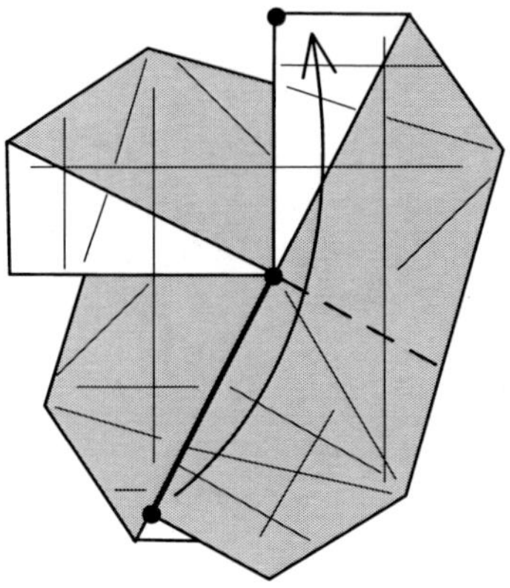

Puff out at the center dot. The other two dots will meet at the top. Rotate the center dot to the bottom.

20

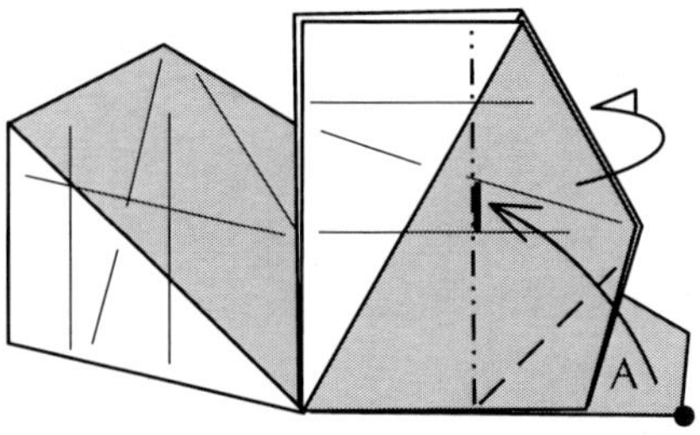

Fold behind along the creases and flatten inside. The dot on flap A will meet the bold line.

21

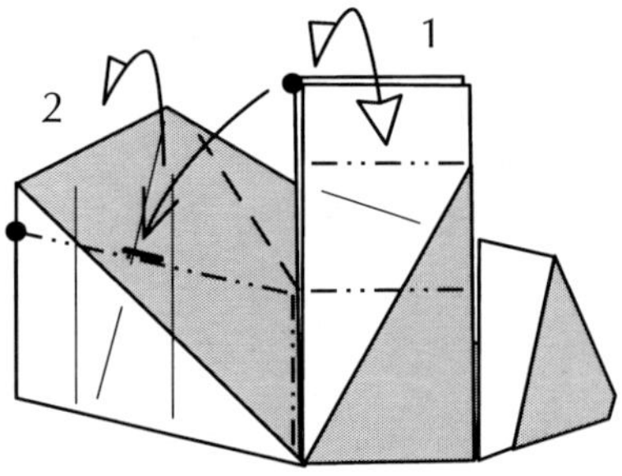

1. Fold and unfold.
2. Repeat step 20 on the left. Rotate so the dot on the left goes to the top.

22

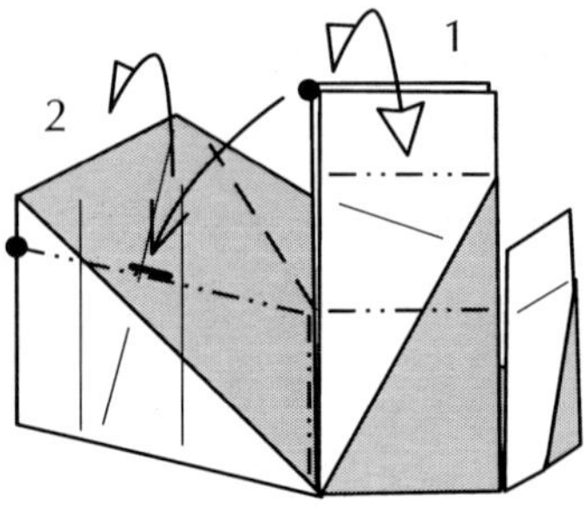

Repeat step 21.

23

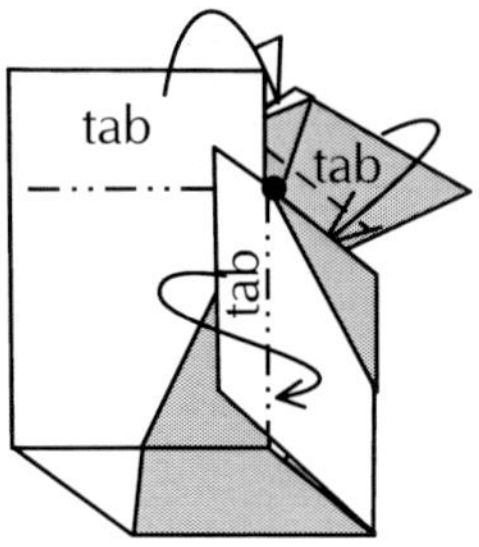

Fold each tab inside at a 90° angle. The tabs meet at the dot and the model closes with a three-way twist lock.

24

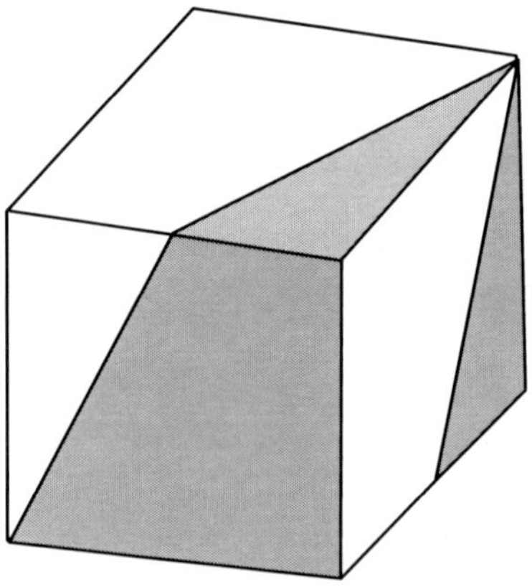

Marble Cube

Plump Dipyramids

Three-dimensional diamonds show the geometric side of origami. Each diamond is composed of identical pyramids joined at the base. The faces are identical isosceles triangles with varying apex angles (angles at the top). For these plump diamonds, the apex angle is 180/(n - 1), where n is the number of sides of the polygon base. This collection ranges from triangular to nonagonal diamonds.

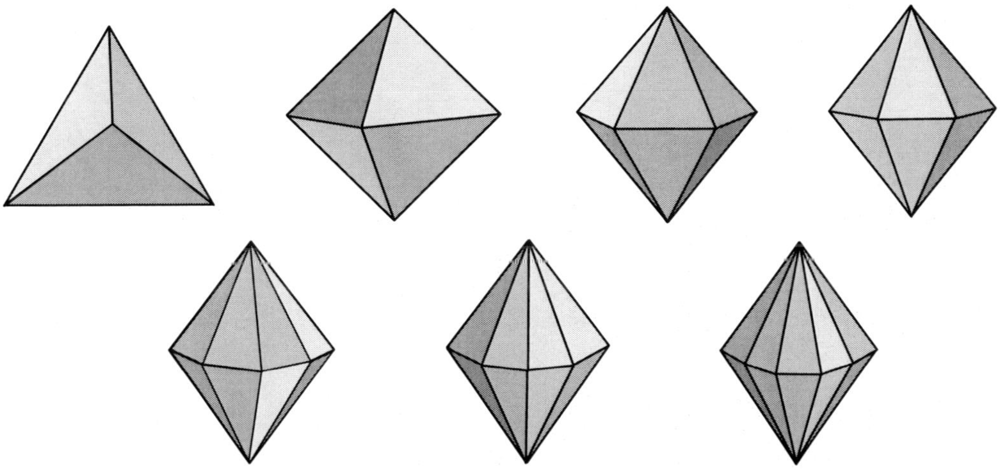

Diamond	Apex Angle
Triangular	90°
Square	60°
Pentagonal	45°
Hexagonal	36°
Heptagonal	30°
Octagonal	25.7°
Nonagonal	22.5°

Triangular Dipyramid 90°

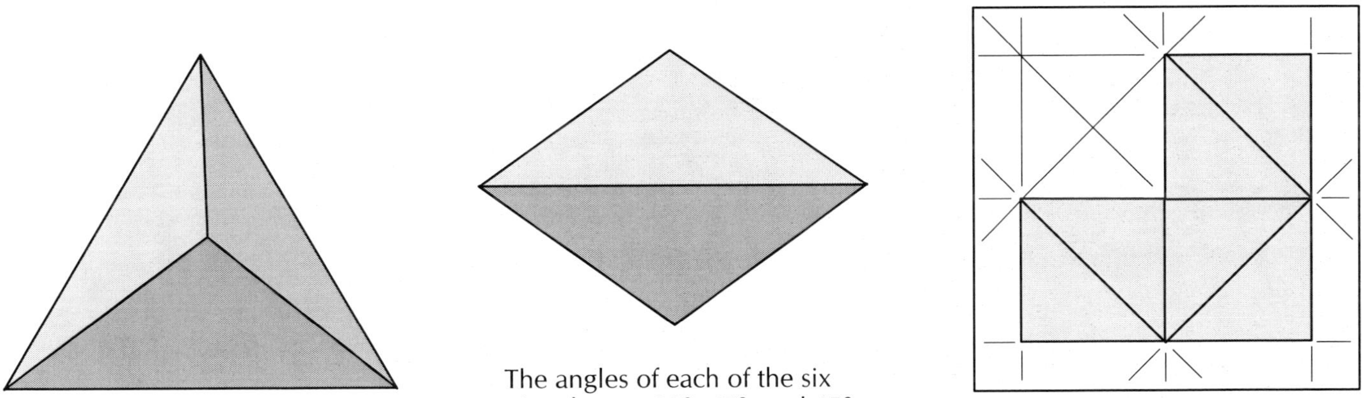

The angles of each of the six triangles are 90°, 45°, and 45°. The layout show 3/4 symmetry.

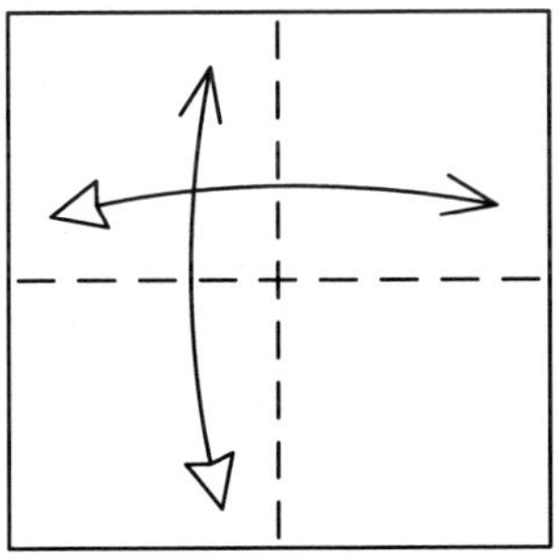

1 Fold and unfold.

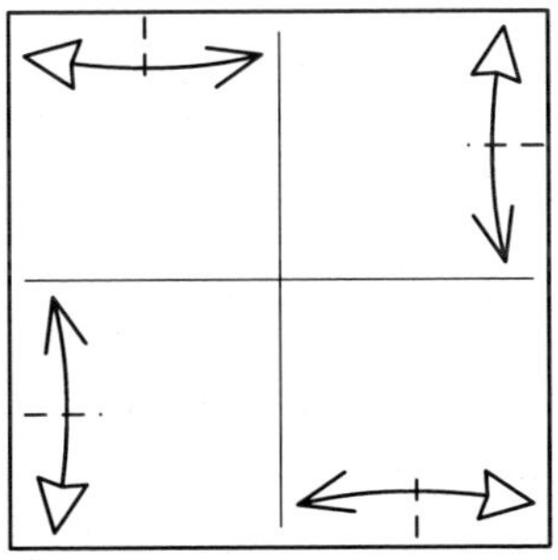

2 Fold and unfold to find the quarter mark on each side.

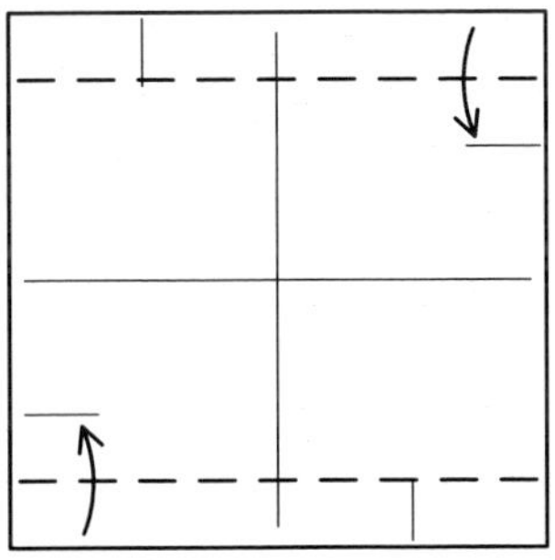

3 Rotate 90°.

4

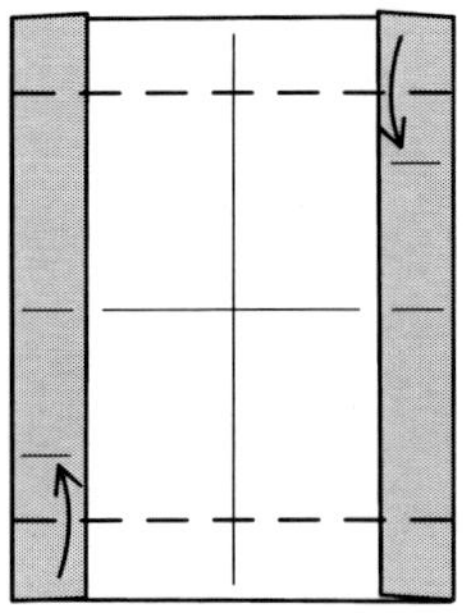

5

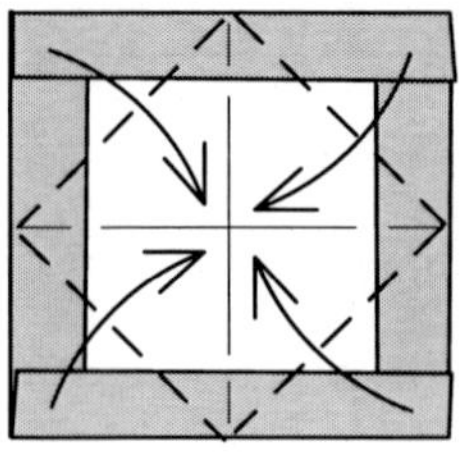

Fold to the center.

6

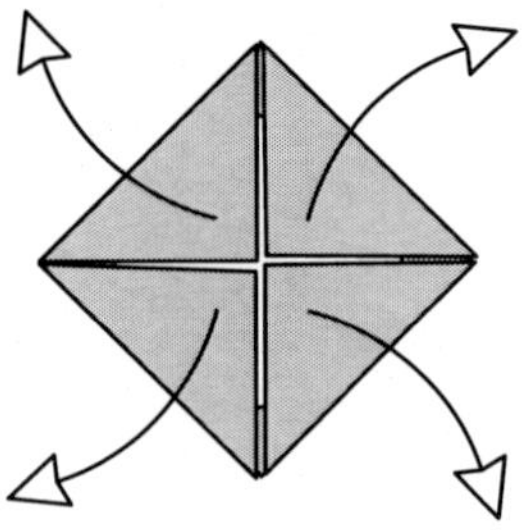

Unfold everything.

7

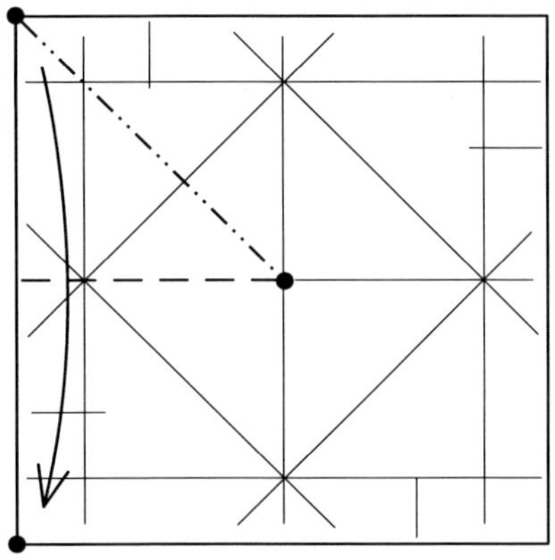

Push in at the center dot. The other two dots will meet.

8

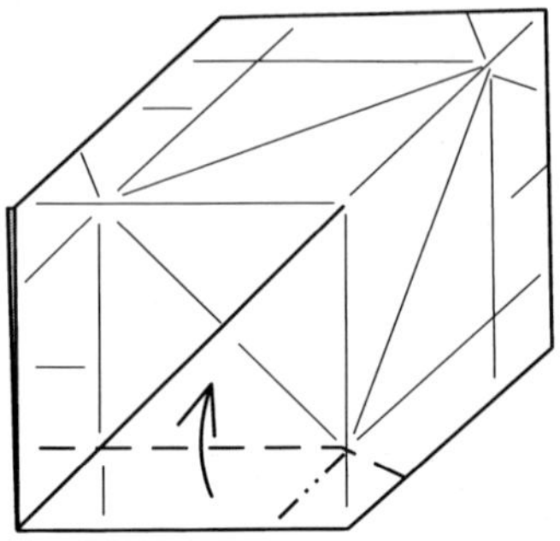

Fold along the creases.

9

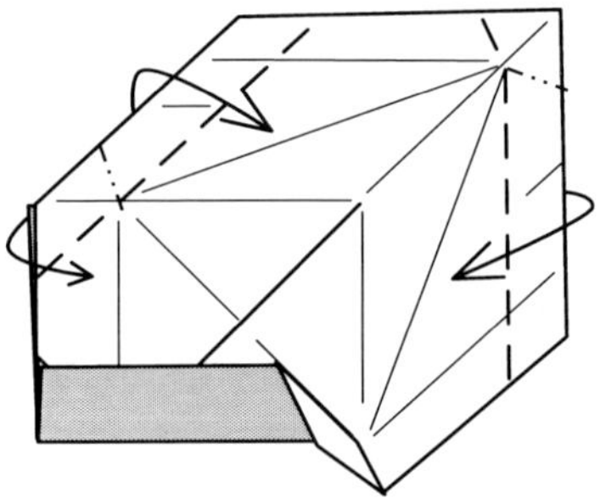

Fold along the creases.

10

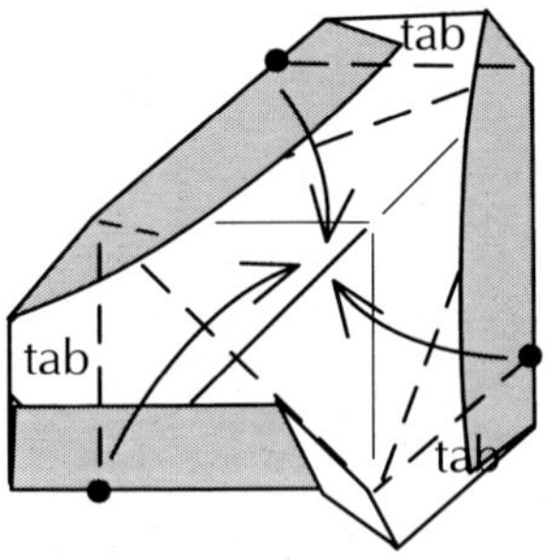

Bring the dots together. The tabs will lock the model.

11

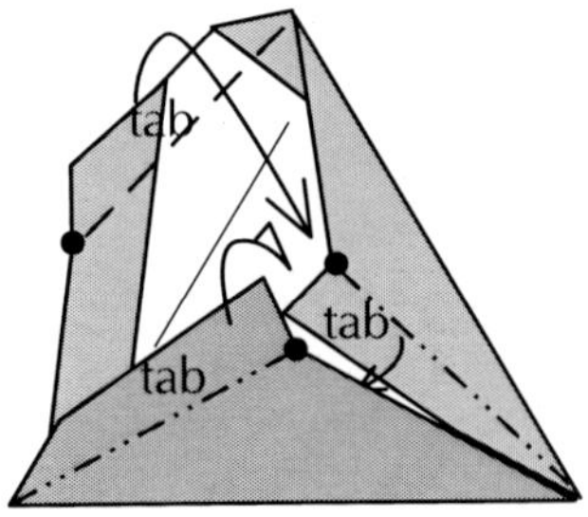

The fold is in progress. Bring the three dots together and close the model with a three-way twist lock.

12

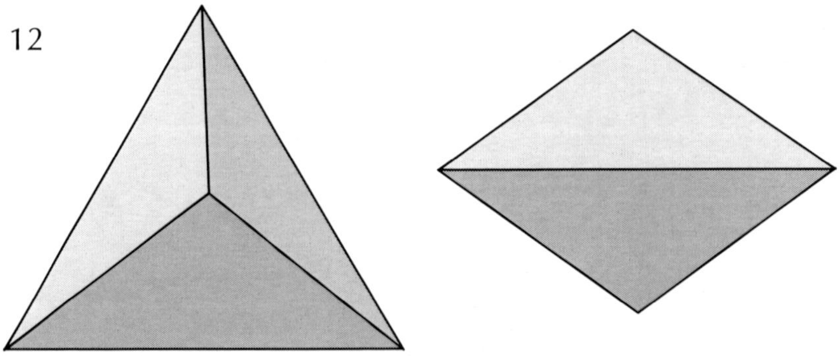

Triangular Dipyramid 90°

Octahedron

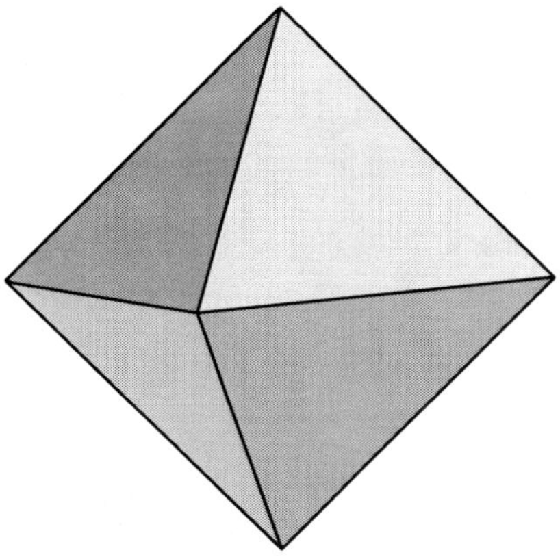

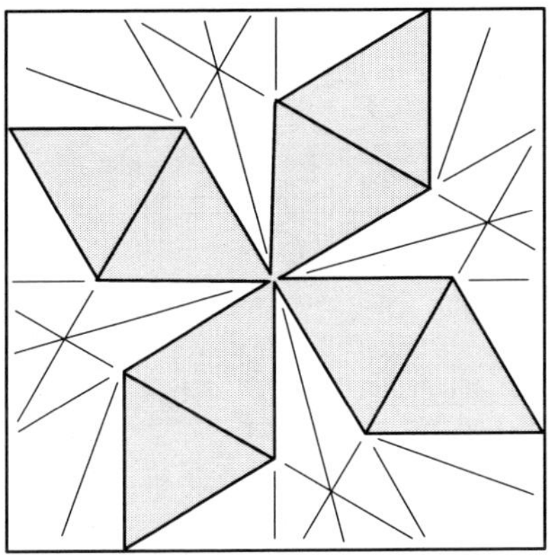

The octahedron is composed of eight equilateral triangles. It can be used as a base to create more polyhedra. According to Plato, the octahedron represented air because it appears to be suspended.

This model uses square symmetry and closes with a four-way twist lock.

1

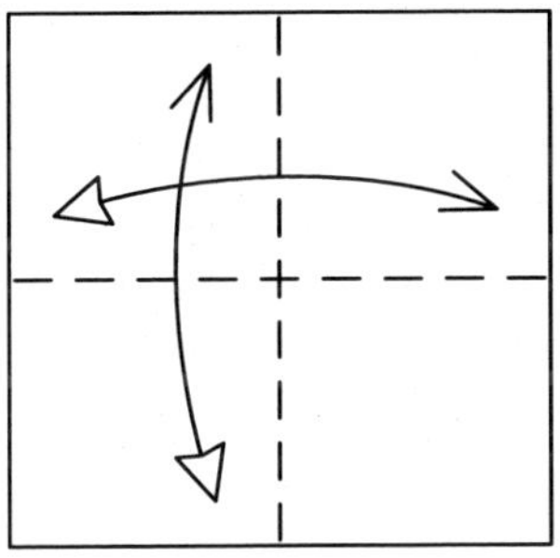

Fold and unfold.

2

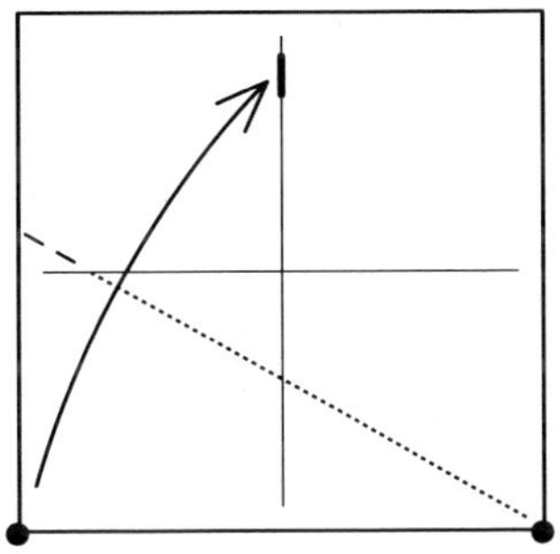

Bring the left dot to the line. Crease on the left.

3

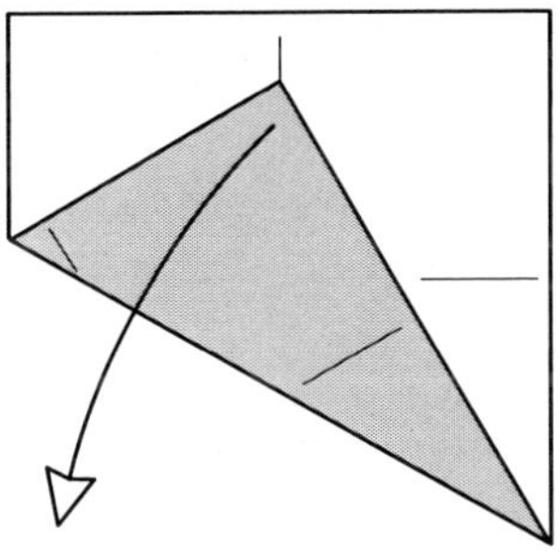

Unfold and rotate 90°.

4

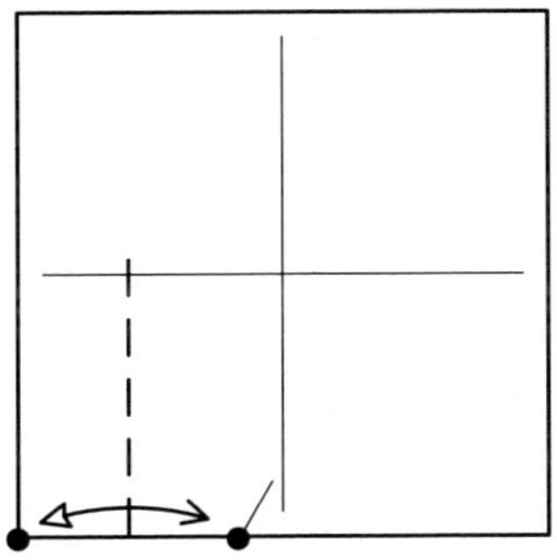

Fold and unfold.

5

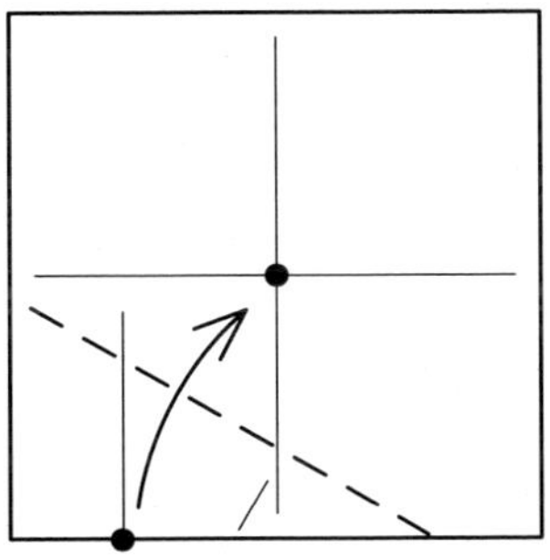

6

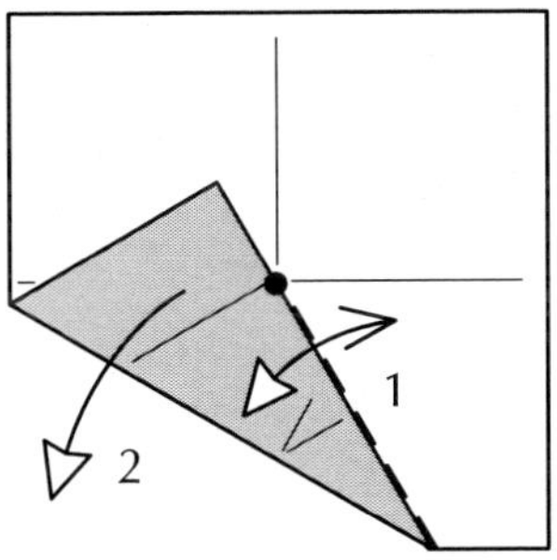

1. Fold and unfold along the edge.
2. Unfold.

7

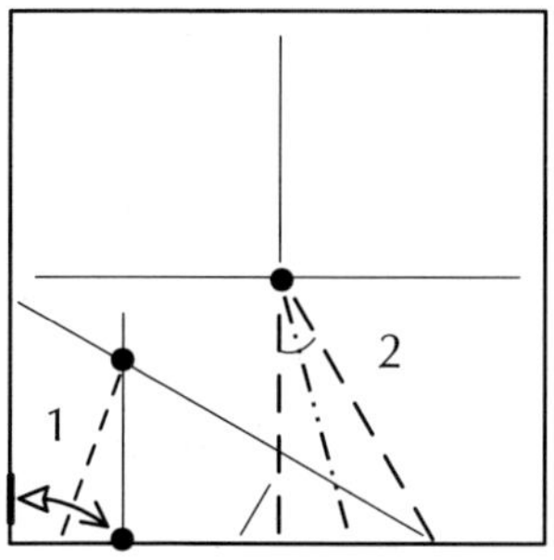

1. Bring the edge to the dot and unfold.
2. Fold and unfold to bisect the angle. It is easier to turn the paper over.

8

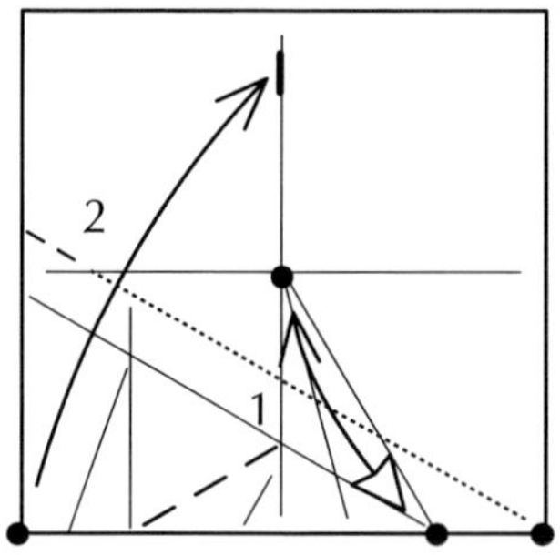

1. Fold and unfold.
2. Repeat steps 2–8 three times.

9

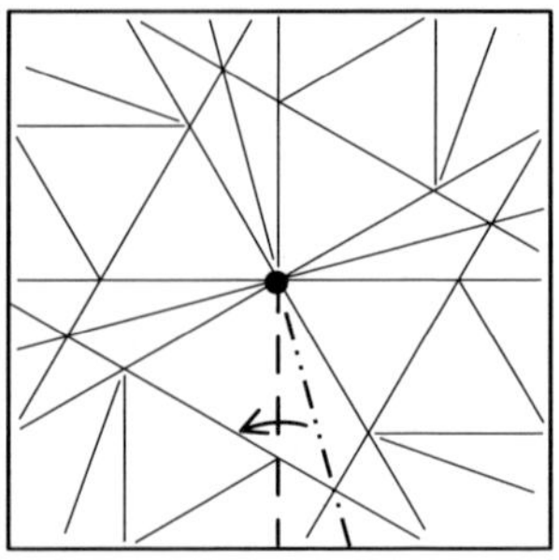

Fold along the creases and push in at the dot.

10

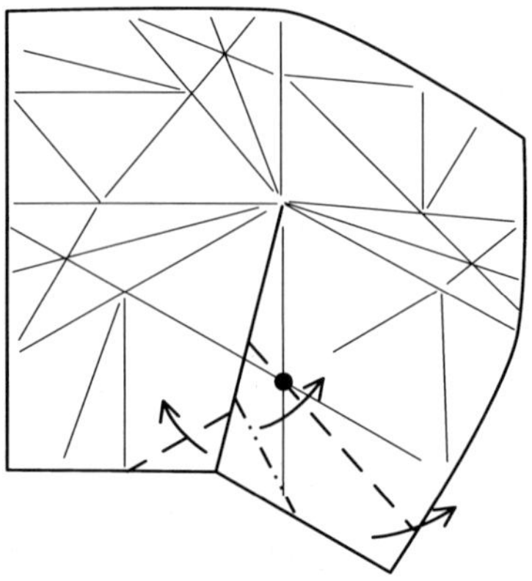

Squash-fold and flatten. Valley-fold along the creases. Rotate 90°.

11

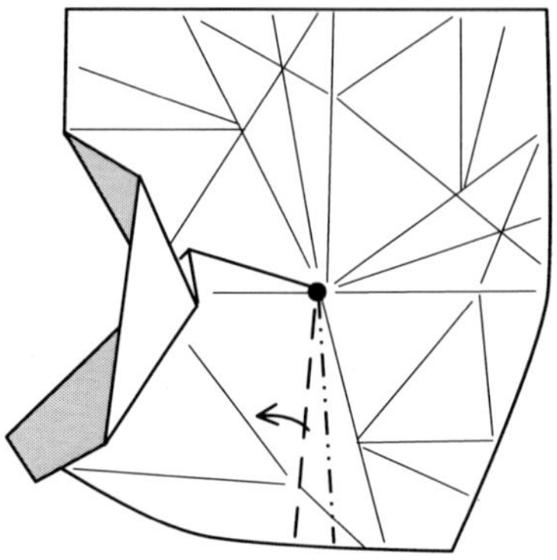

Repeat steps 9–10 three times. Rotate to view the outside.

12

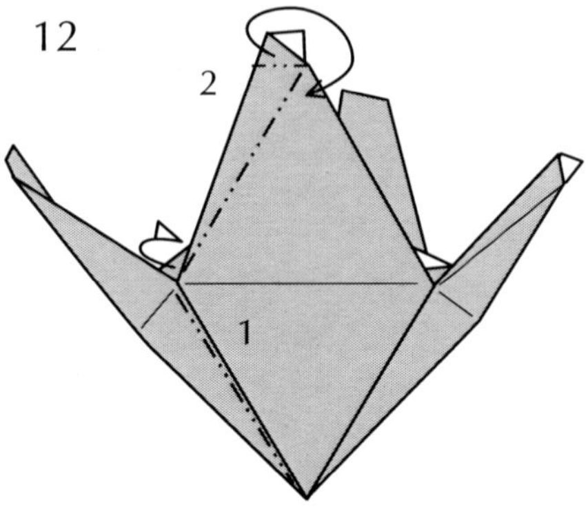

1. Fold and unfold.
2. Wrap all around.
Repeat all around.

13

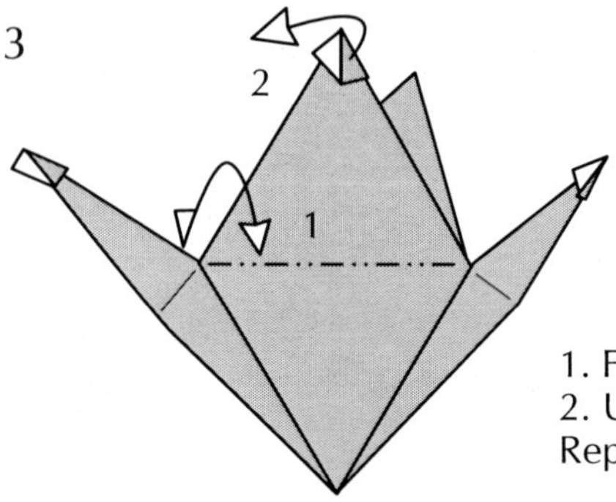

1. Fold and unfold.
2. Unfold.
Repeat all around.

14

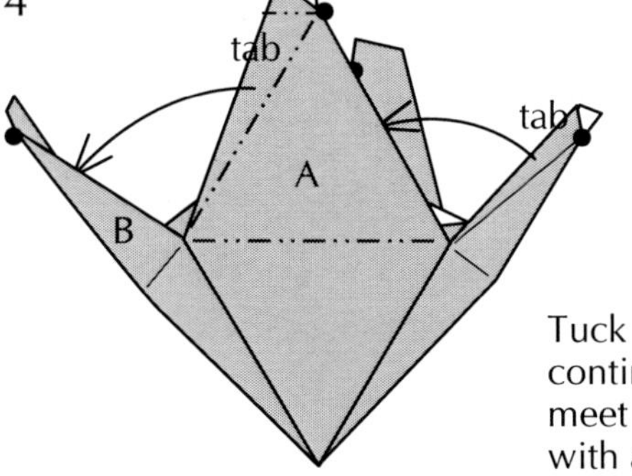

Tuck the tabs under A and B and continue all around. The dots will meet at the top. The model closes with a four-way twist lock.

15

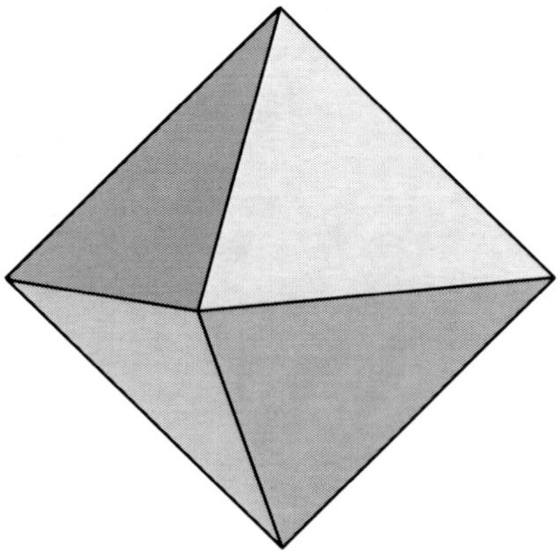

Octahedron

Pentagonal Dipyramid 45°

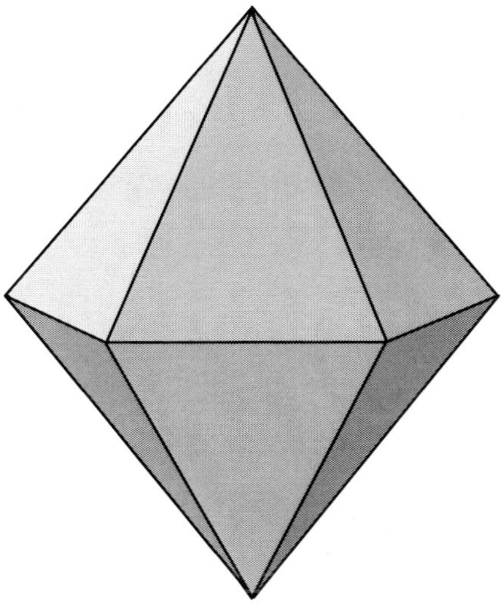

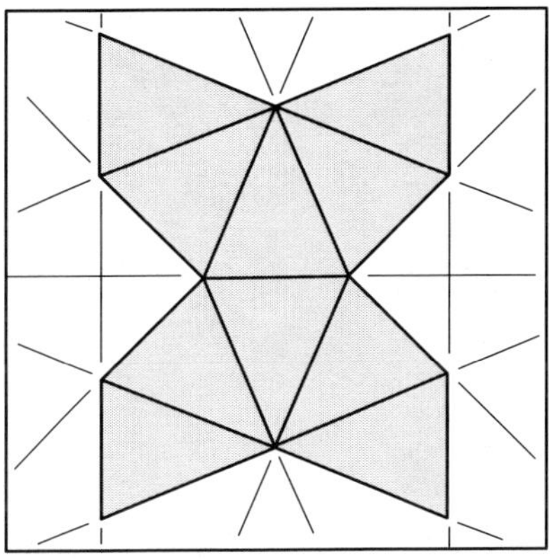

The angles of each of the ten triangles are 45°, 67.5°, and 67.5°. The model uses odd symmetry.

1

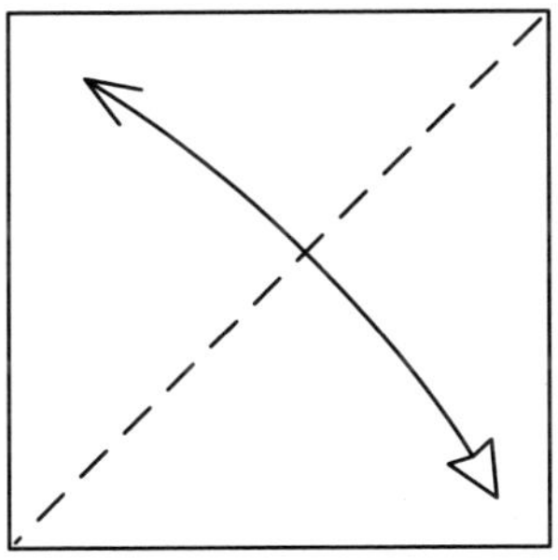

Fold and unfold.

2

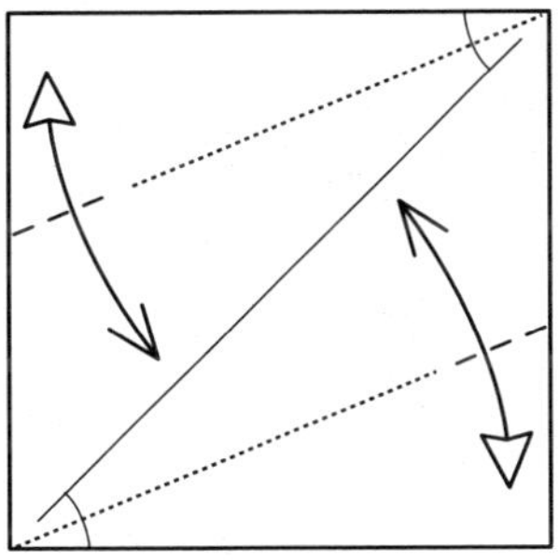

Fold and unfold along the edges.

3

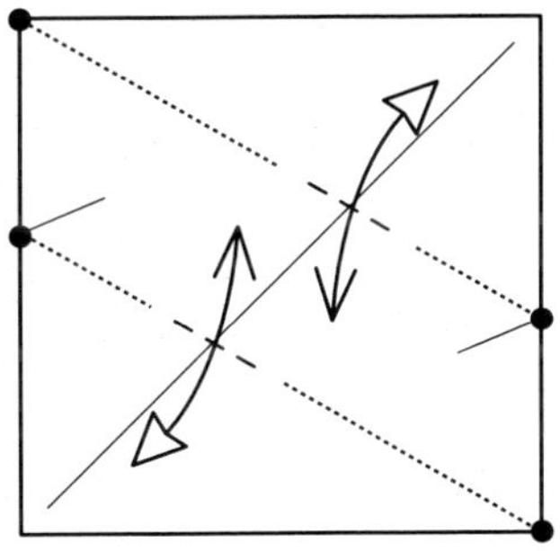

Fold and unfold on the diagonal.

4

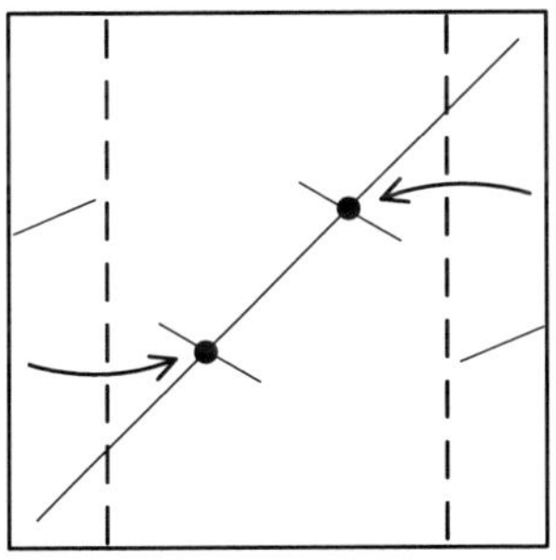

5

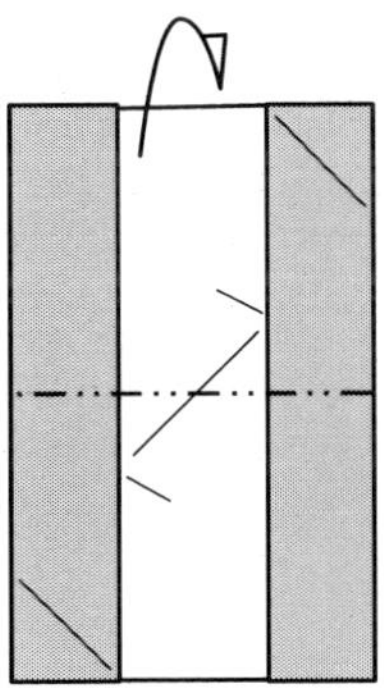

Fold in half.

6

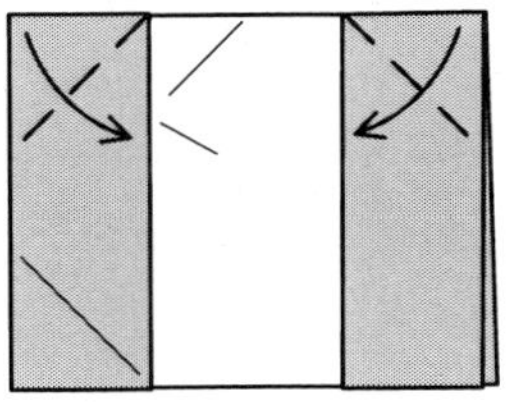

7

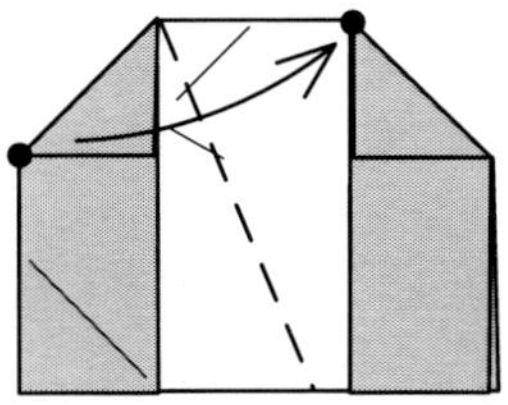

8

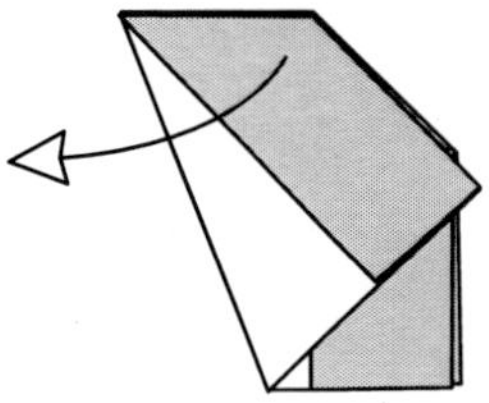

Unfold.

9

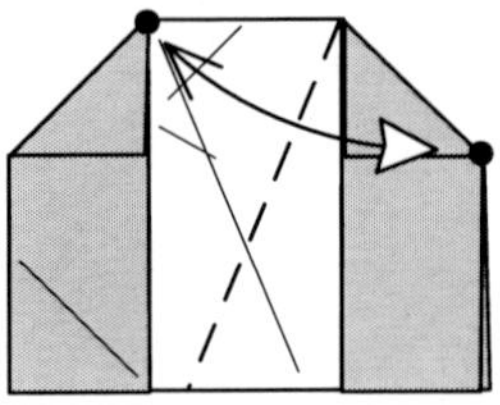

Fold and unfold.

10

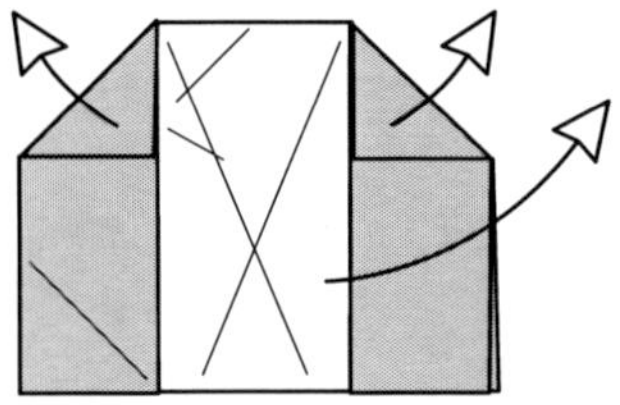

Unfold everything.

11

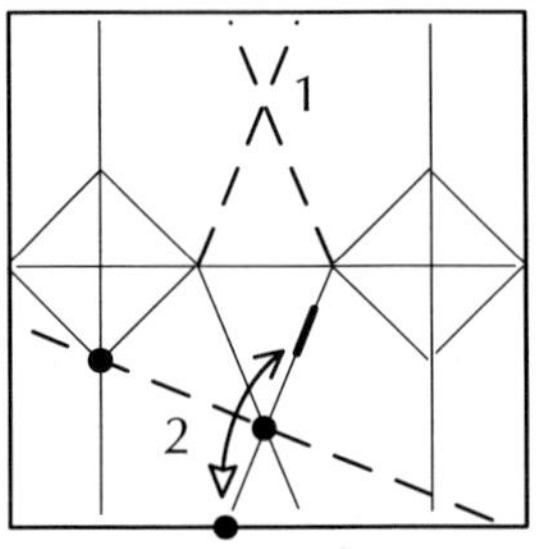

1. Fold and unfold along the creases.
2. Fold and unfold.

12

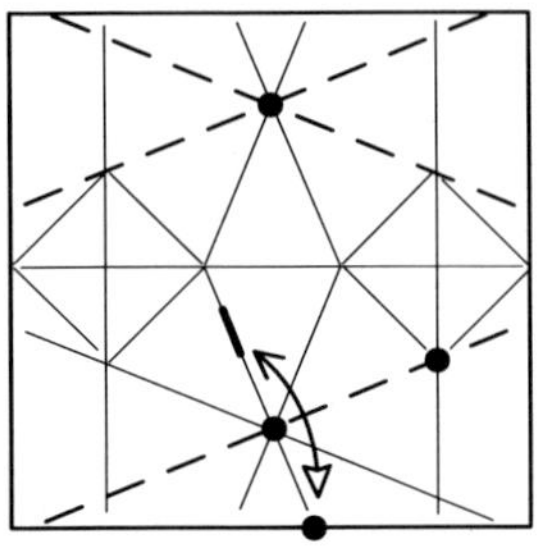

Fold and unfold three times.

13

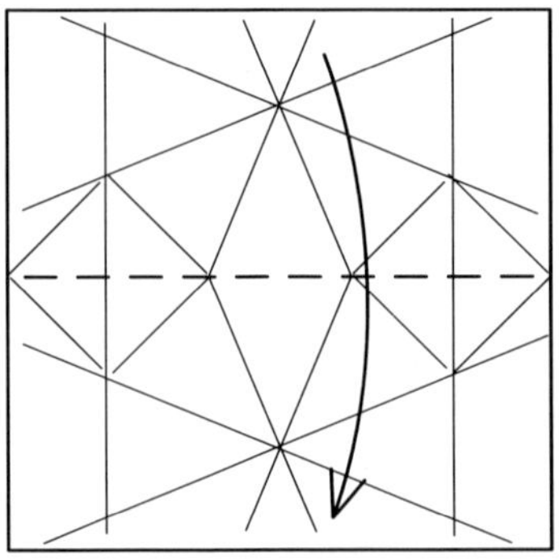

14

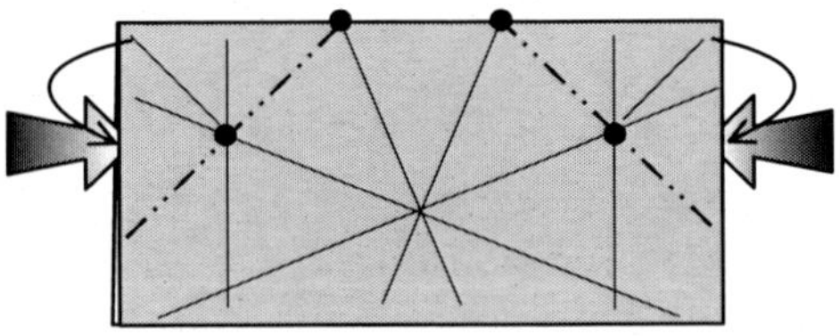

Make reverse folds along the creases by the dots.

15

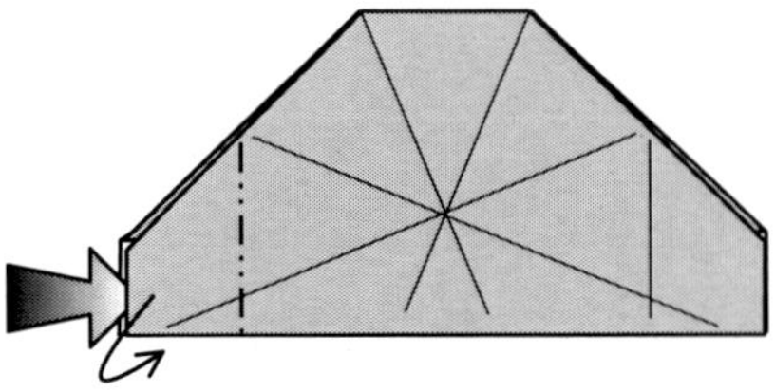

Reverse-fold. Turn over and repeat.

16

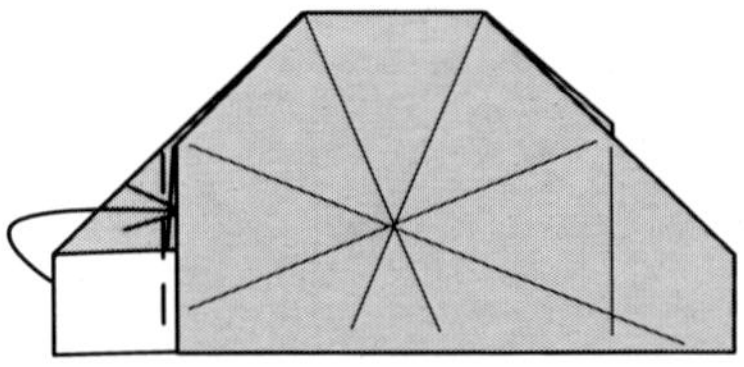

Tuck inside. Turn over and repeat.

17

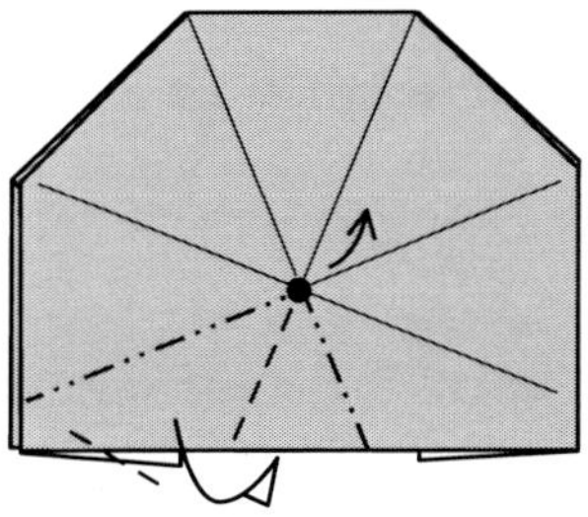

Puff out at the dot and fold along the creases. Turn over and repeat.

18

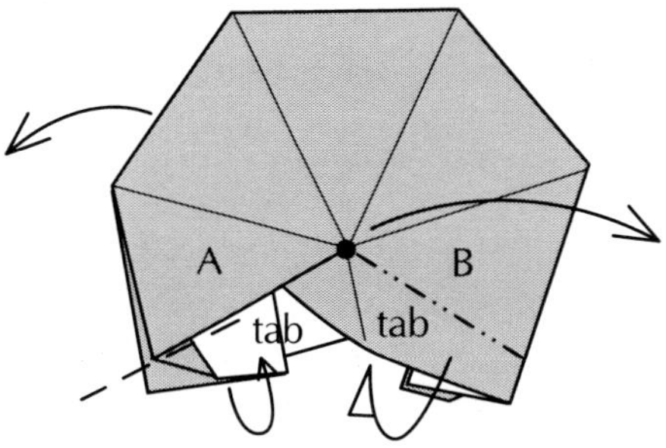

Bring the dot to the right and the one behind to the left. Follow regions A and B into the next step. Fold all the layers of each tab.

19

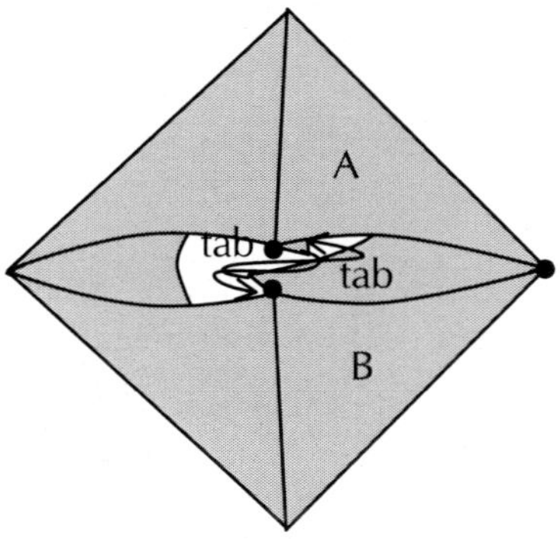

Tuck the two tabs into each other so the center dots meet.

20

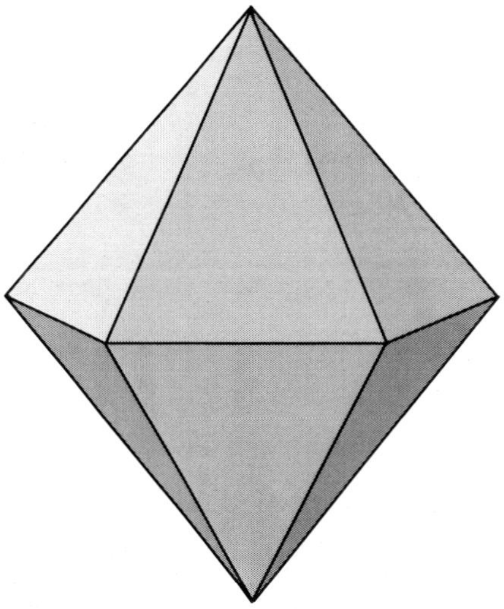

Pentagonal Dipyramid 45°

Hexagonal Dipyramid 36°

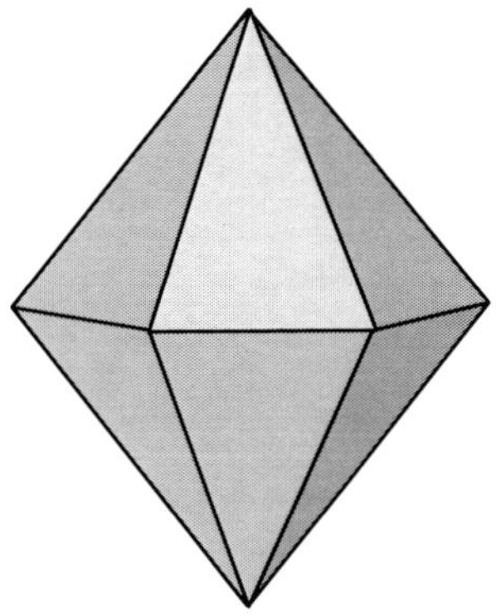

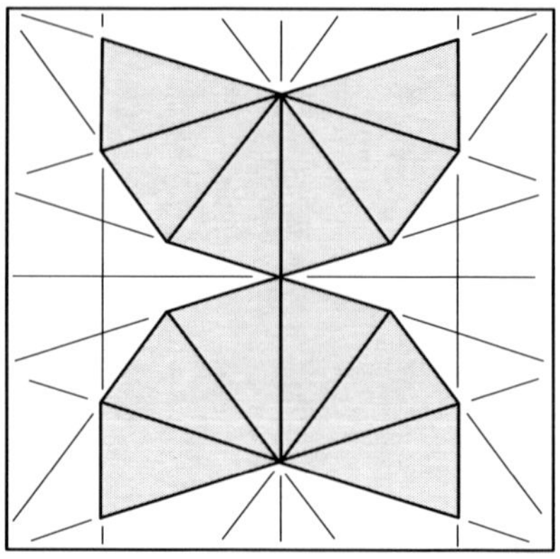

The angles of each of the twelve triangles are 36°, 72°, and 72°. Odd symmetry is used.

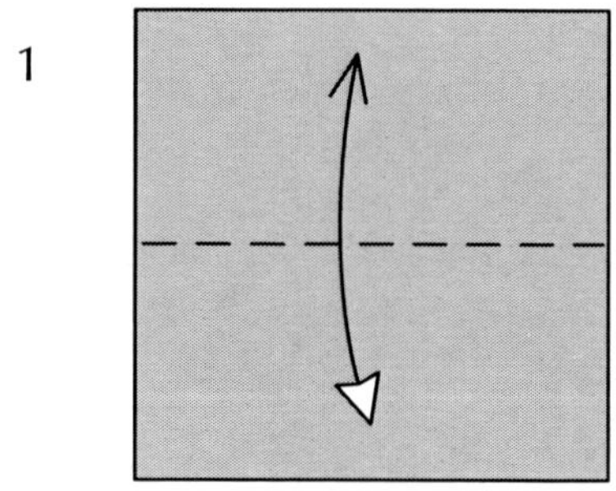

Fold and unfold.

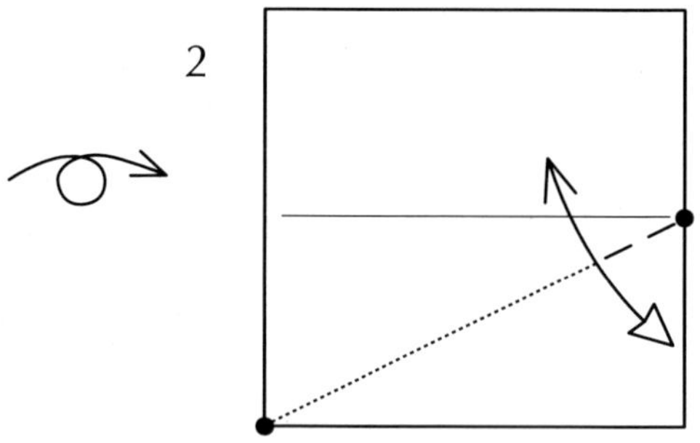

Fold and unfold by the right.

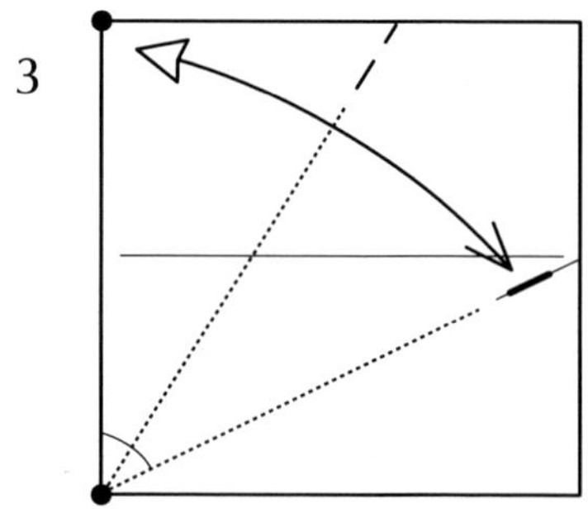

Fold and unfold on the top to bisect the angle.

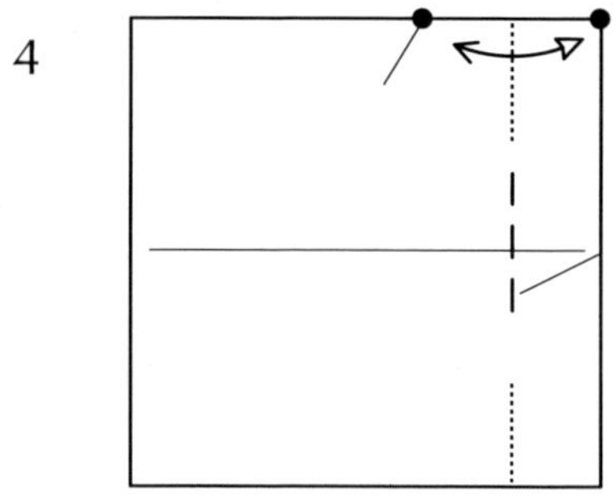

Fold and unfold in the center.

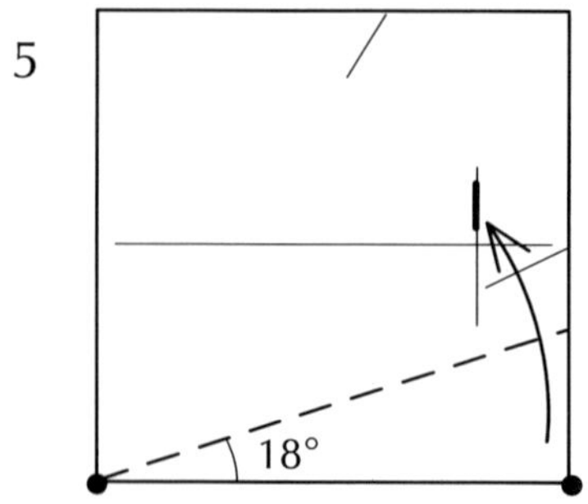

Bring the corner to the crease. The 18° angle is exact.

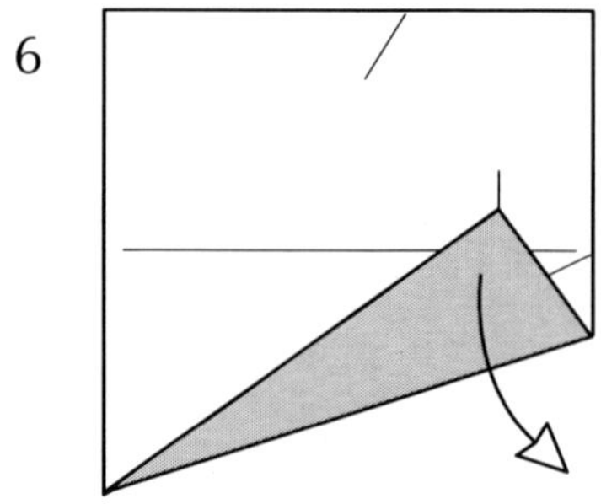

Unfold.

7

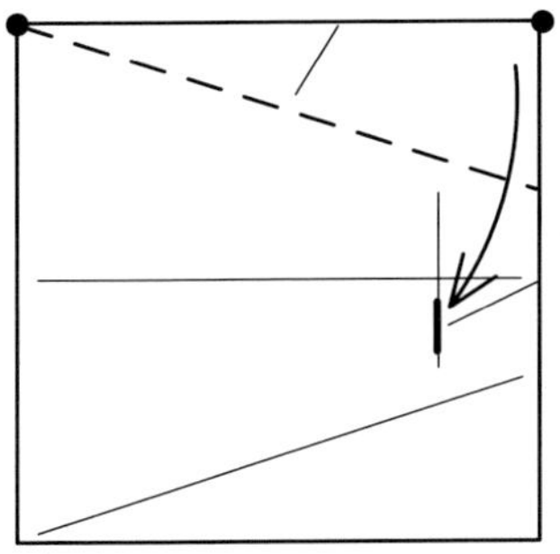

Repeat steps 5–6
at the top.

8

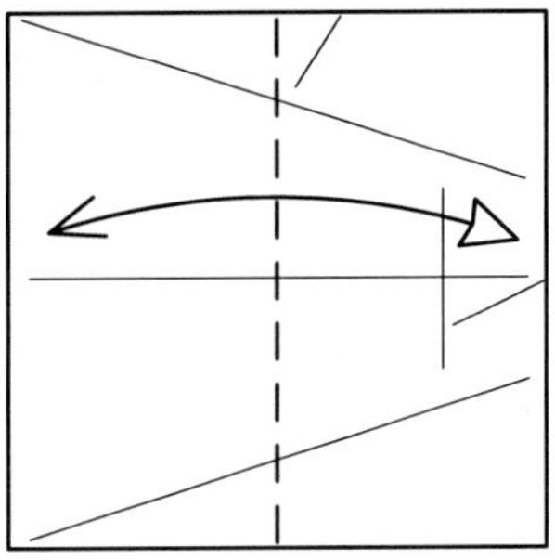

Fold and unfold.

9

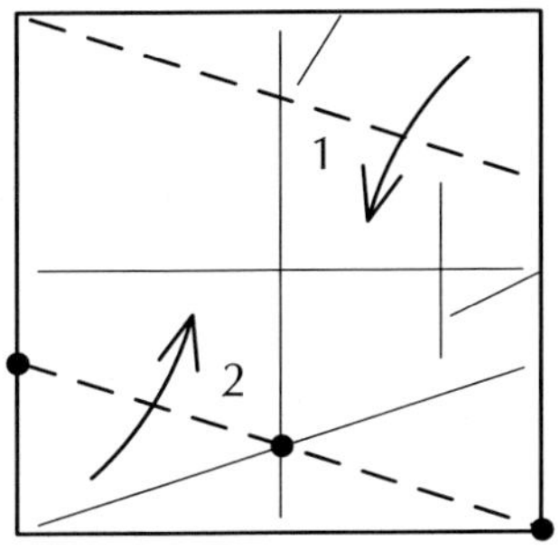

1. Fold along the crease.
2. Valley-fold.

10

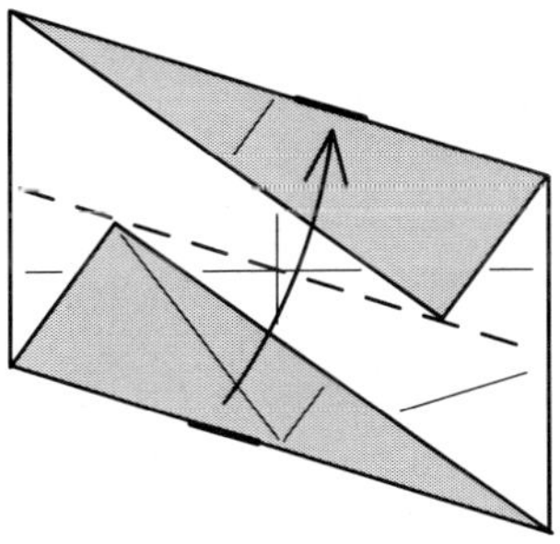

11

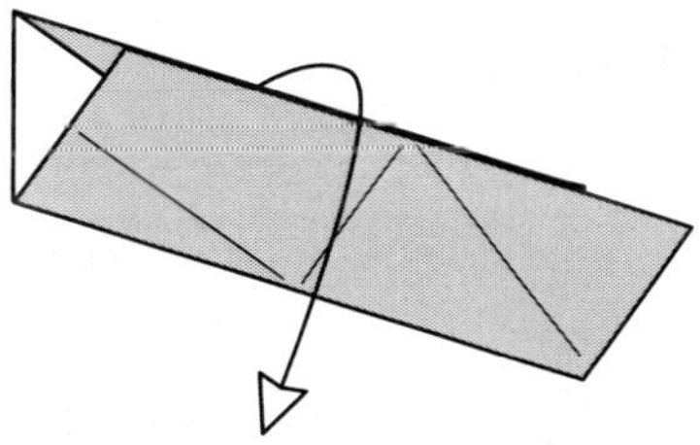

Unfold everything.

12

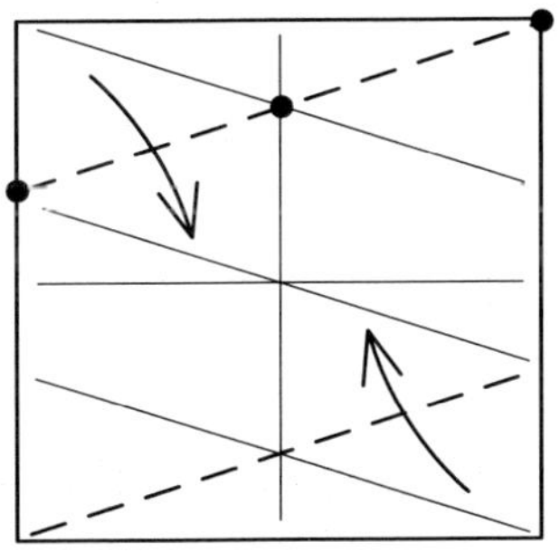

Repeat steps 9–11 in
the other direction.

13

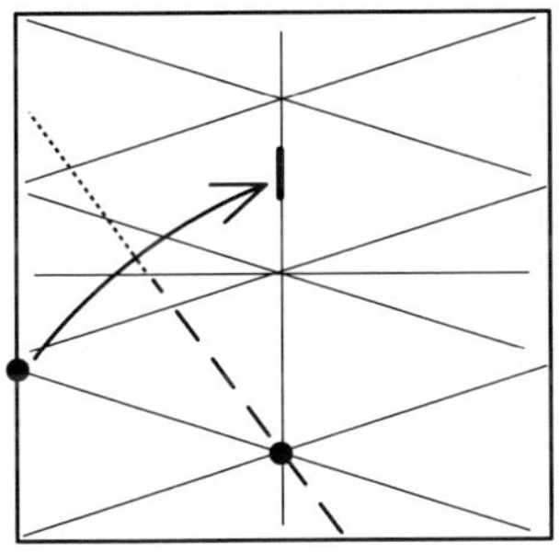

14

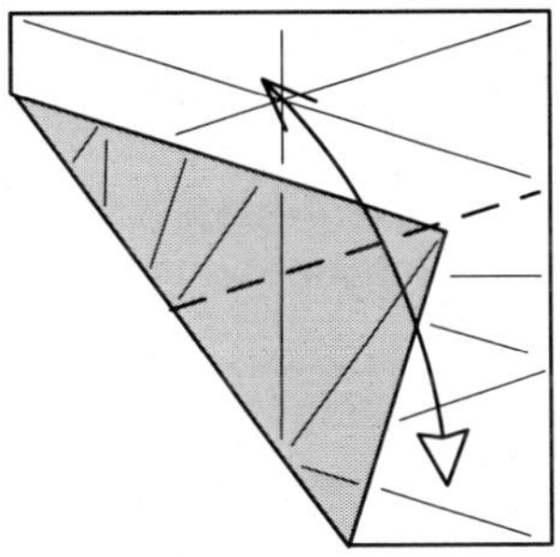

Fold and unfold along a
partially hidden crease.

15

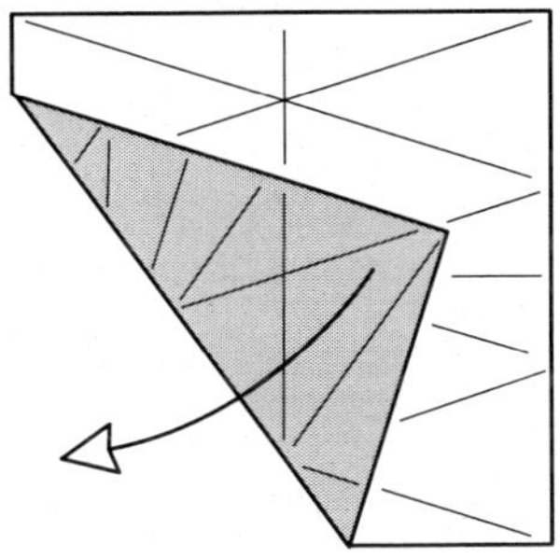

Unfold.

16

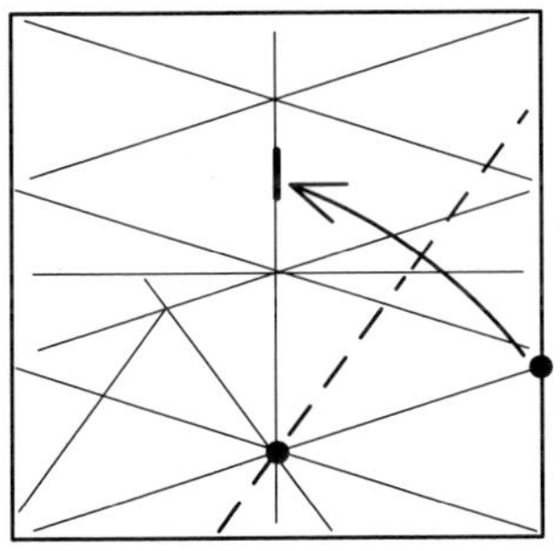

Repeat steps 13–15
three times, on the
right and at the top.

17

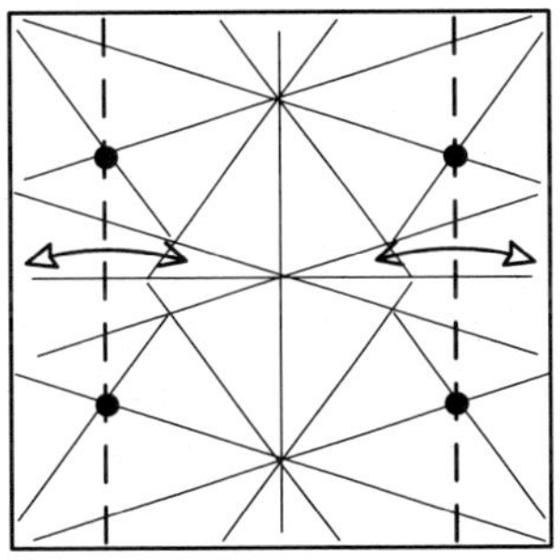

Fold and unfold.

18

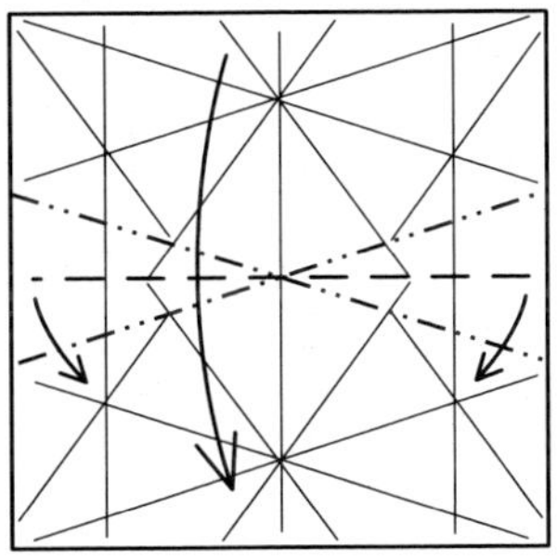

Fold along the creases.

19

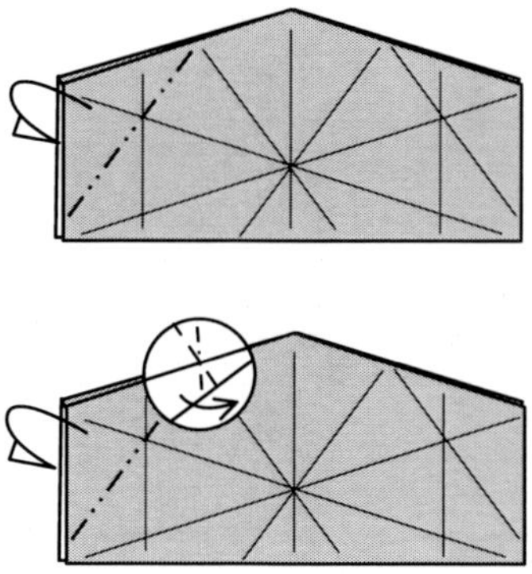

Inside view. Fold the inside layers together for this spine-lock fold. Turn over and repeat.

20

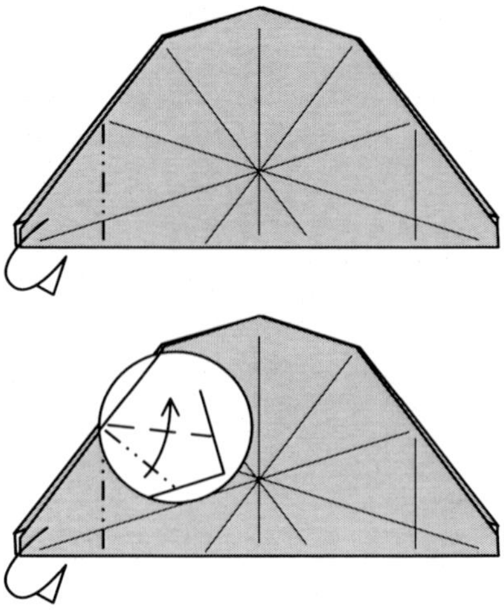

Inside view. Fold the inside layers together for this spine-lock fold. Turn over and repeat.

21

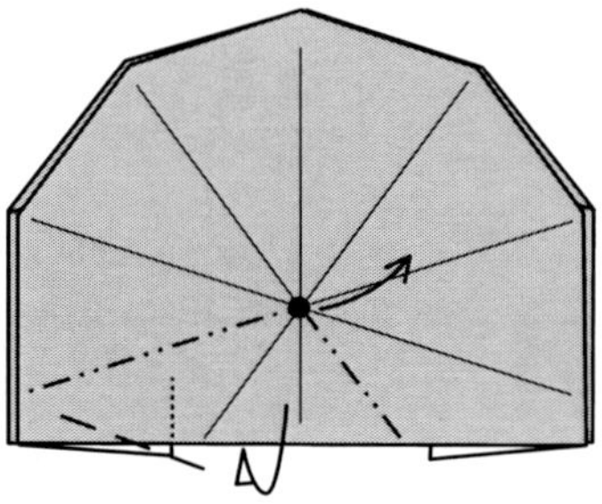

Puff out at the dot and fold along the creases. Turn over and repeat.

22

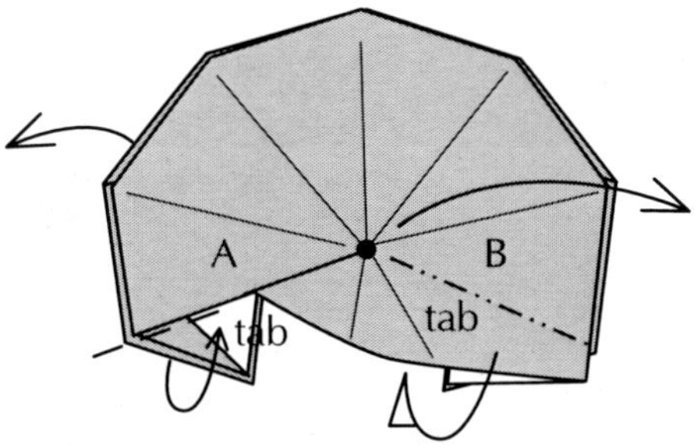

Bring the dot to the right and the one behind to the left. Follow regions A and B into the next step. Fold all the layers of each tab.

23

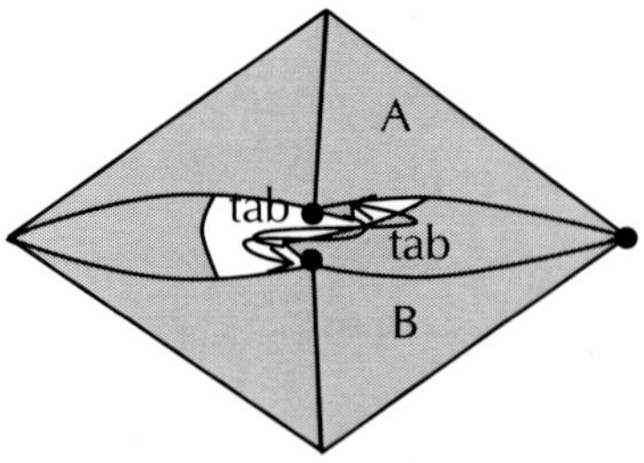

Tuck the two tabs into each other so the center dots meet.

24

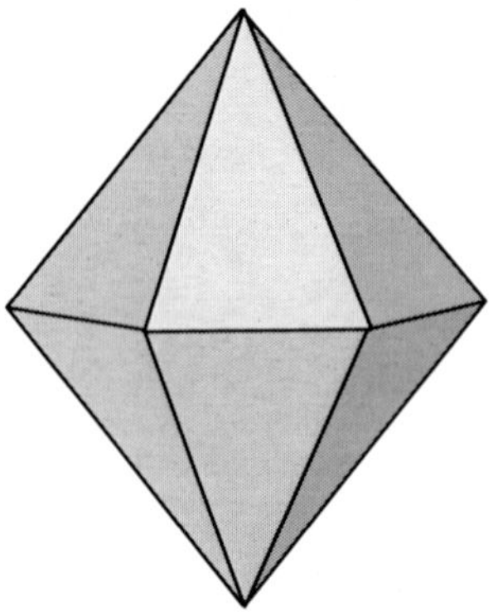

Hexagonal Dipyramid 36°

Heptagonal Dipyramid 30°

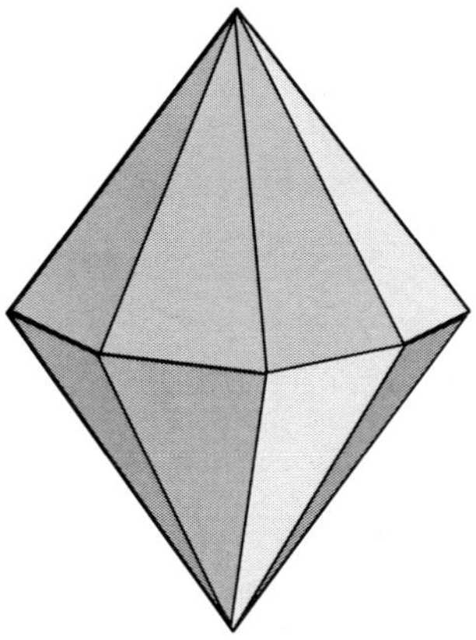

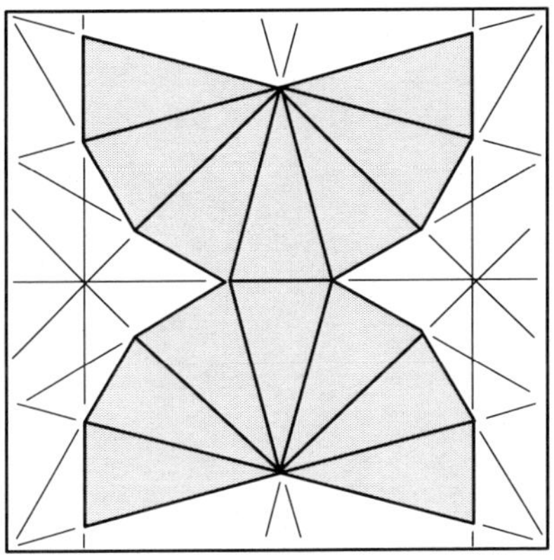

The angles of each of the twelve triangles are 30°, 75°, and 75°. Odd symmetry is used.

1
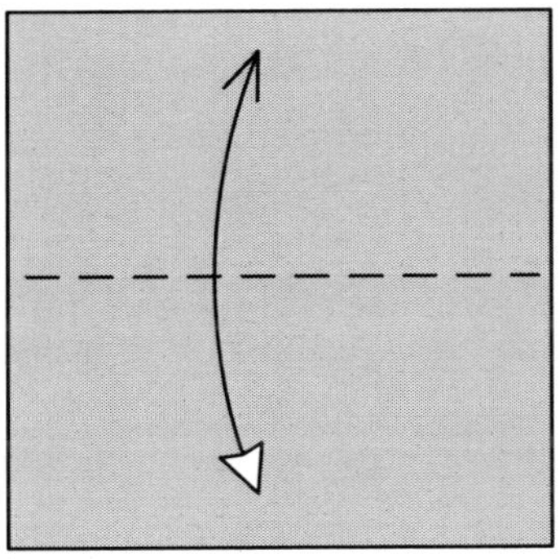

Fold and unfold.

2
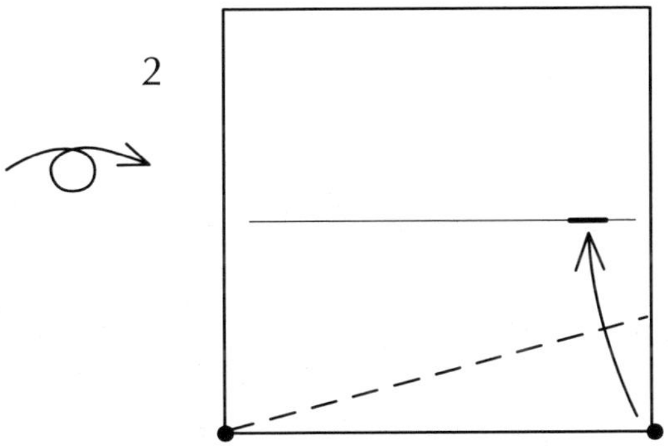

Bring the corner to the line.

3
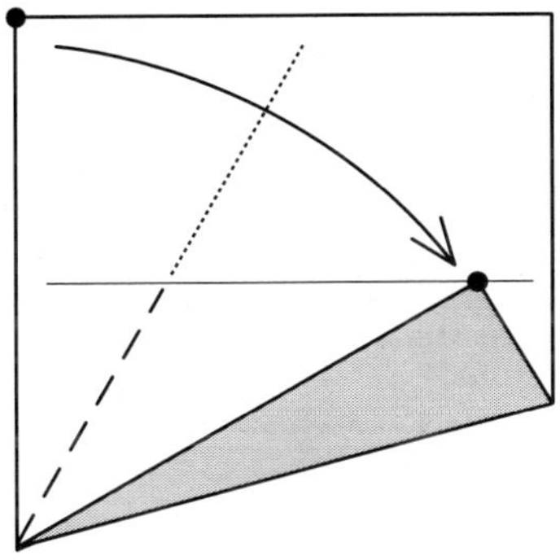

Fold on the lower half.

4
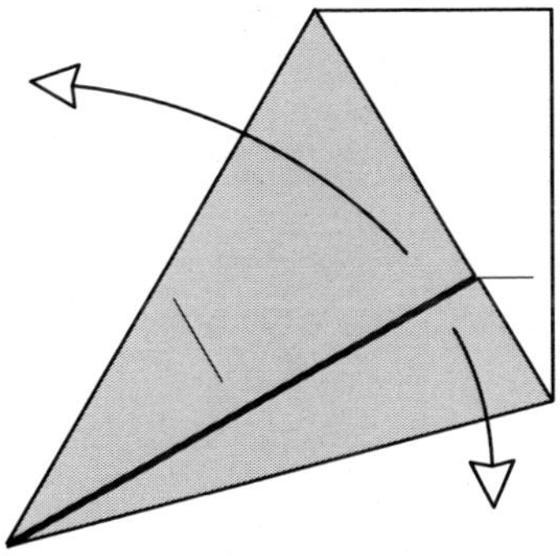

Unfold.

5
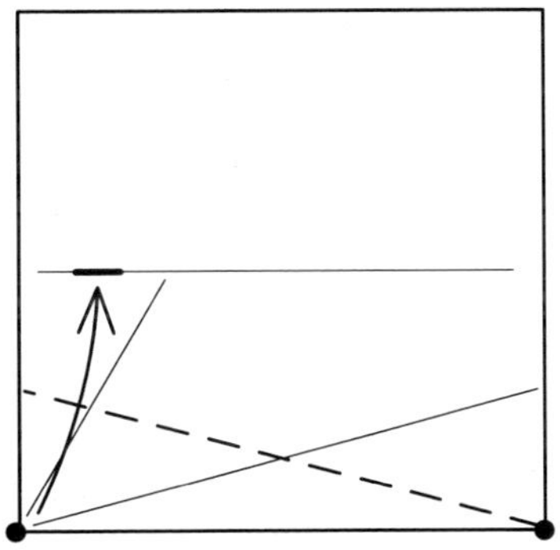

Repeat steps 2–4 in the other direction. Rotate 180°.

6
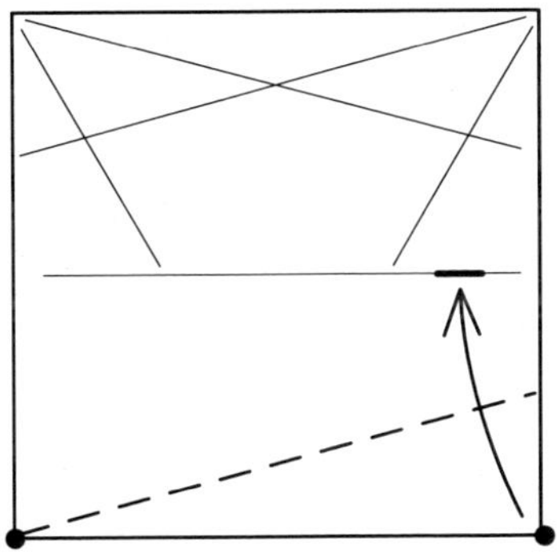

Repeat steps 2–5.

7

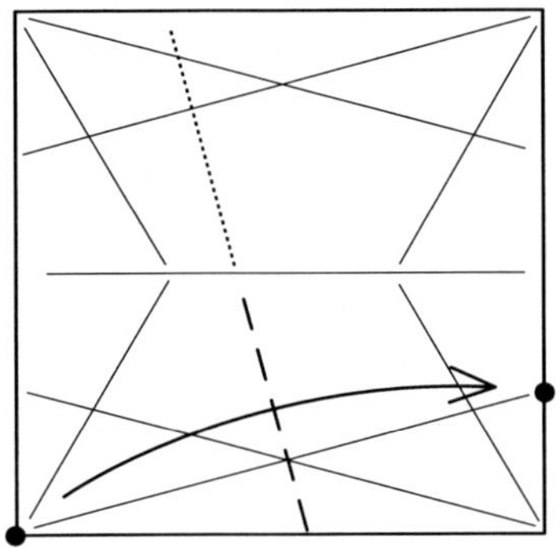

8

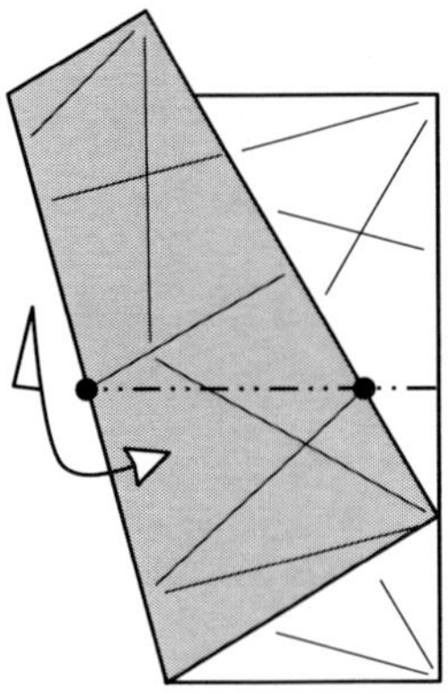

Fold and unfold all the layers along a partially hidden crease.

9

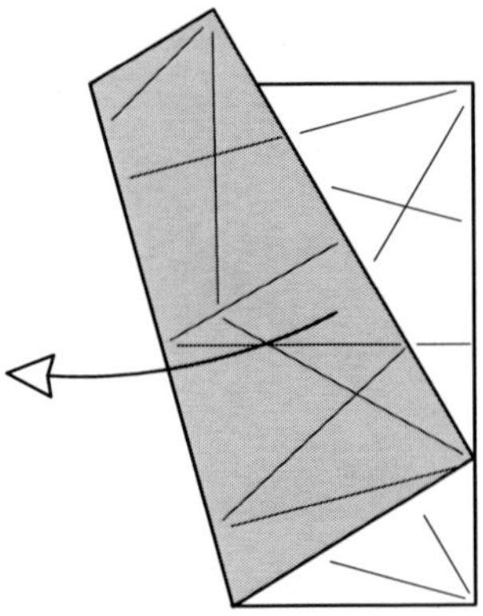

Unfold.

10

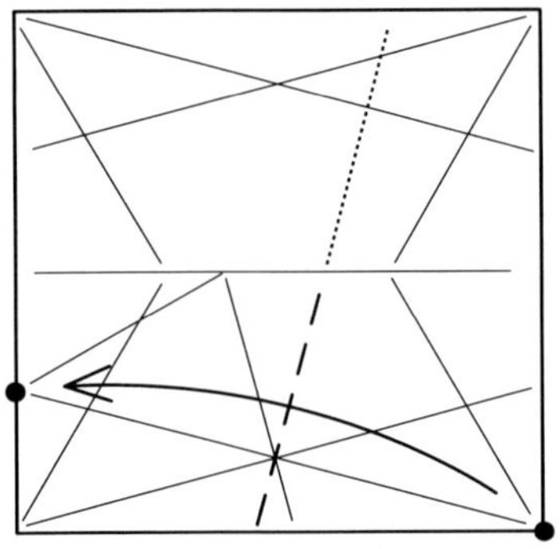

Repeat steps 7–9 three times, on the right and top.

11

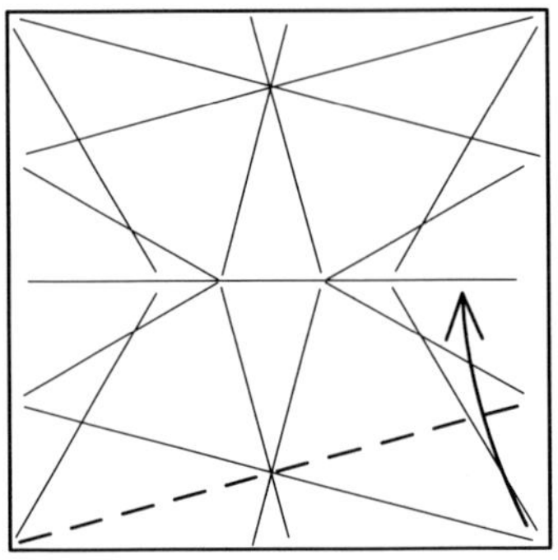

Fold along the crease.

12

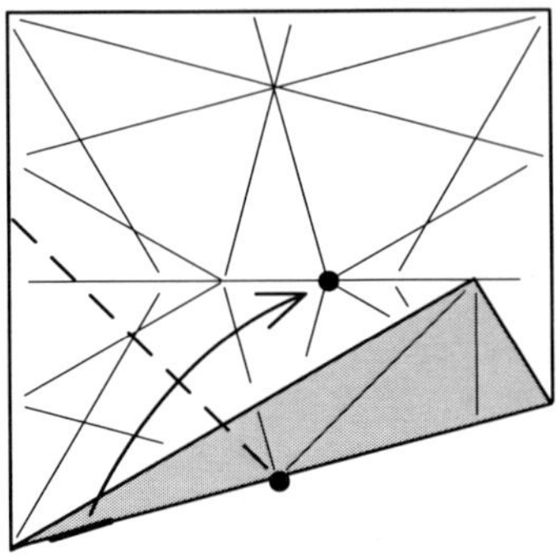

13

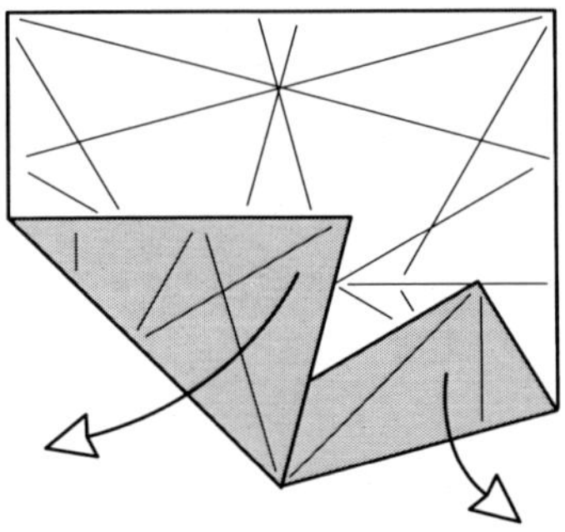

Unfold.

14

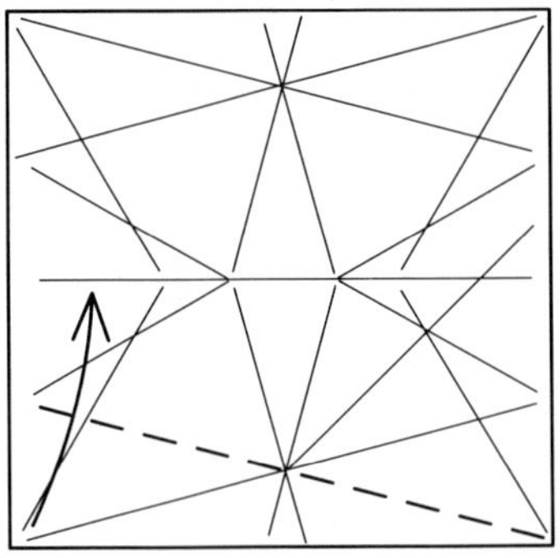

Repeat steps 11–13 three times, in the other direction and on top.

15

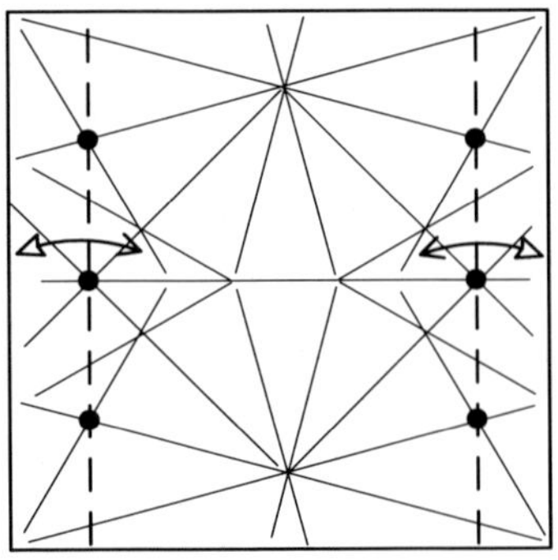

Fold and unfold.

16

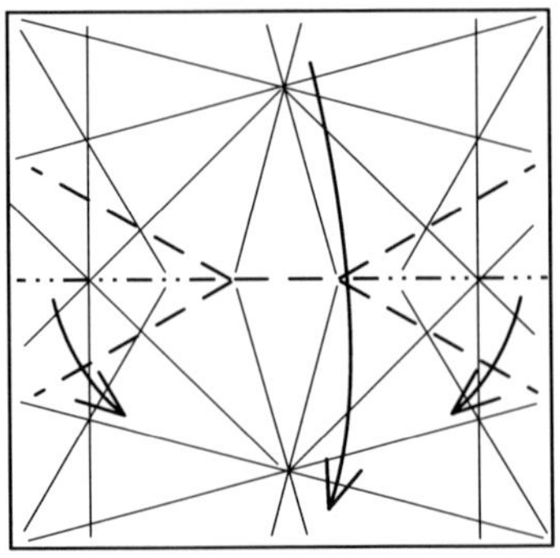

Fold along the creases.

17

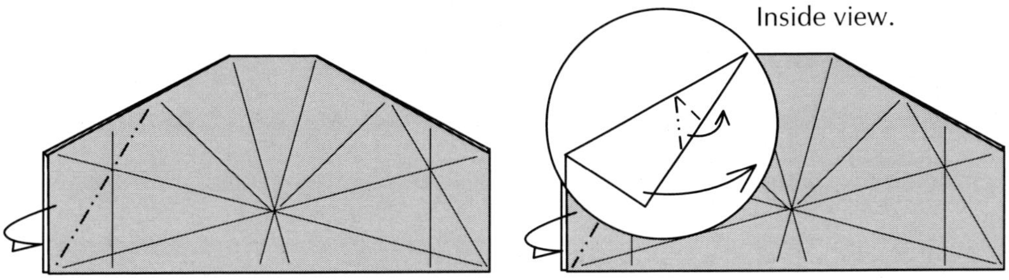

Fold the layers together at the dot for this spine-lock fold. Turn over and repeat.

18

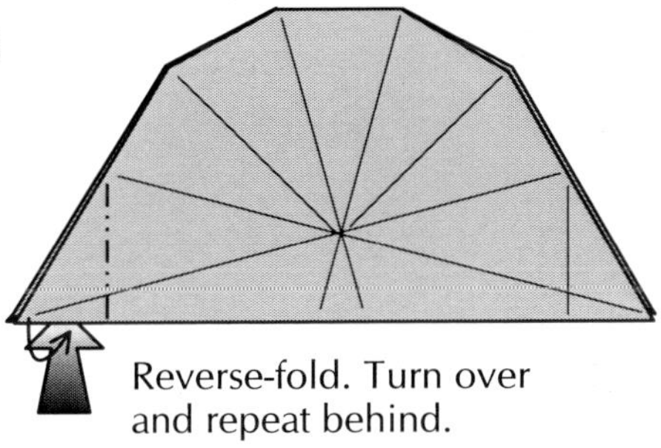

Reverse-fold. Turn over and repeat behind.

19

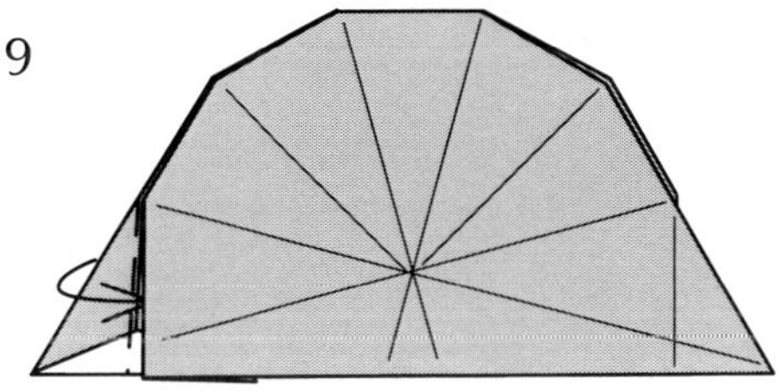

Tuck inside. Turn over and repeat behind.

20

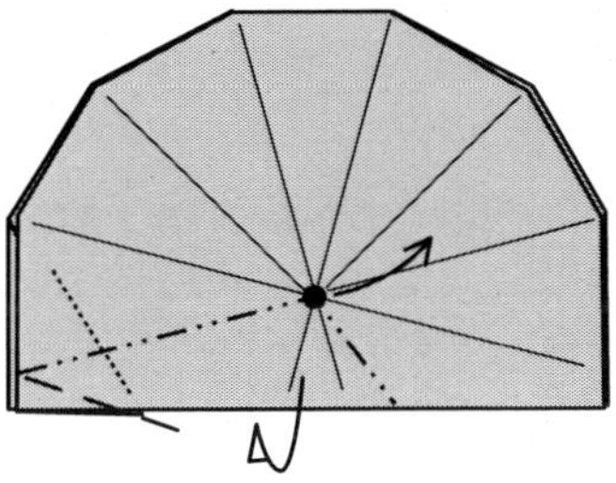

Puff out at the dot and fold along the creases. Turn over and repeat.

21

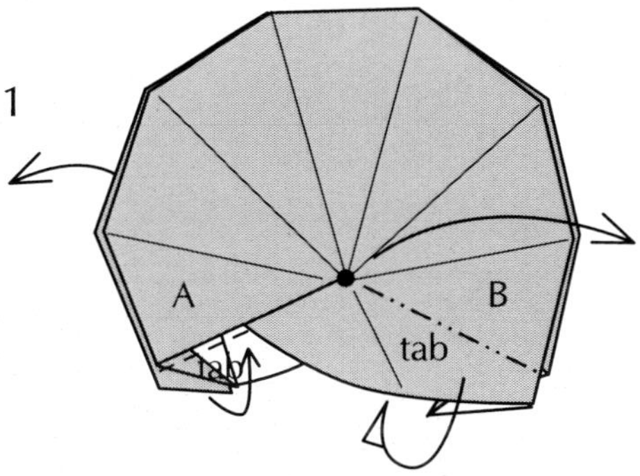

Bring the dot to the right and the one behind to the left. Follow regions A and B into the next step. Fold all the layers of each tab.

22

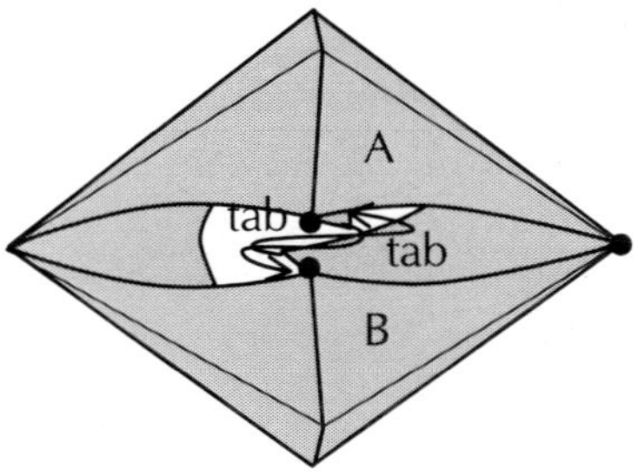

Tuck the two tabs into each other so the center dots meet.

23

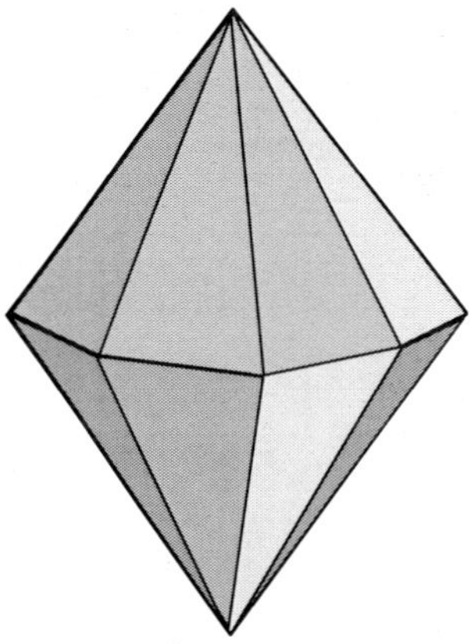

Heptagonal Dipyramid 30°

Octagonal Dipyramid 26°

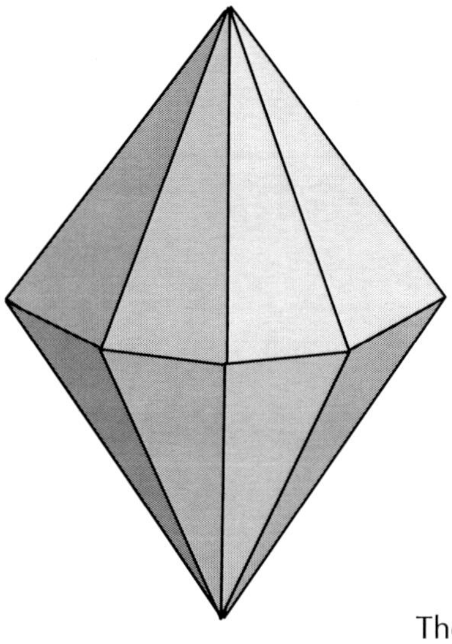

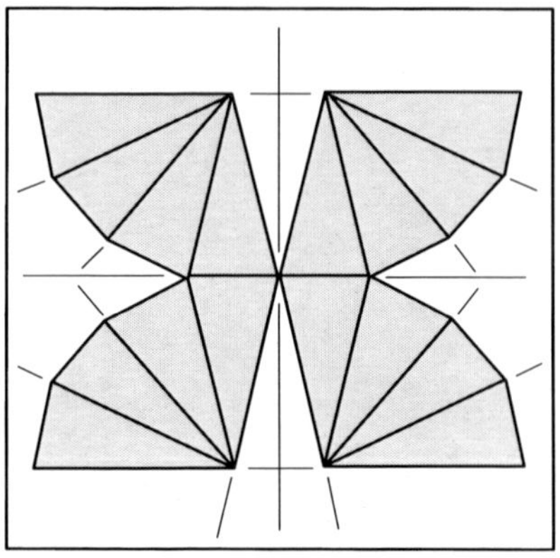

The folding of this 16-sided dipyramid is similar to the Octagonal Dipyramid. The small angle in each triangle is about 25.7°. That angle makes the geometry convenient since the horizontal lines are part of the crease pattern.

1

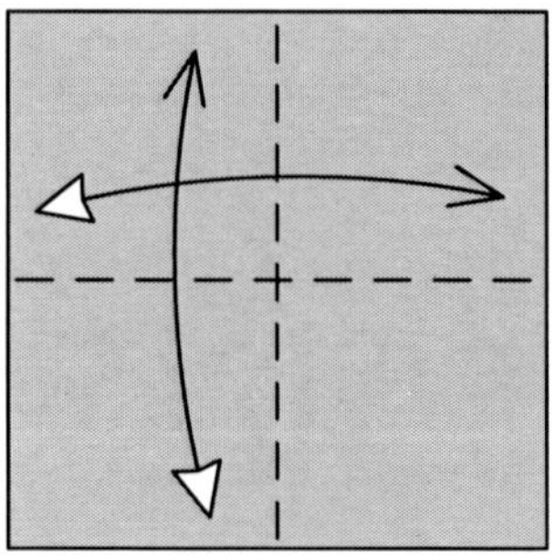

Fold and unfold.

2

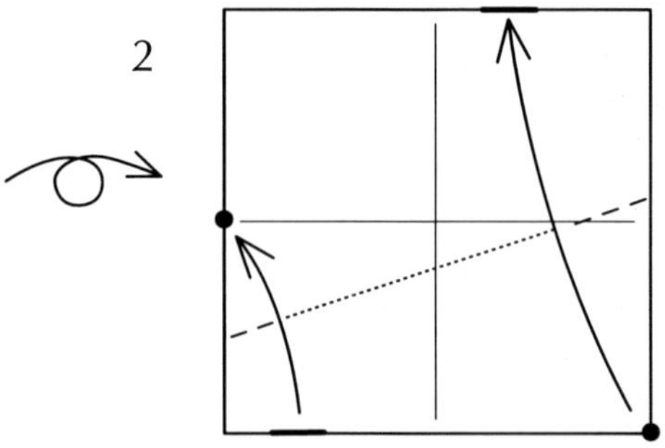

Bring the lower right corner to the top edge and the bottom edge to the left center. Crease on the left and right.

3

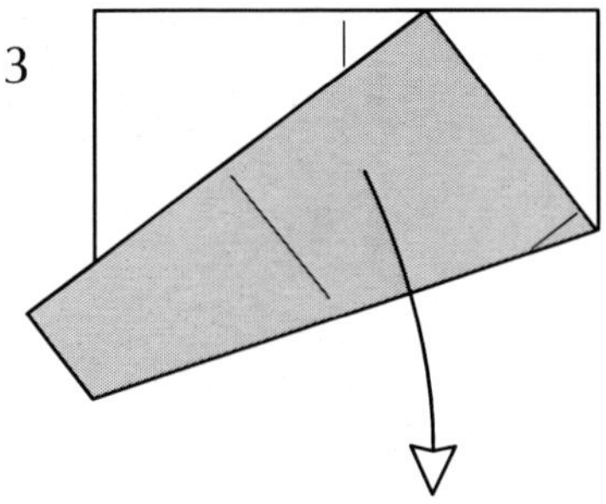

Unfold.

4

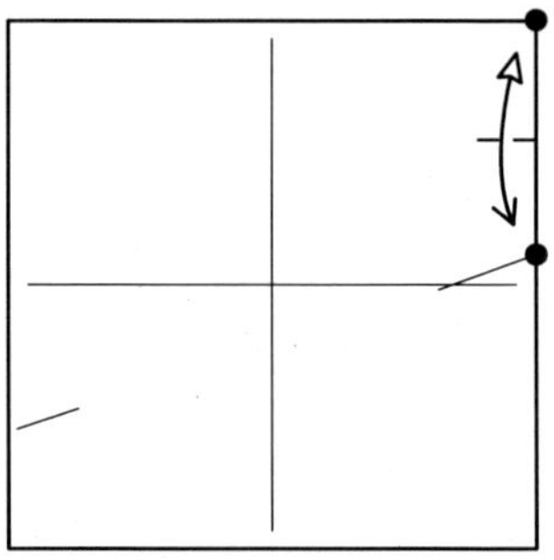

Fold and unfold on the right.

5

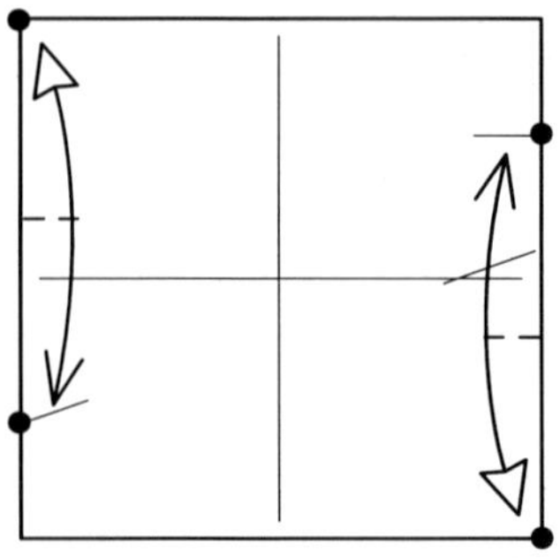

Fold and unfold on the edges. Rotate 90°.

6

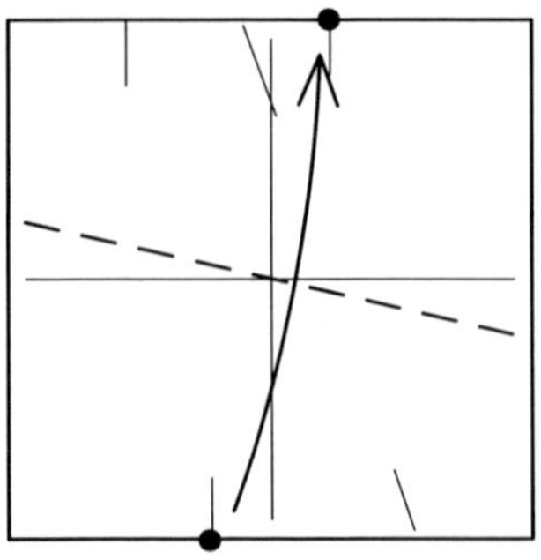

7

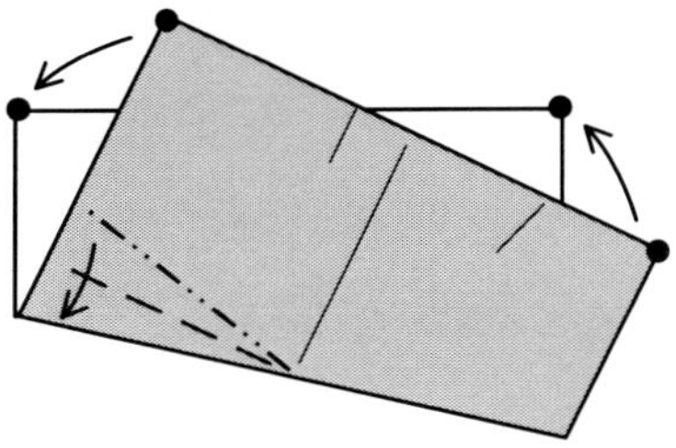

Valley-fold along the crease.
Turn over and repeat.

8

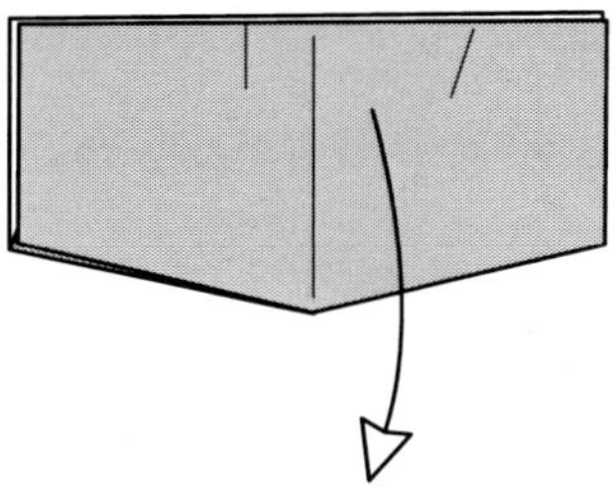

Unfold and rotate 90°.

9

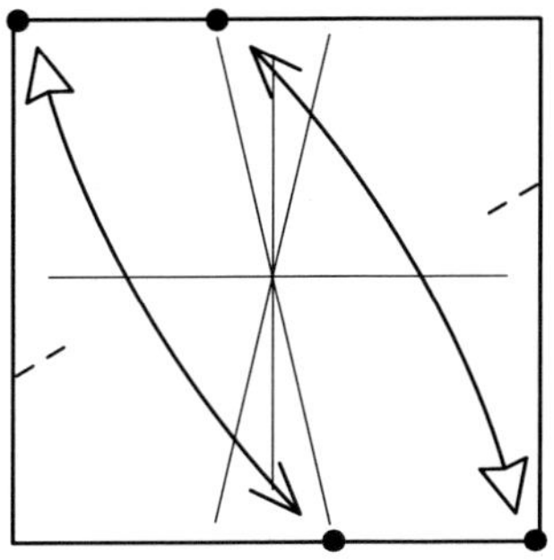

Fold and unfold
on the edges.

10

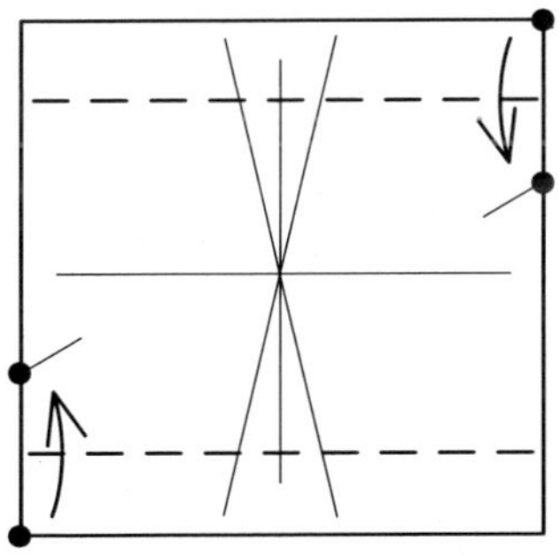

11

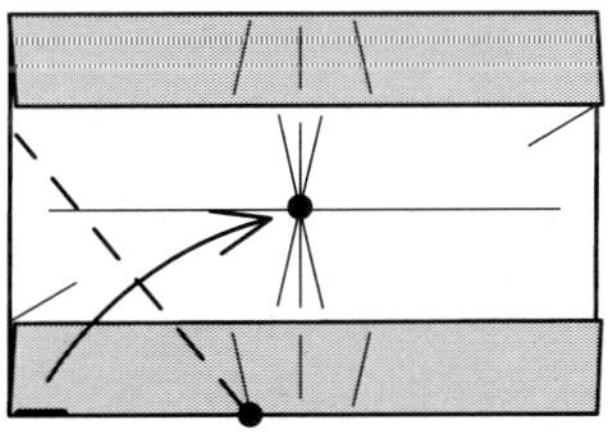

12

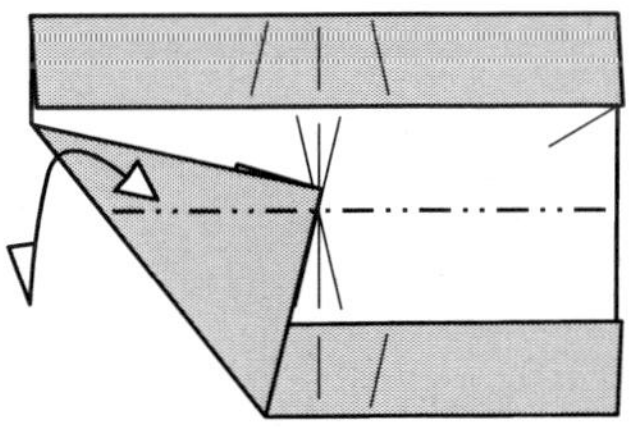

Fold and unfold.

13

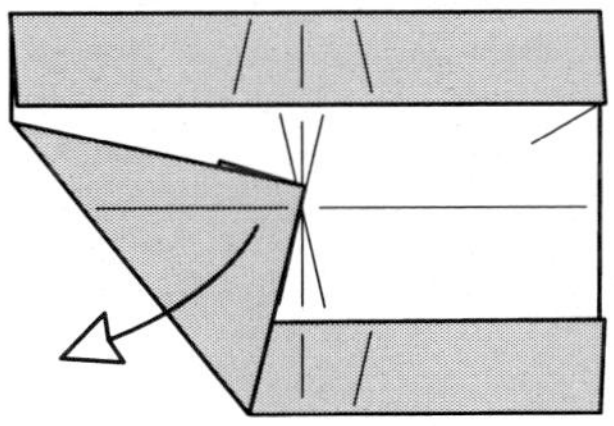

Unfold.

14

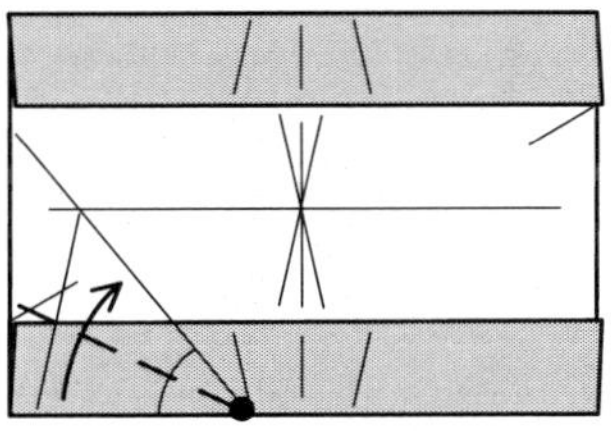

15

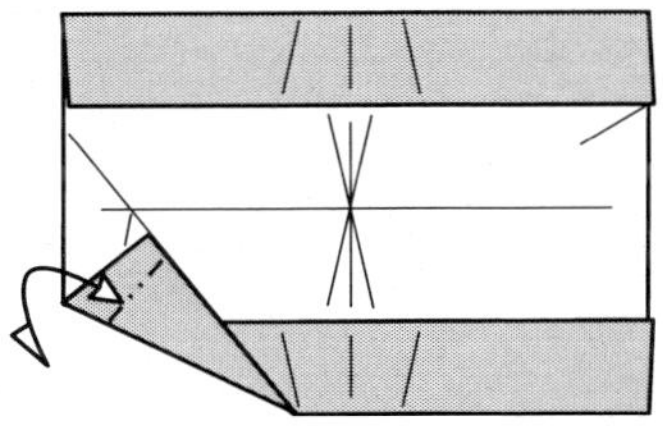

Fold and unfold.

16

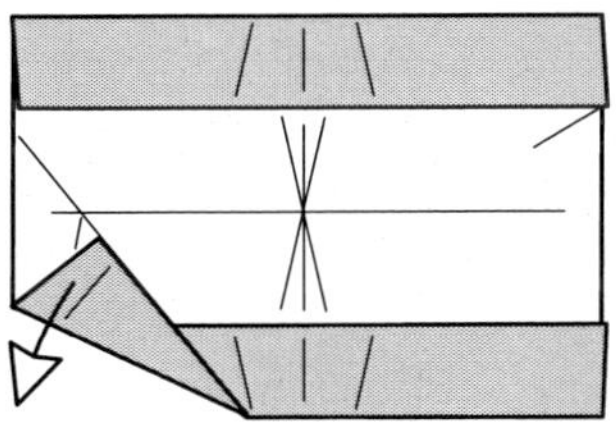

Unfold.

17

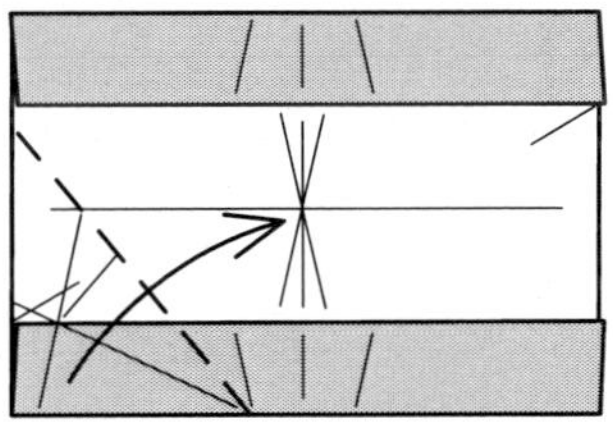

Fold along the crease.

18

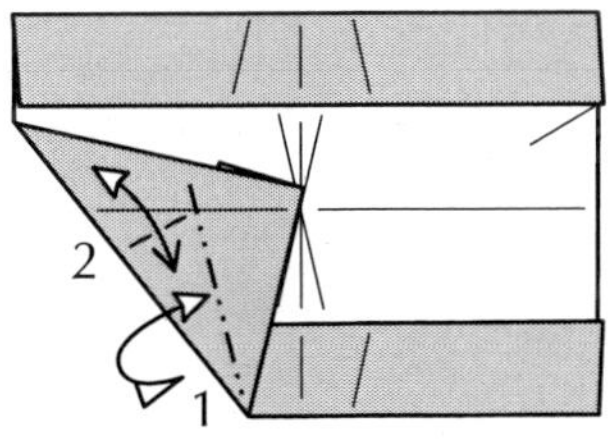

Fold and unfold
along the creases.

19

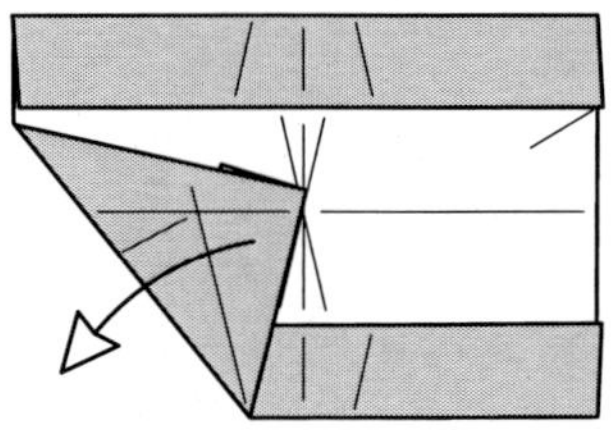

Unfold.

20

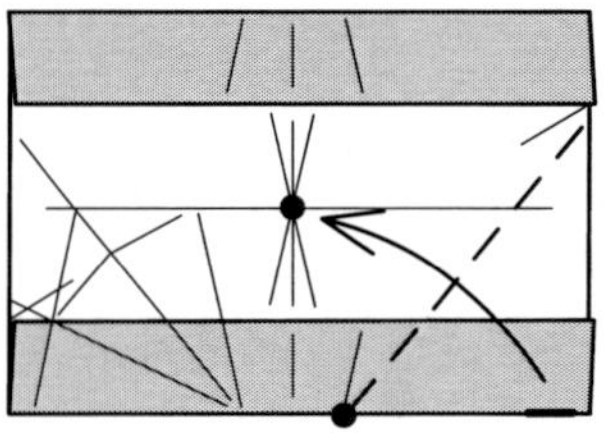

Repeat steps 11–19 three times, on the right and at the top.

21

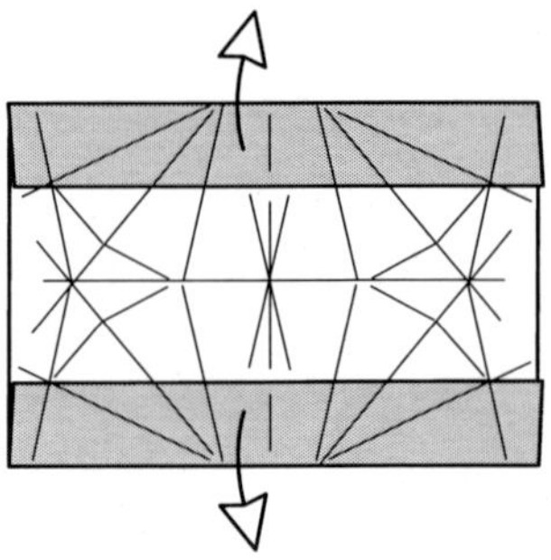

Unfold.

22

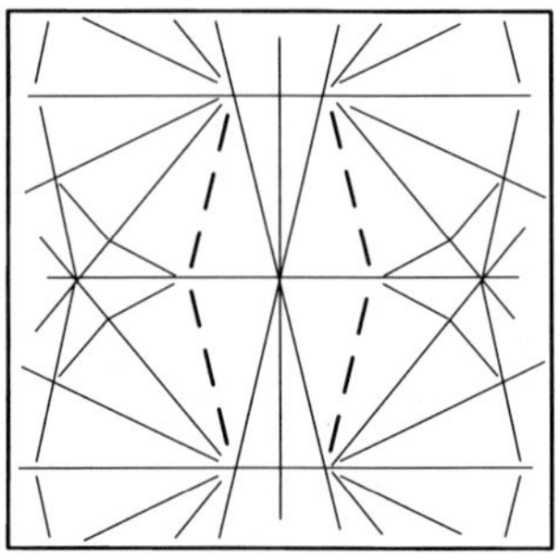

Fold and unfold along the creases.

23

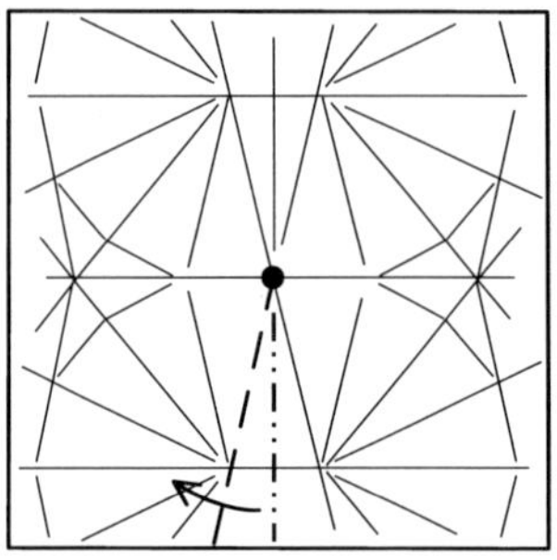

Push in at the dot.

24

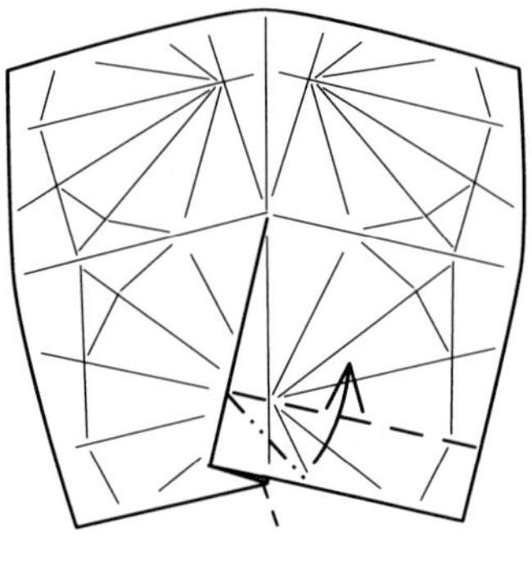

Squash-fold and rotate 180°.

25

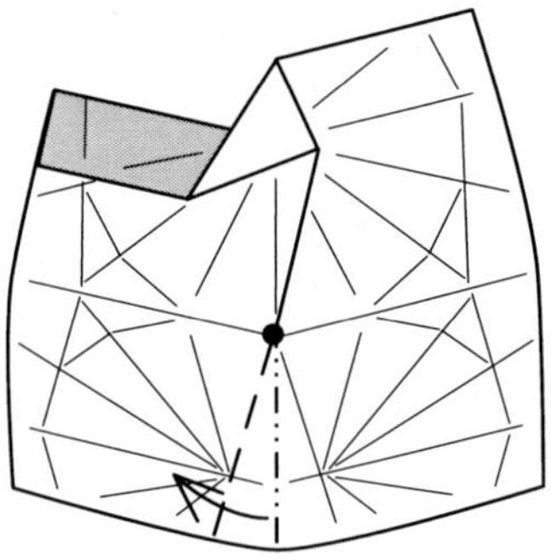

Repeat steps 23–24. Then flatten.

26

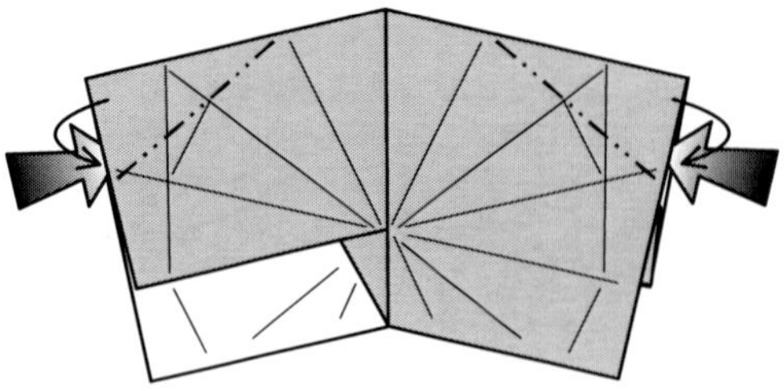

Make reverse folds.

27

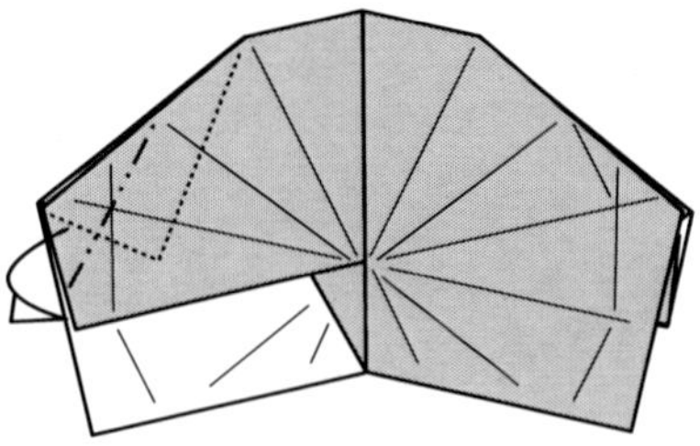

Fold the inside layers together for this spine-lock fold. Turn over and repeat.

28

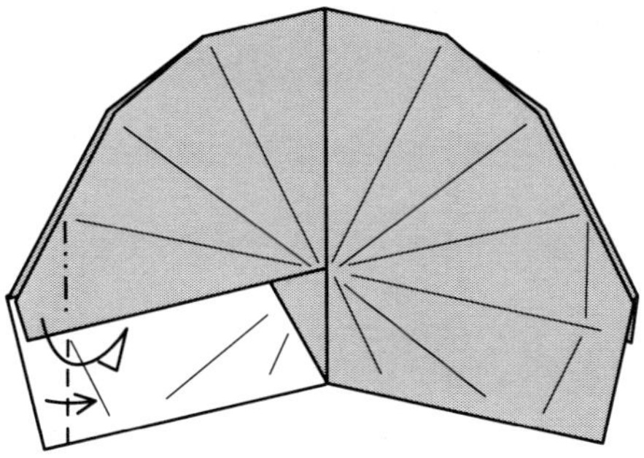

Fold the inside layers together for this spine-lock fold. Turn over and repeat.

29

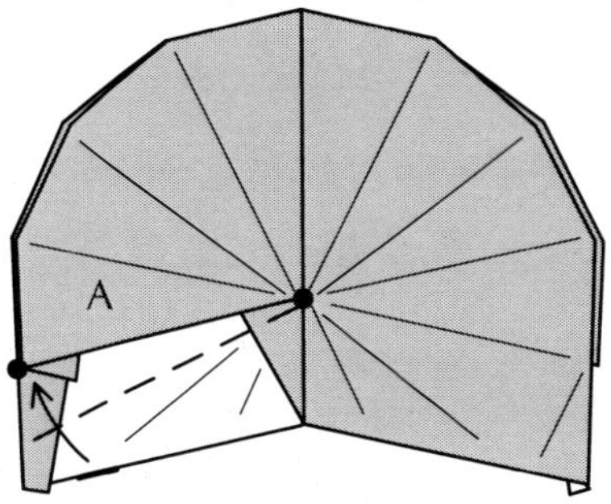

Bring the edge to the dot. All of the folds will be under region A. Turn over and repeat.

30

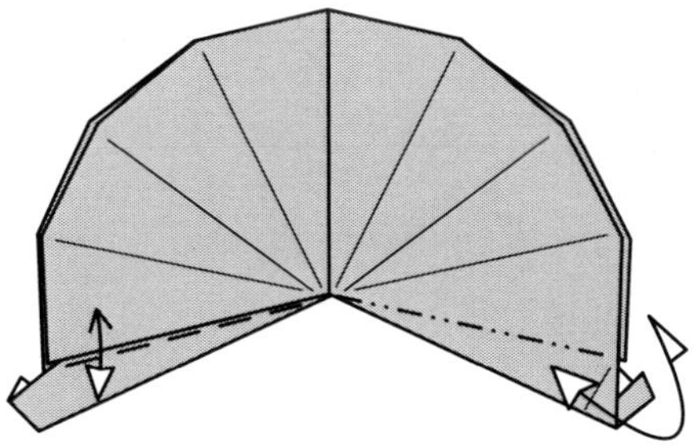

Fold and unfold.

31

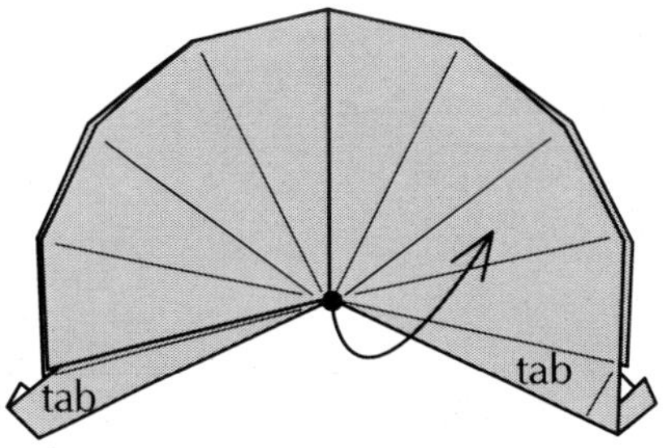

Lift up at the dot to open the model. Tuck and interlock the tabs.

32

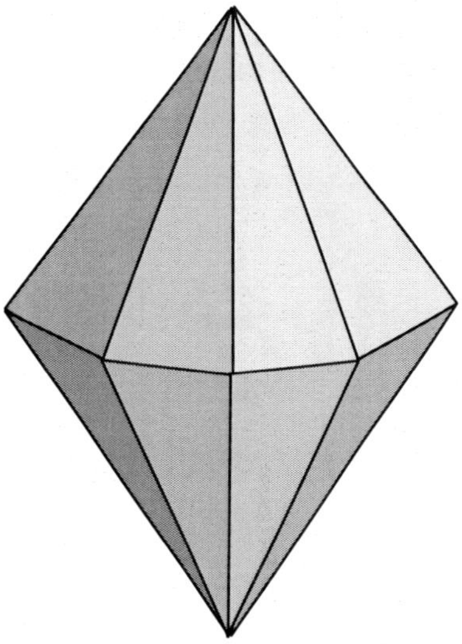

Octagonal
Dipyramid 26°

Nonagonal Dipyramid 22.5°

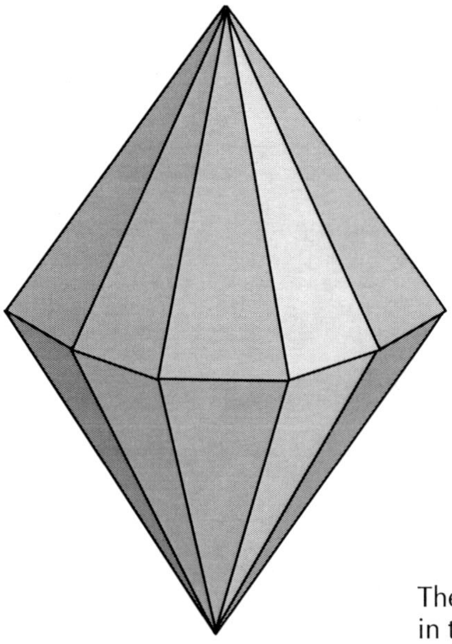

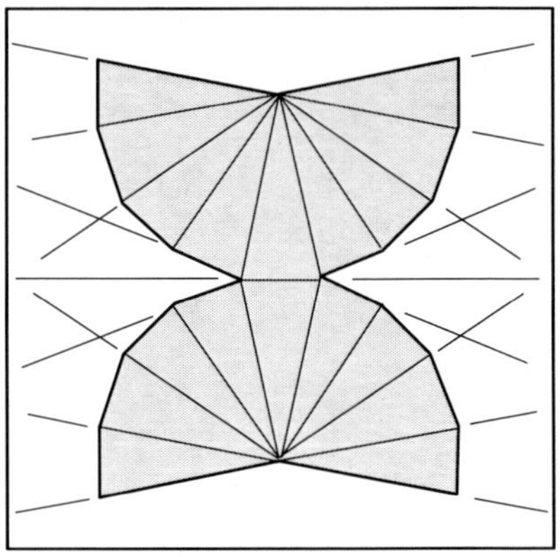

The angle at the top of the triangle in this 18-sided dipyramid is 22.5°.

1

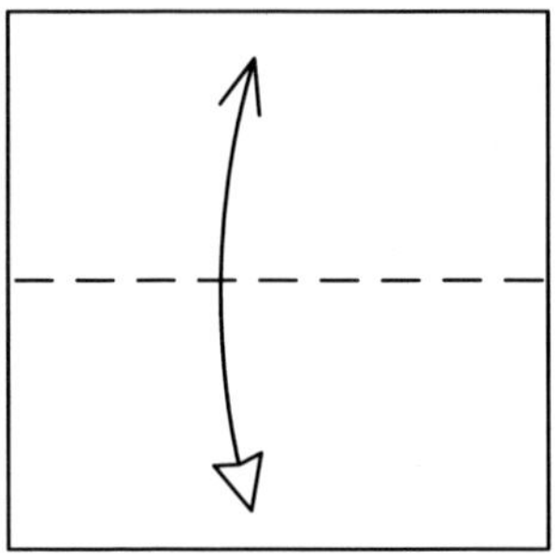

Fold and unfold.

2

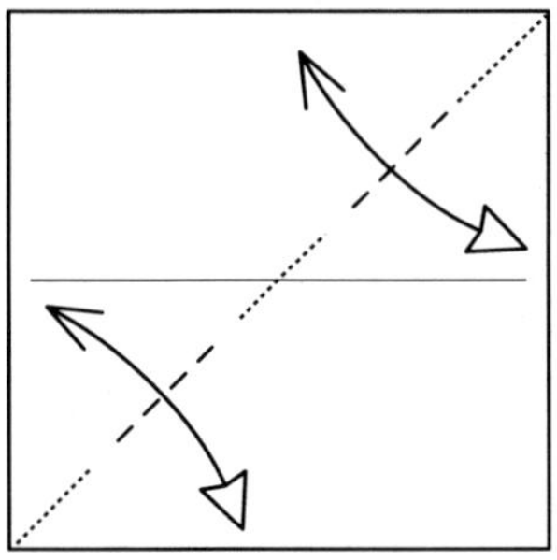

Fold and unfold along the diagonal.

3

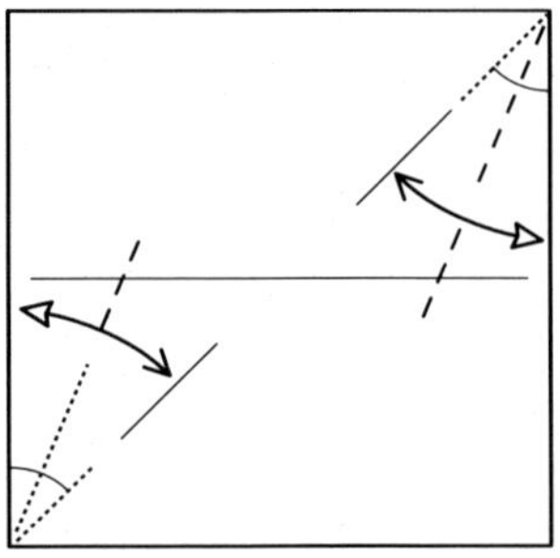

Fold and unfold on the center line.

4

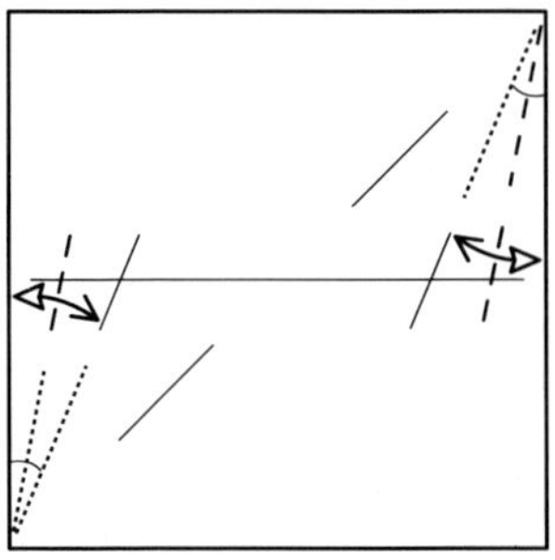

Fold and unfold on the center line.

5

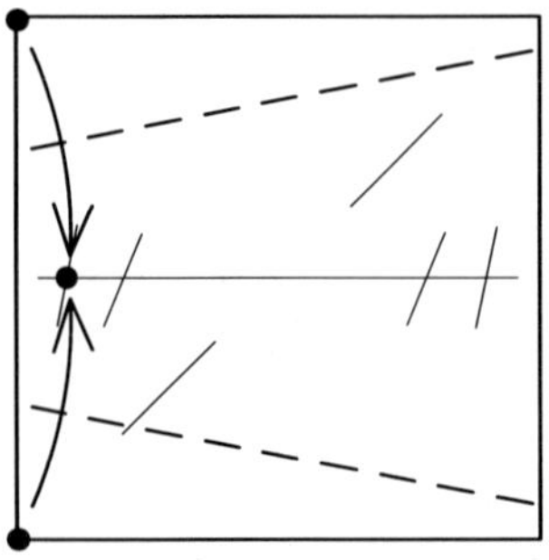

The dots will meet.

6

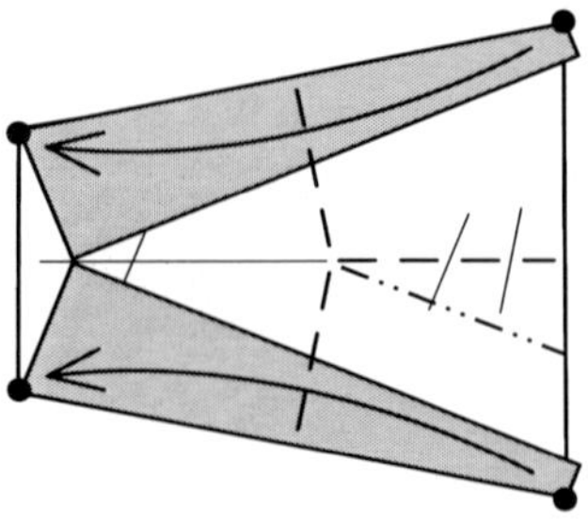

This is similar to a rabbit ear. The dots will meet.

7

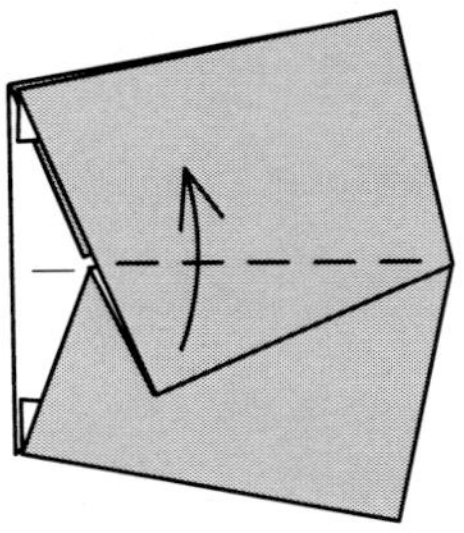

8

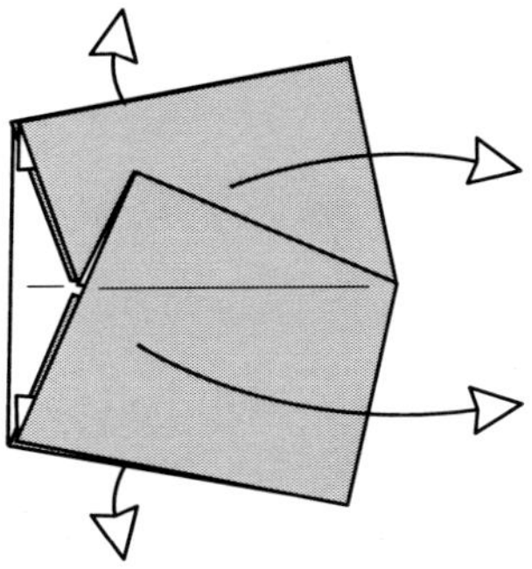

Unfold everything
and rotate 180°.

9

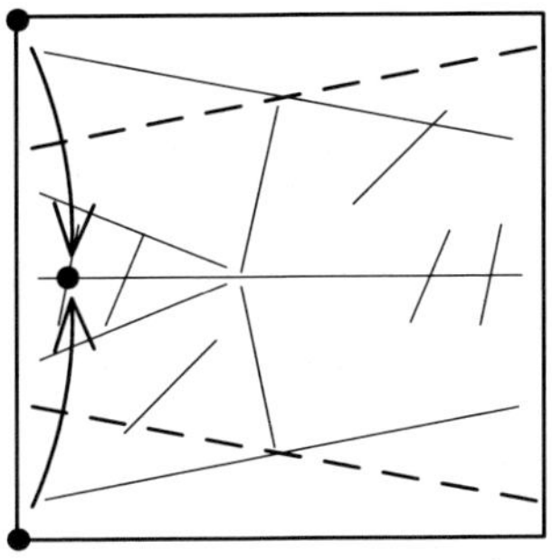

Repeat steps 5–8.

10

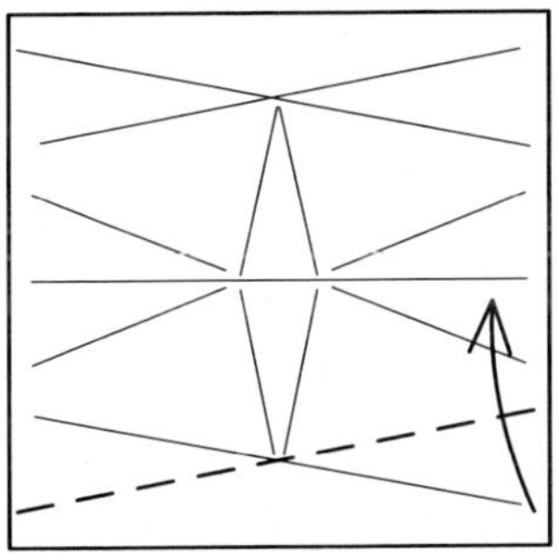

Fold along the crease.

11

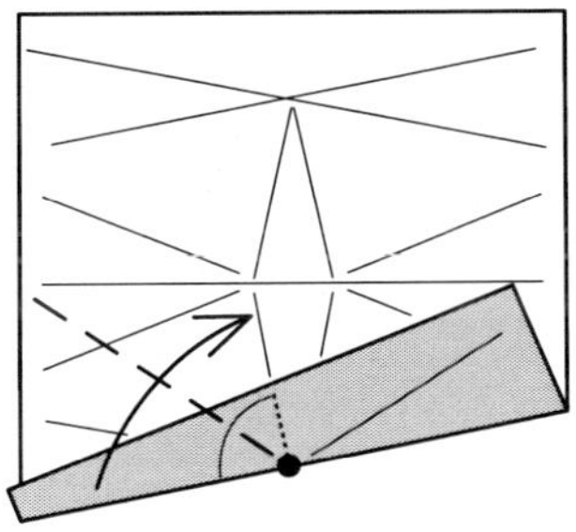

12

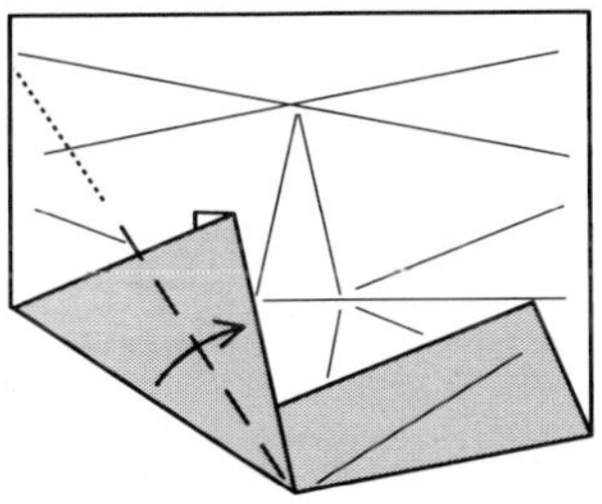

Fold along the crease.

13

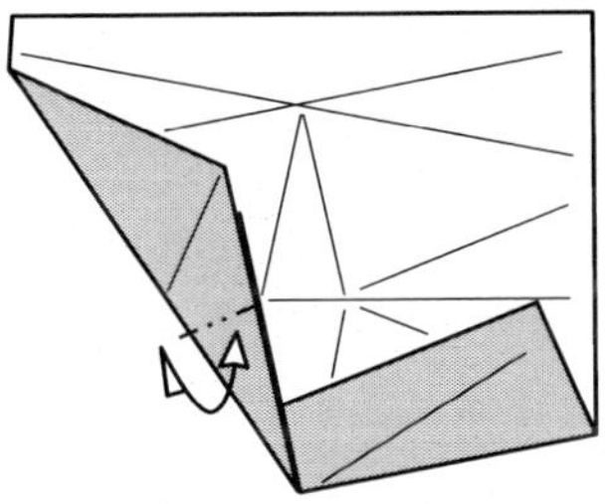

Fold and unfold along
a hidden crease.

14

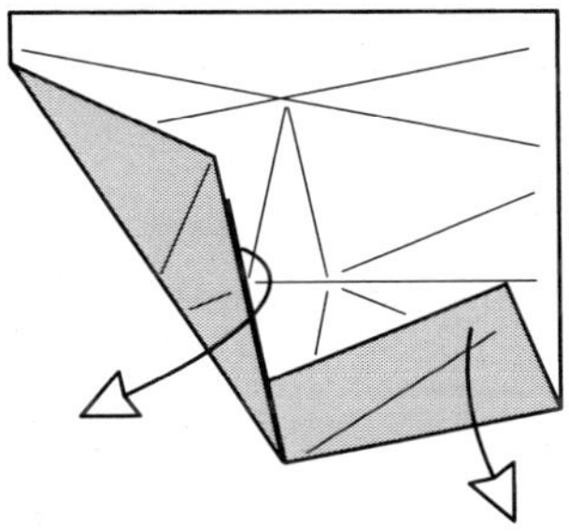

Unfold everything.

15

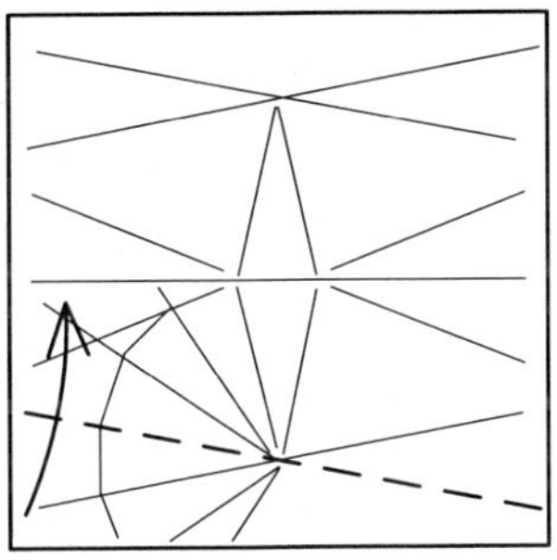

Repeat steps 10–14 in
the opposite direction.
Rotate 180°.

16

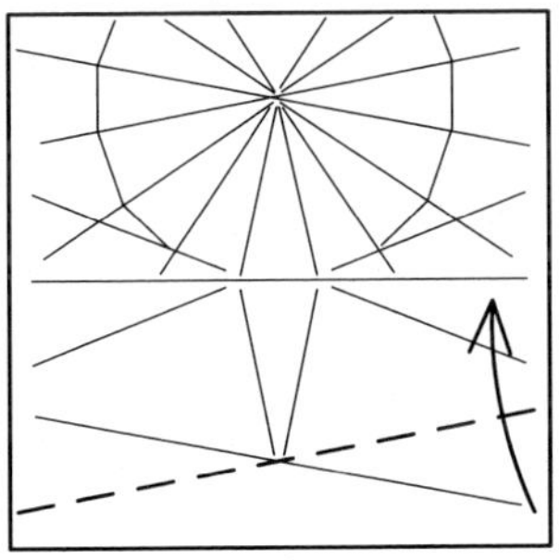

Repeat steps 10–15.

17

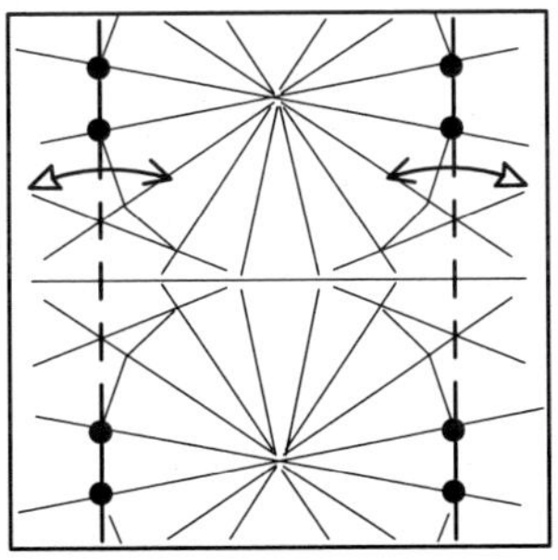

Fold and unfold along
creases at the dots.

18

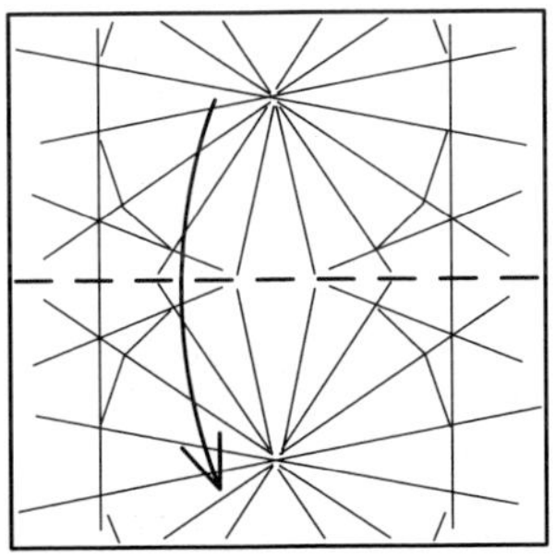

Fold in half.

19

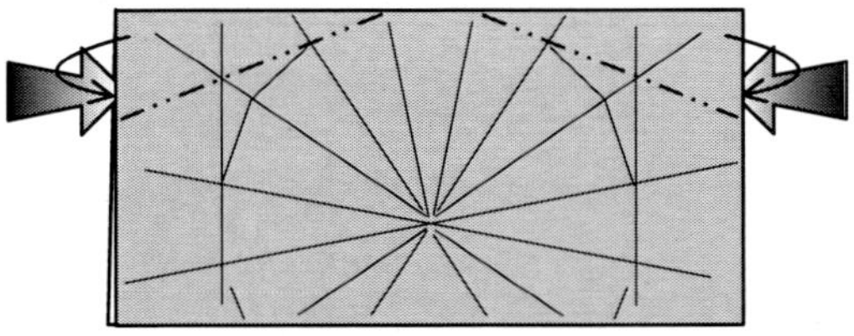

Make revese folds along the creases.

20

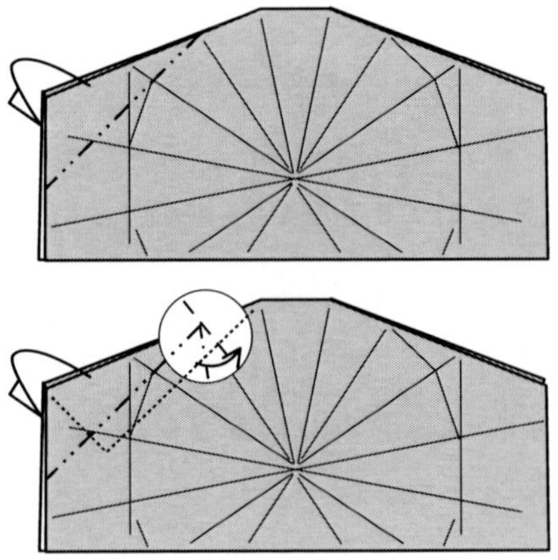

Inside view. Fold the inside layers together for this spine-lock fold. Turn over and repeat.

21

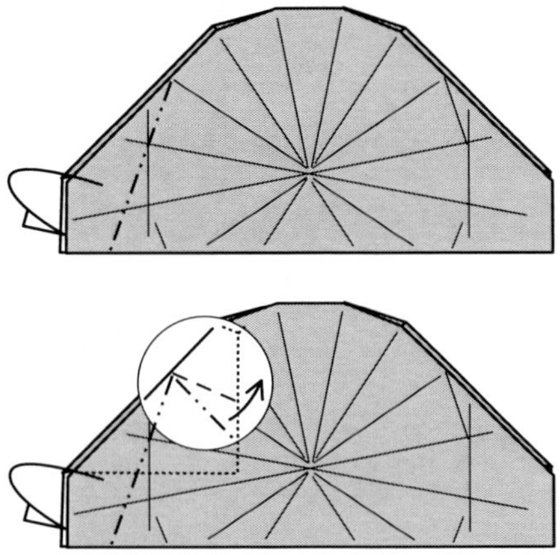

Inside view. Fold the inside layers together for this spine-lock fold. Turn over and repeat.

22

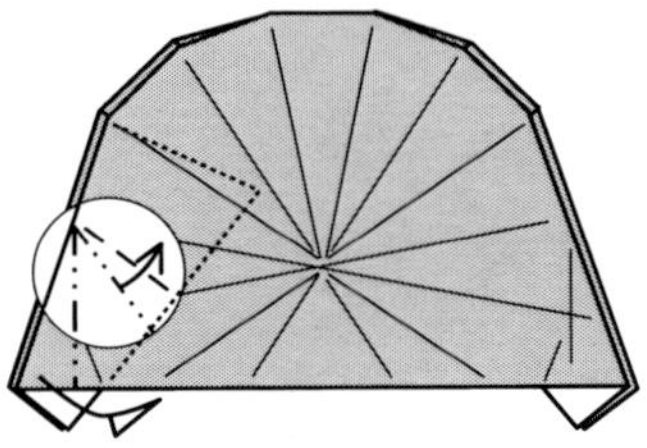

Inside view. Fold the inside layers together for this spine-lock fold. Turn over and repeat.

23

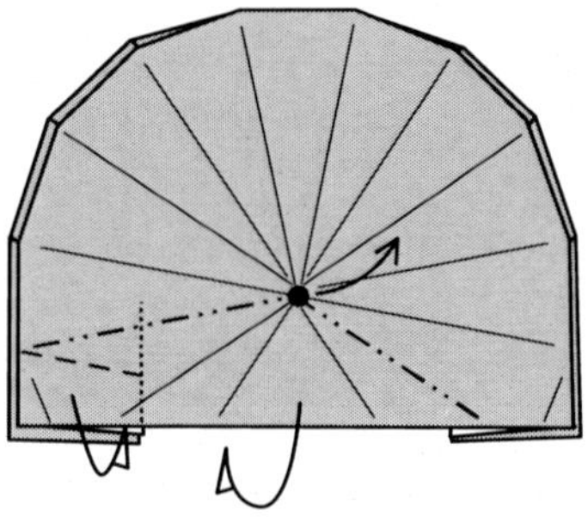

Puff out at the dot and fold along the creases. Valley-fold along an inner flap. Turn over and repeat.

24

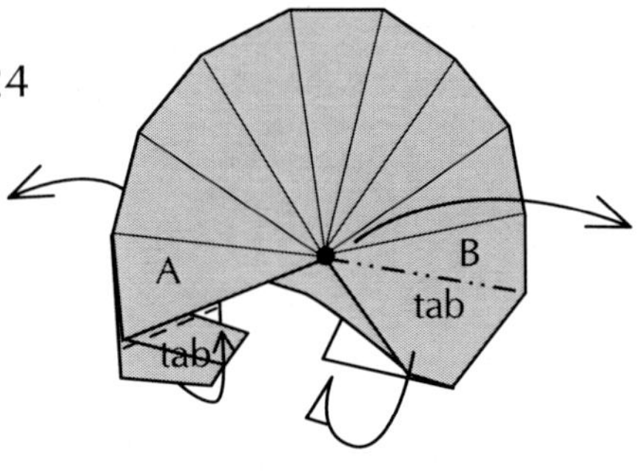

Bring the dot to the right and the one behind to the left. Follow regions A and B into the next step. Fold all the layers of each tab.

25

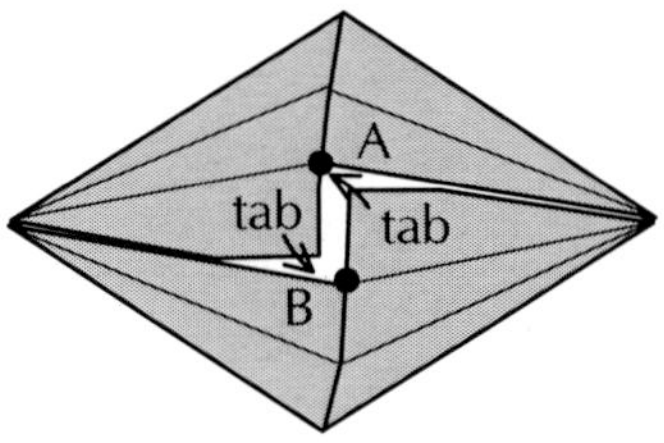

Tuck the two tabs inside to lock the model. The dots will meet.

26

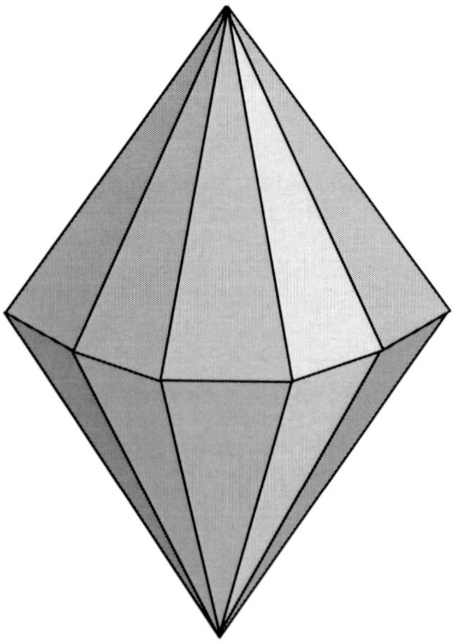

Nonagonal Dipyramid

Colorful Octahedra

Adding colorful patterns to an octahedron creates a stunning model. The octahedron was shown in the previous section on Dipyramids. Here are three octahedra with colorful patterns.

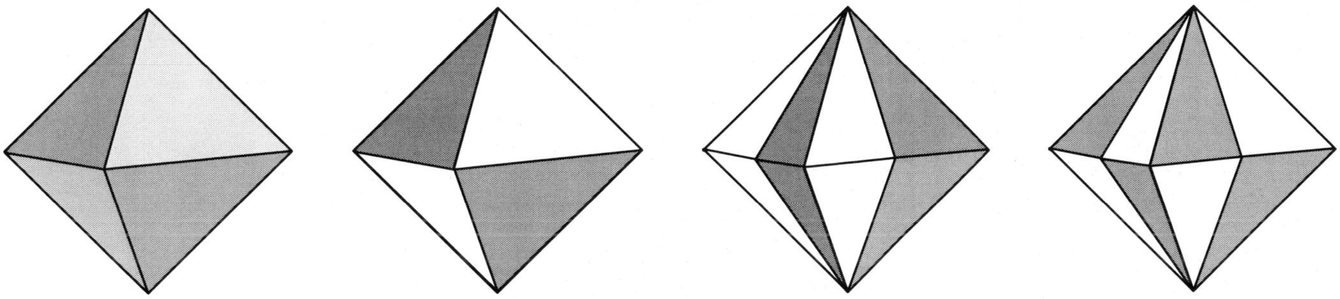

Magic Octahedron

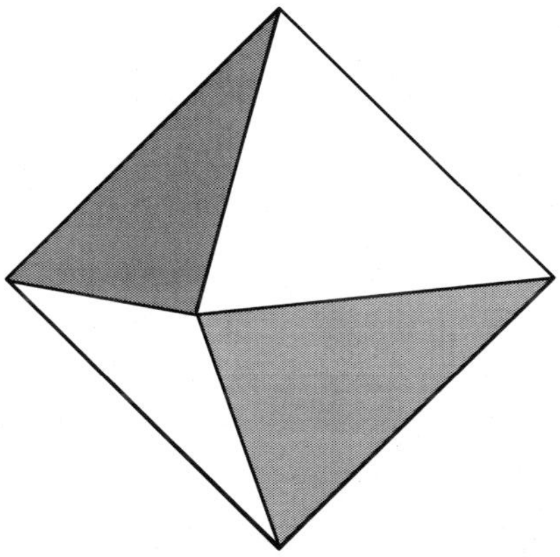

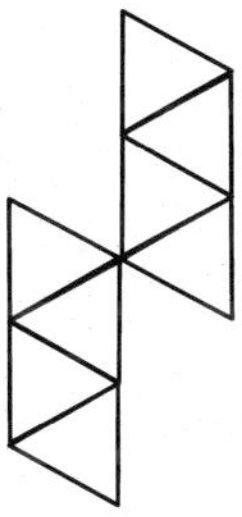

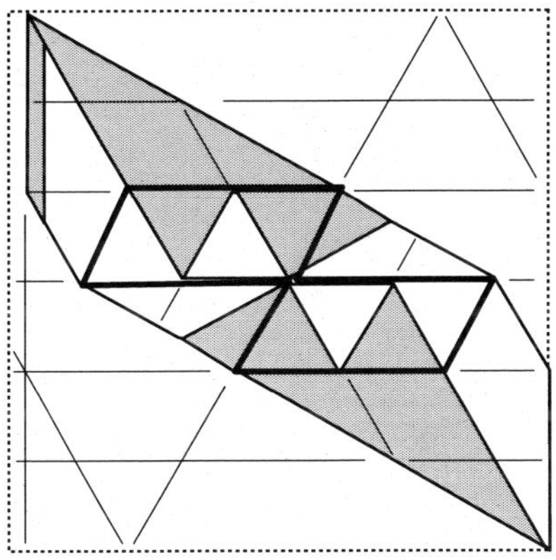

The colors on the faces alternate between a color and white. Odd symmetry is used. The layout, above, from step 15, shows the triangles with boh sides of the paper.

1

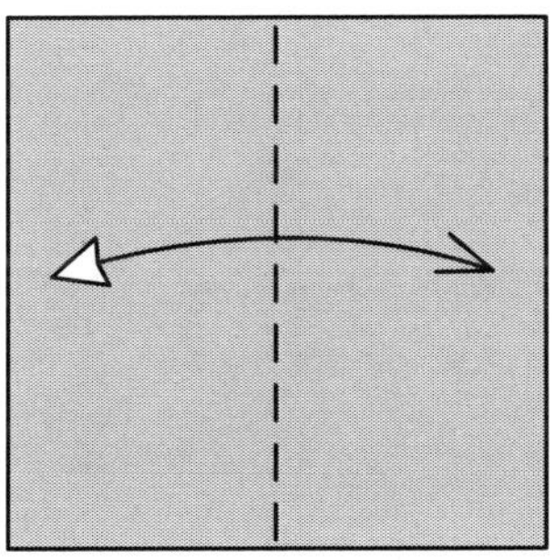

Fold and unfold.

2

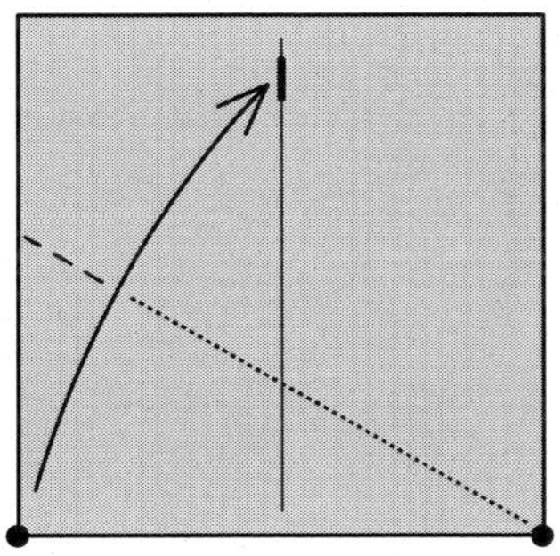

Bring the left dot to the line. Crease on the left.

3

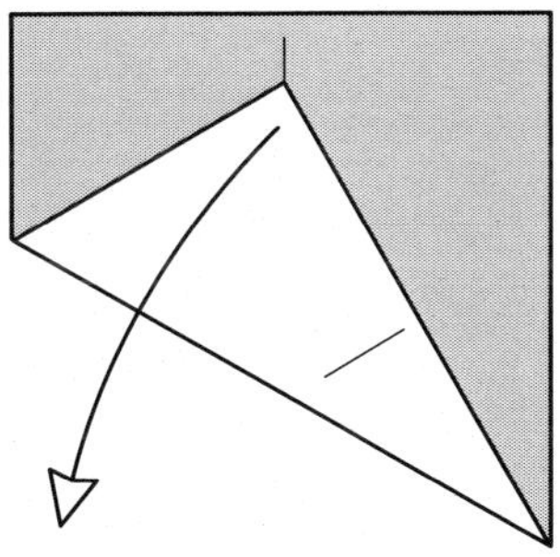

Unfold.

4

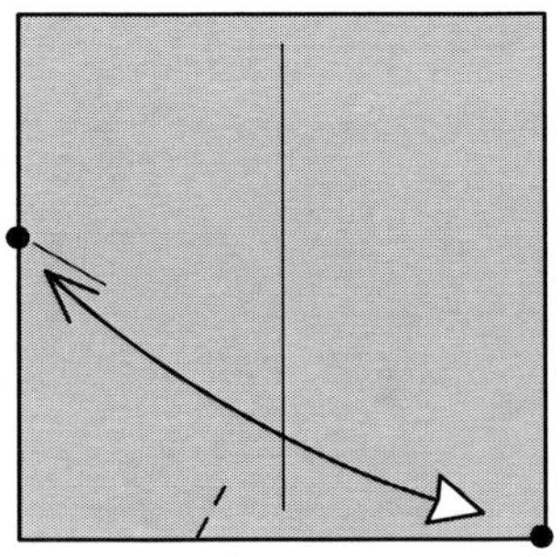

Fold and unfold.

5

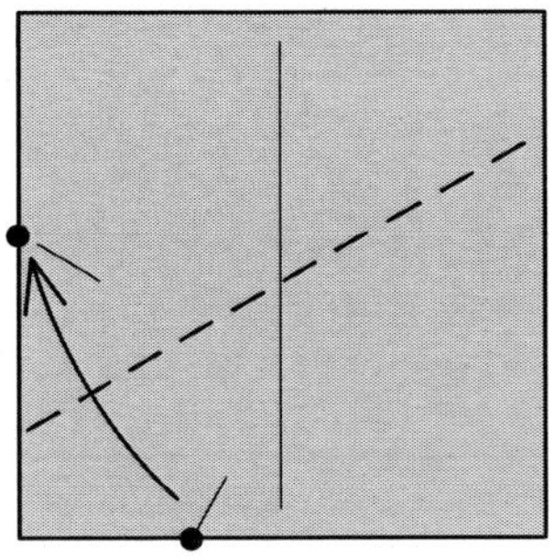

The dots will meet.

6

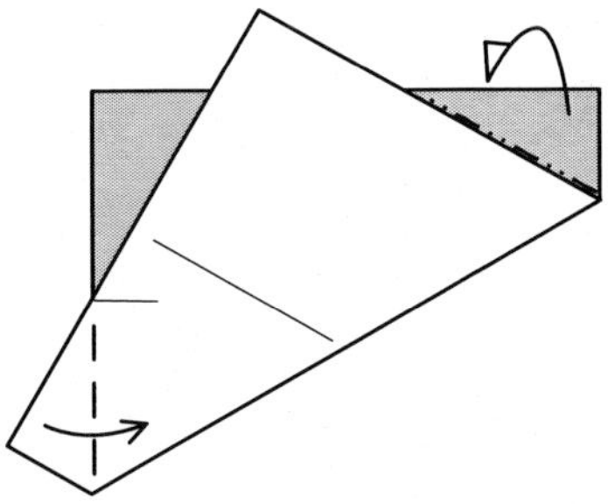

7

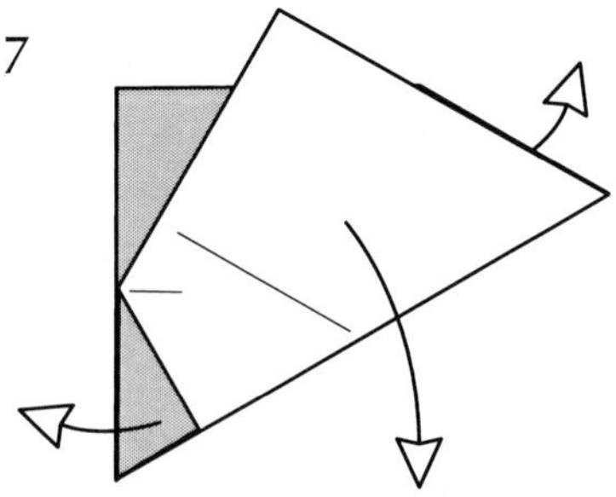

Unfold everything
and rotate 90°.

8

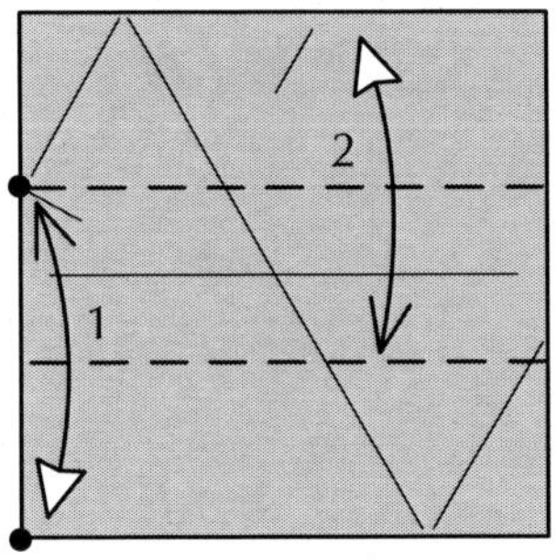

Fold and unfold.

9

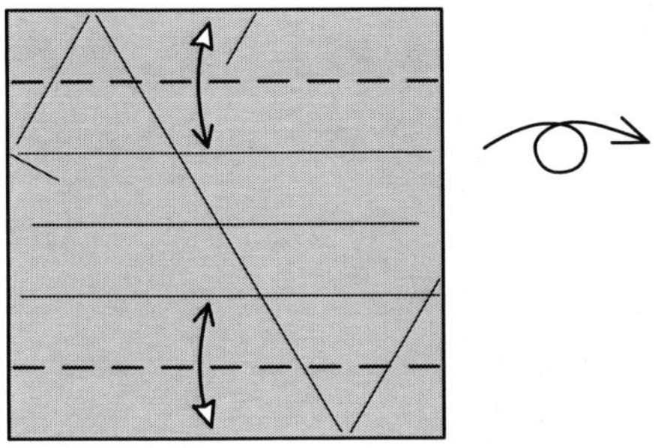

Fold and unfold.

10

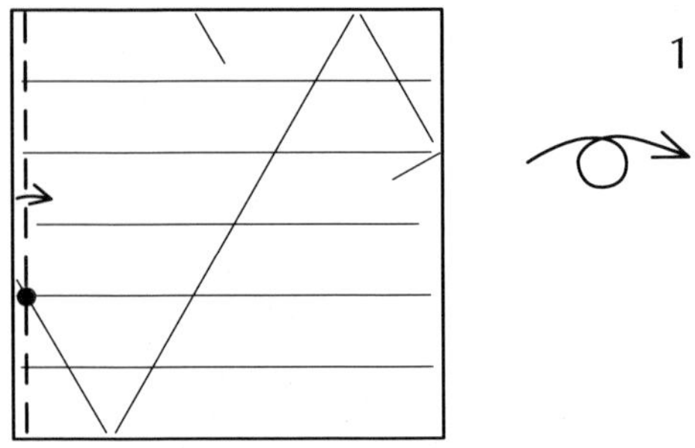

Fold a thin strip at the dot.

11

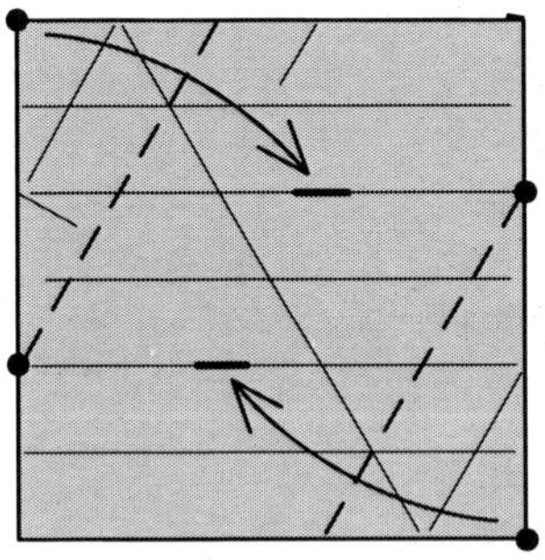

Bring the corners
to the lines.

12

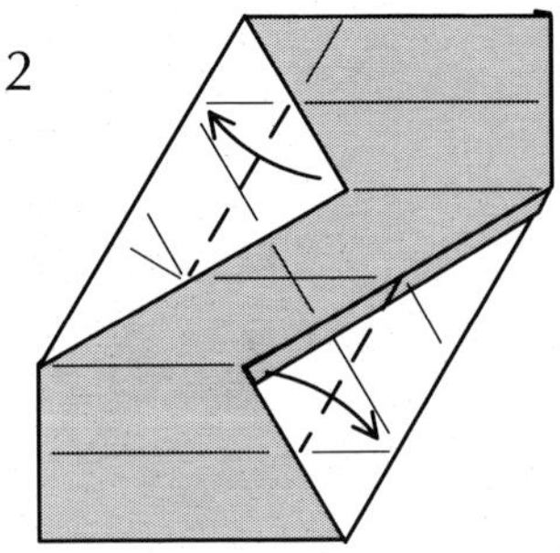

Fold along the creases.

13

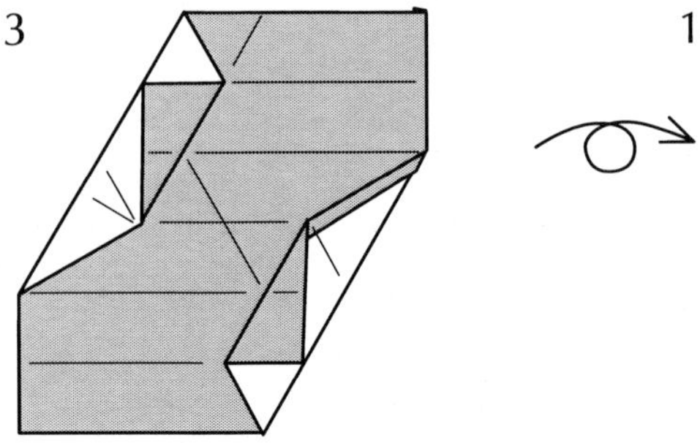

14

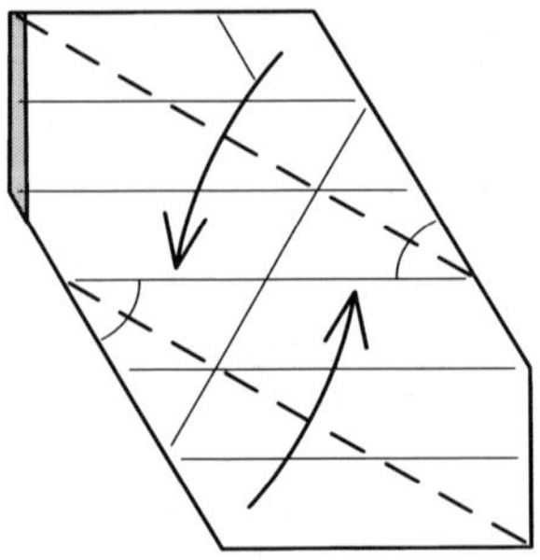

15

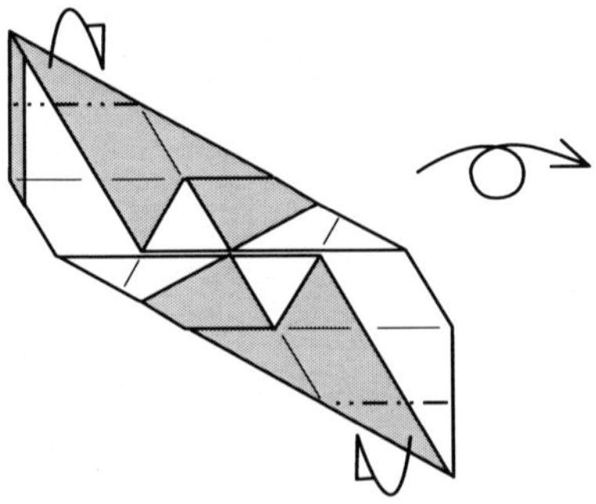

Fold along the
creases. Rotate 90°.

16

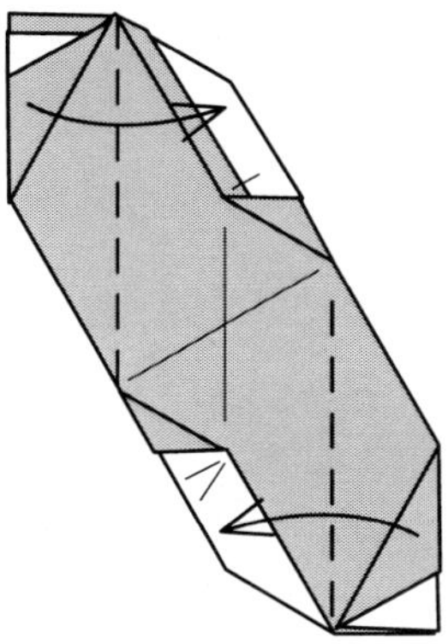

17

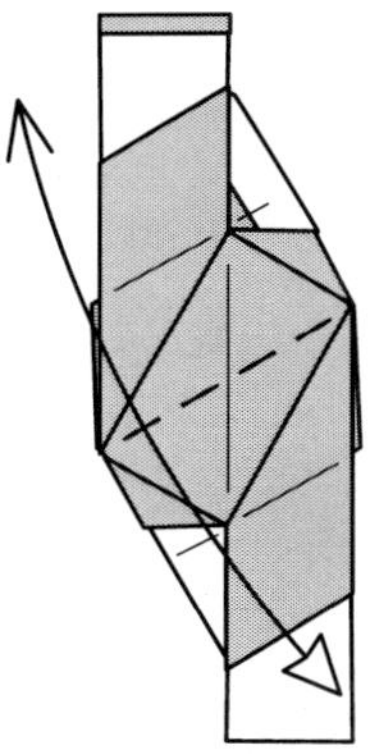

Fold and unfold.

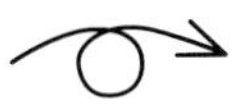

18

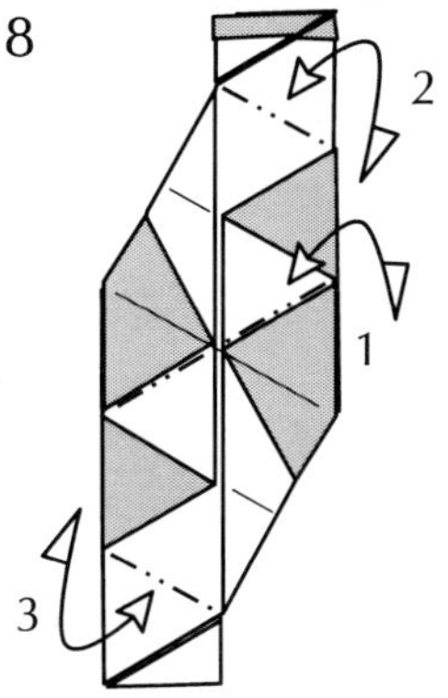

Fold and unfold.

19

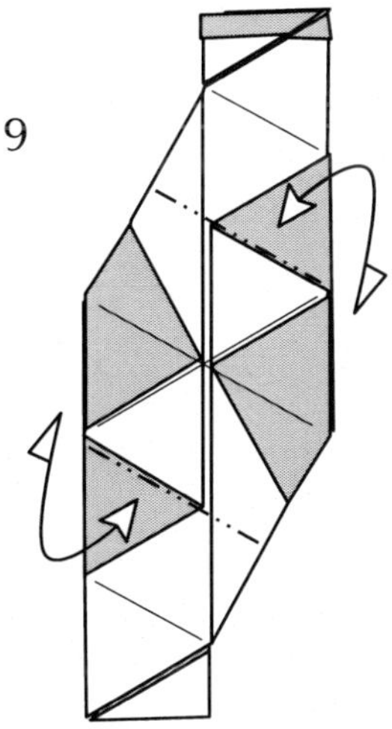

Fold and unfold.

20

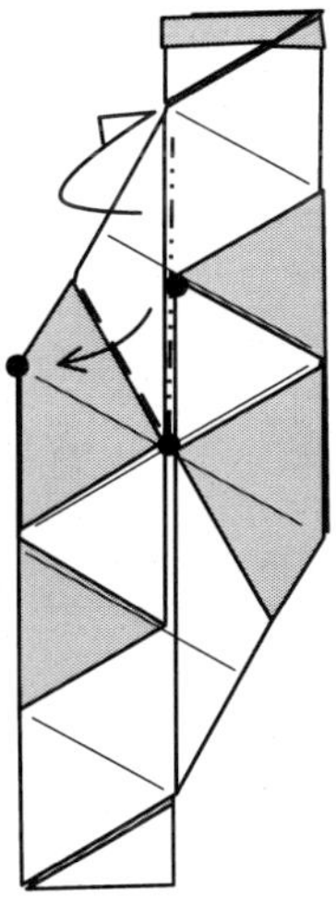

Puff out at the lower dot.

21

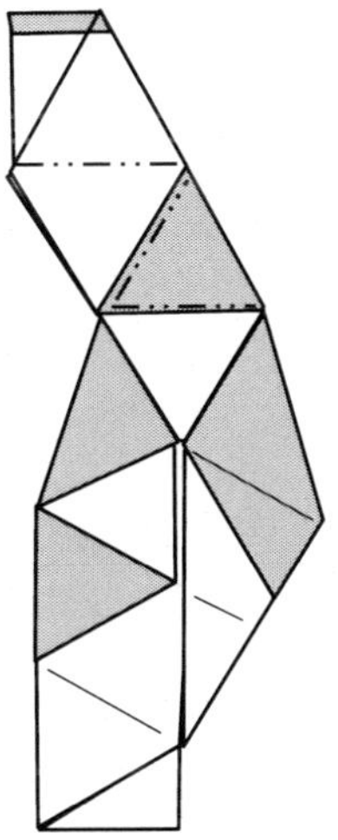

Fold and unfold along the edges. Rotate 180°.

22

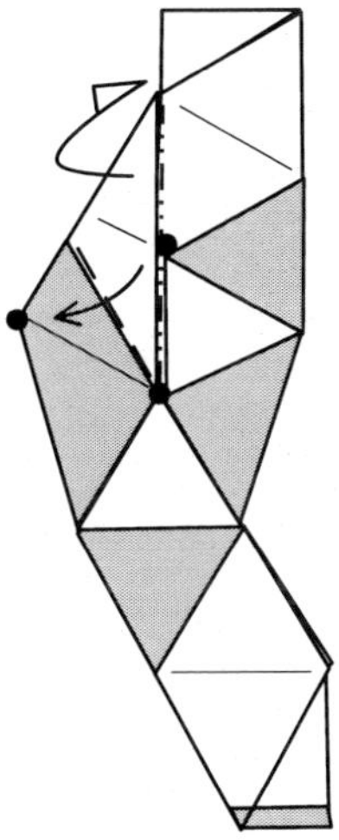

Repeat steps 20–21.

23

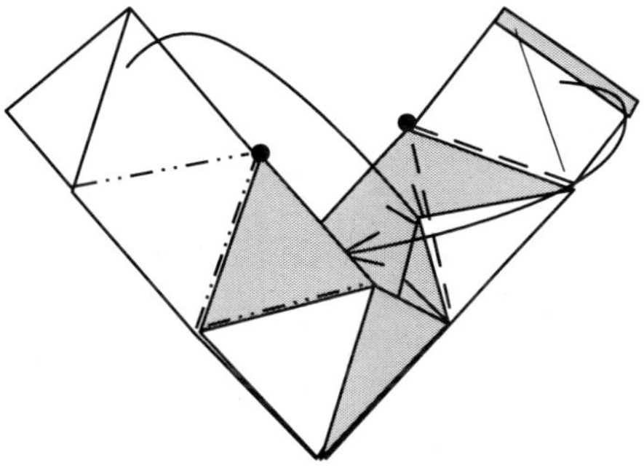

Tuck inside. The dots will meet.

24

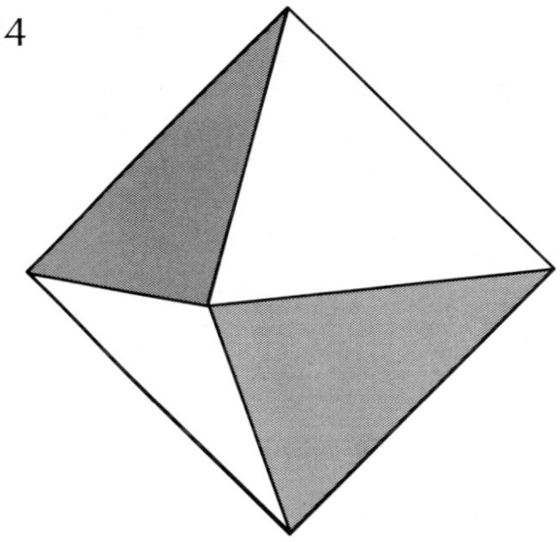

Magic Octahedron

Marble Octahedron

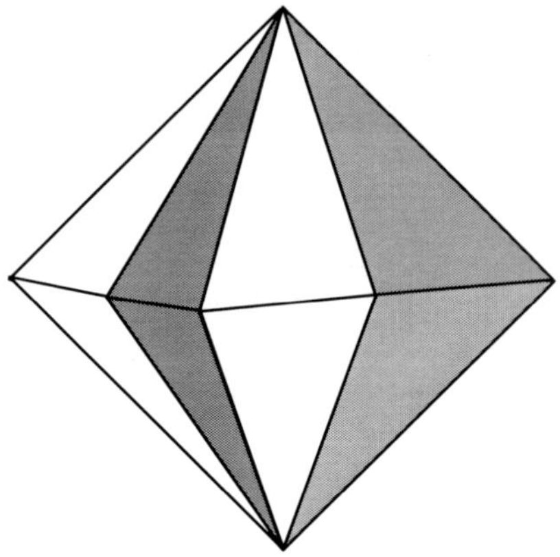

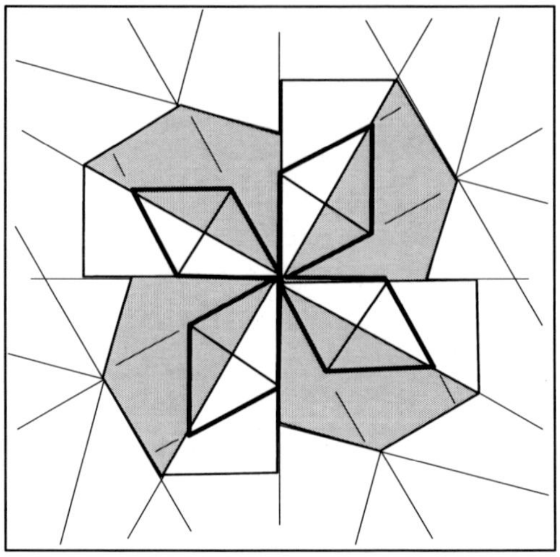

This marble octahedron has two colors on each face. The color pattern is made from the base in step 11, which is then formed into an octahedron. Square symmetry is used and the model closes with a four-way twist lock.

1

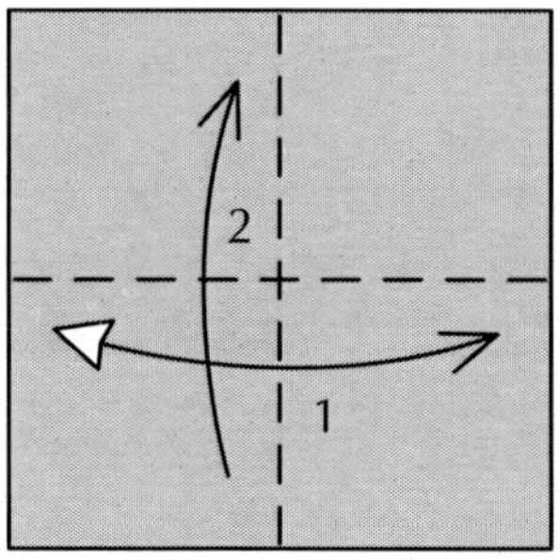

1. Fold and unfold.
2. Fold up.

2

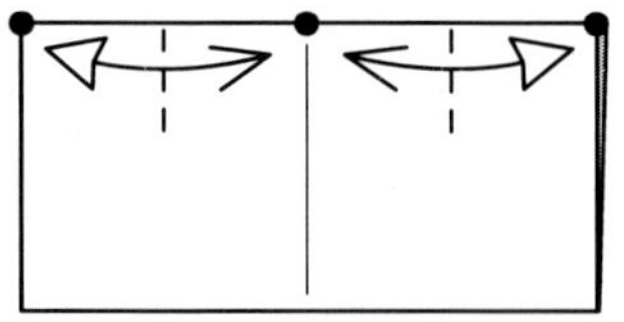

Fold and unfold.

3

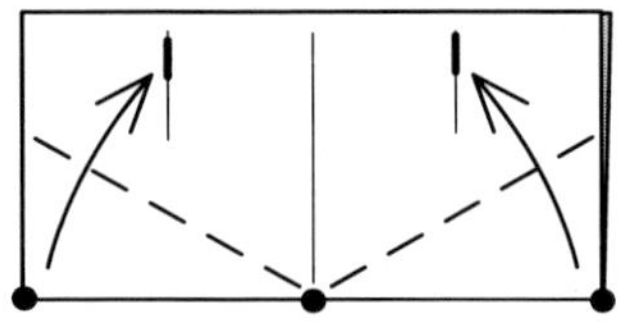

Bring the corners to the lines.

4

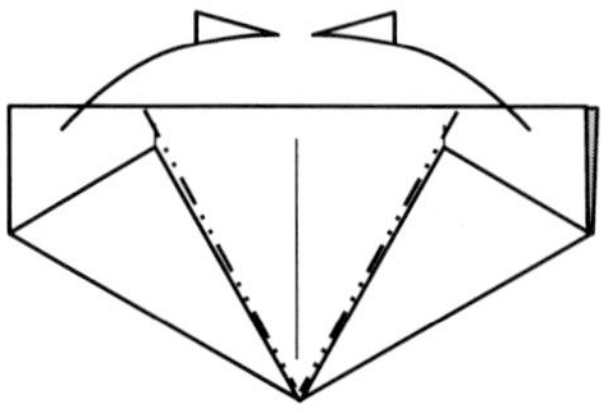

Fold to the center.

5

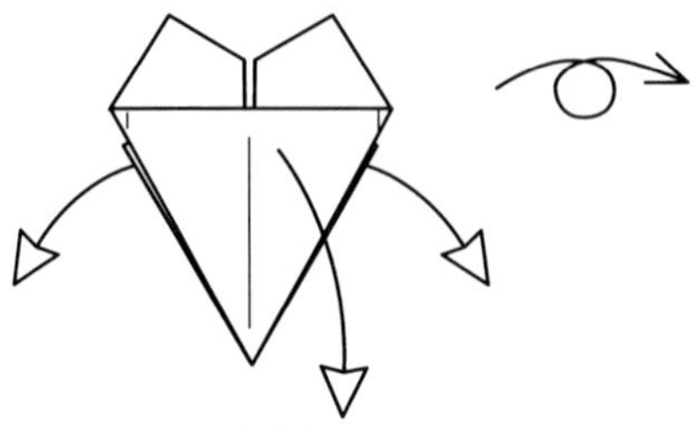

Unfold.

6

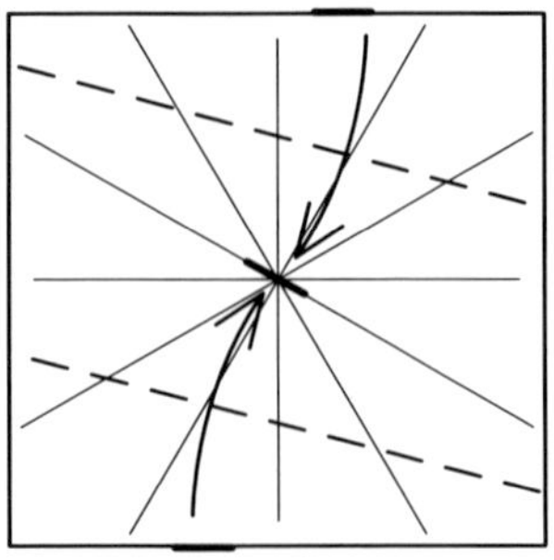

Bring the edges to the line.

7

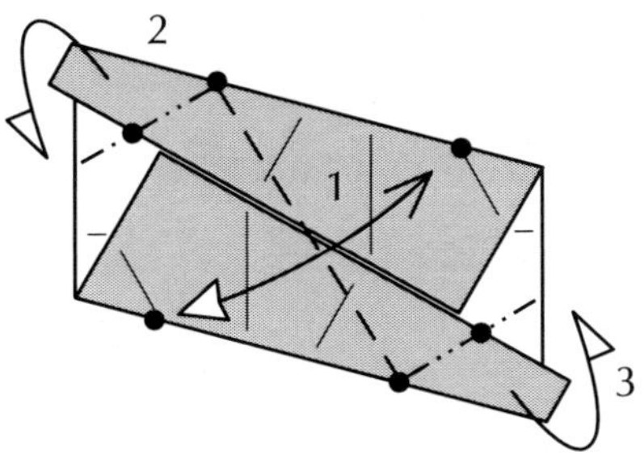

1. Fold and unfold.
2,3. Fold along the creases between the dots.

8

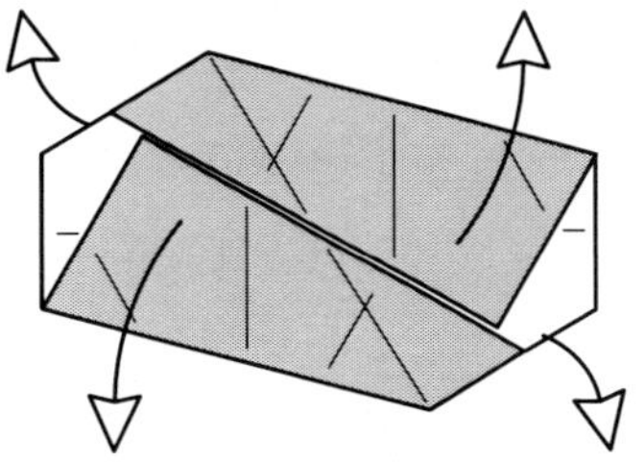

Unfold and rotate 90°.

9

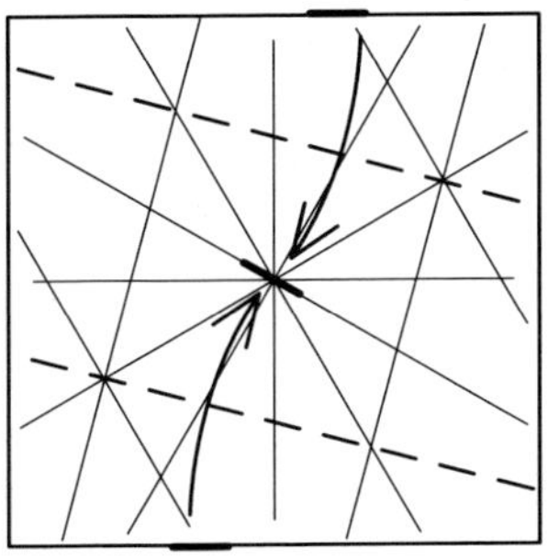

Repeat steps 6–7.

10

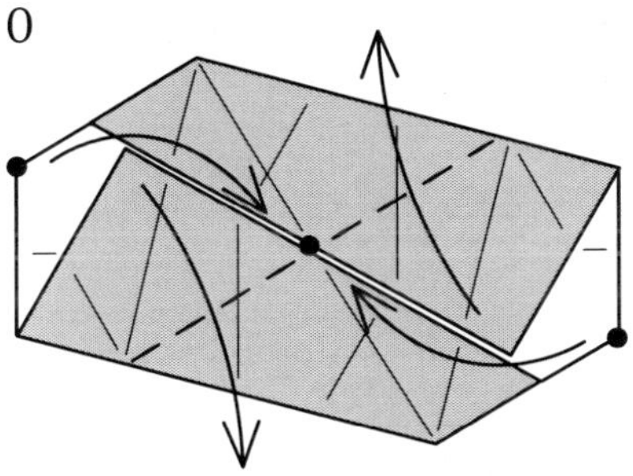

Fold along the creases and bring the outer dots to the center.

11

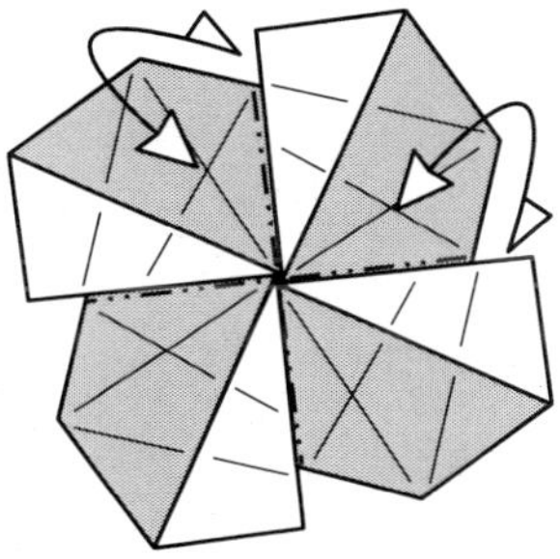

Fold and unfold.

12

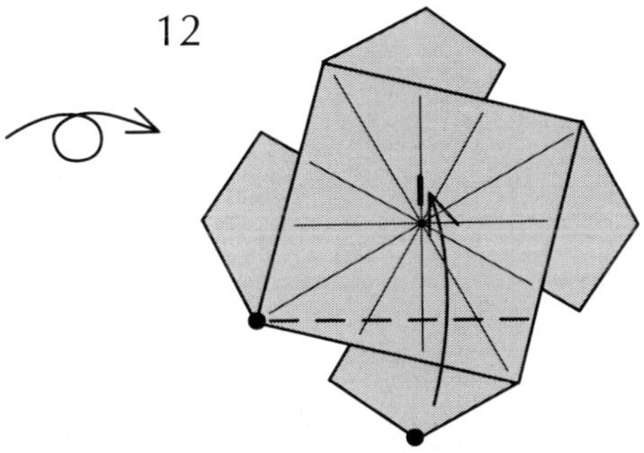

Bring the dot to the line.

13

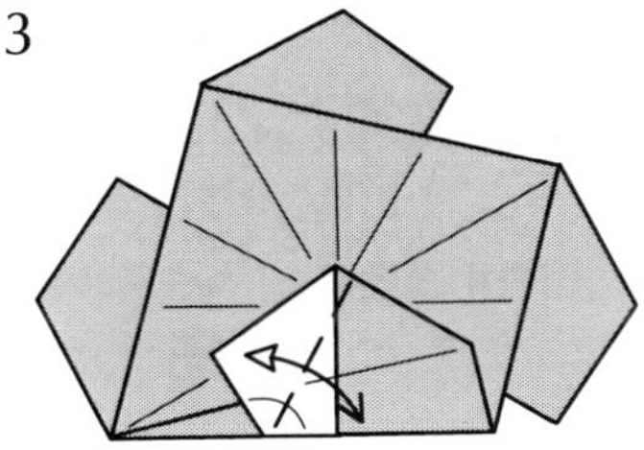

Fold and unfold.

14

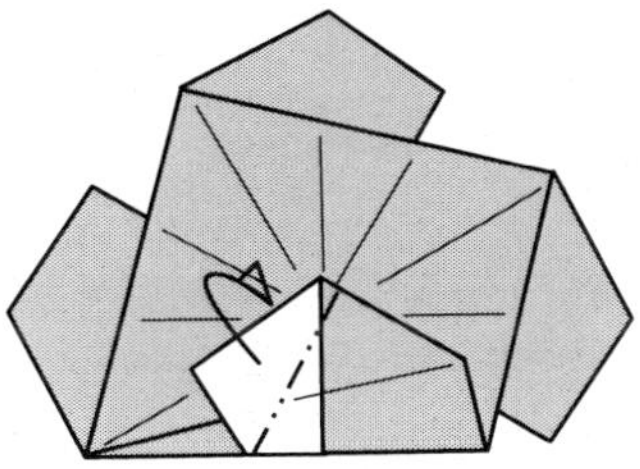

Fold behind along the crease.

15

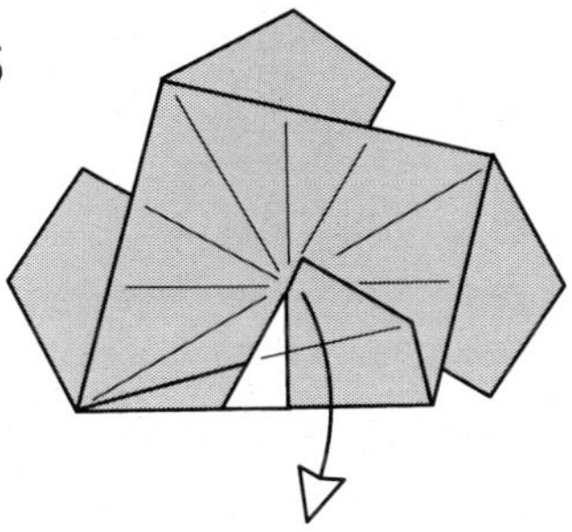

Unfold.

16

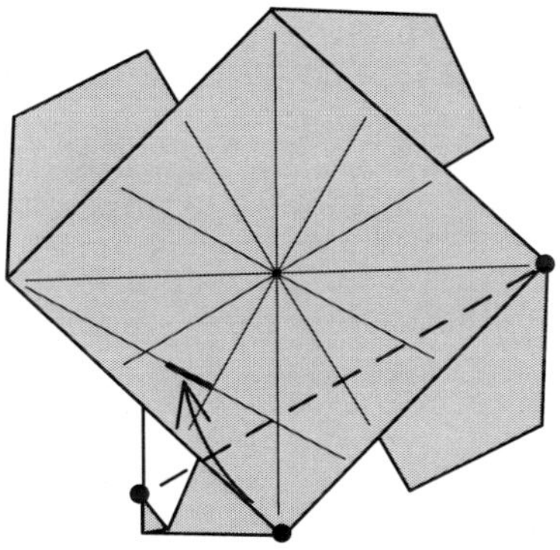

17

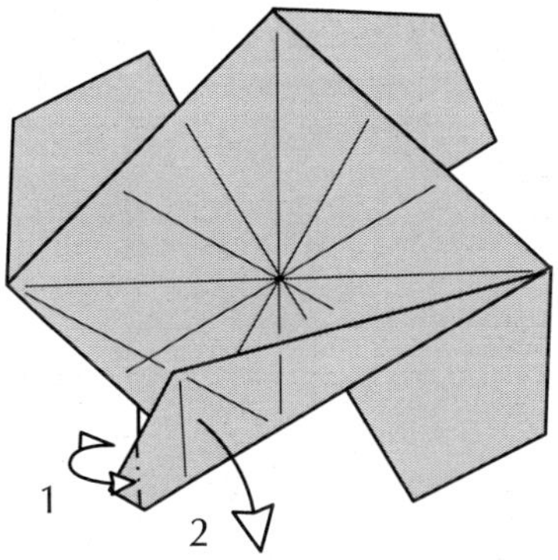

1. Fold and unfold.
2. Unfold.
Rotate.

18

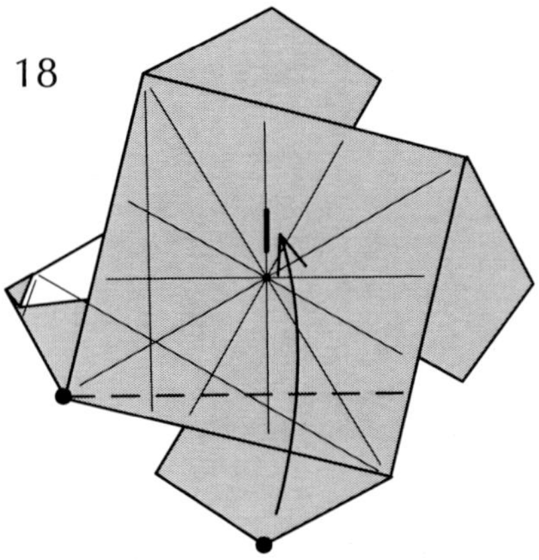

Repeat steps 12–17 three times.

19

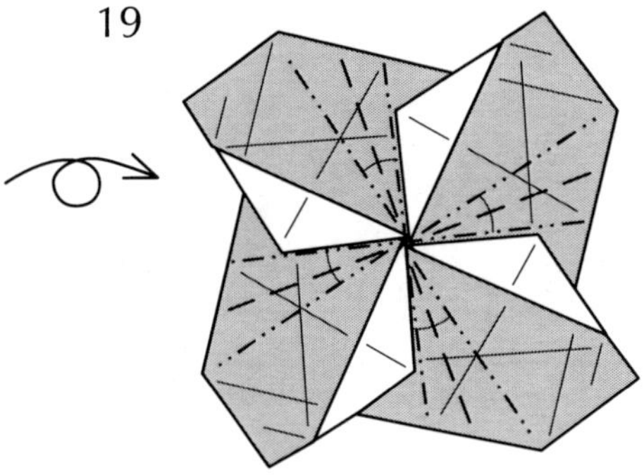

Fold and unfold to bisect the angle.

20

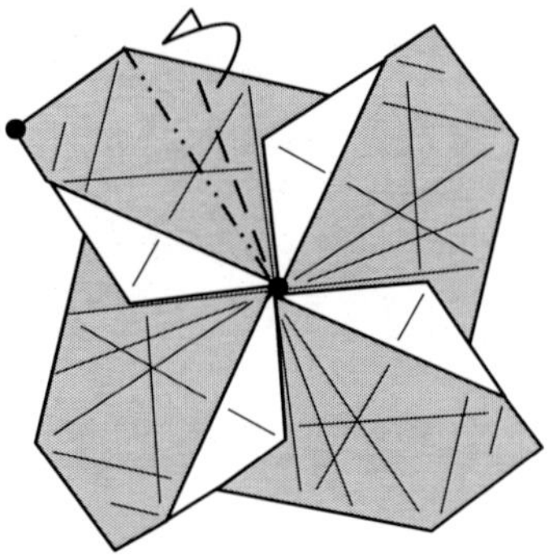

Puff out at the center dot. Rotate so the dot on the left goes to the top.

21

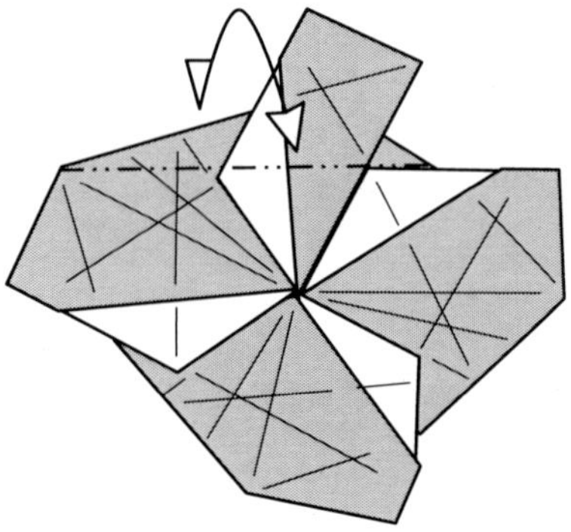

Fold and unfold along the creases. Rotate 90°.

22

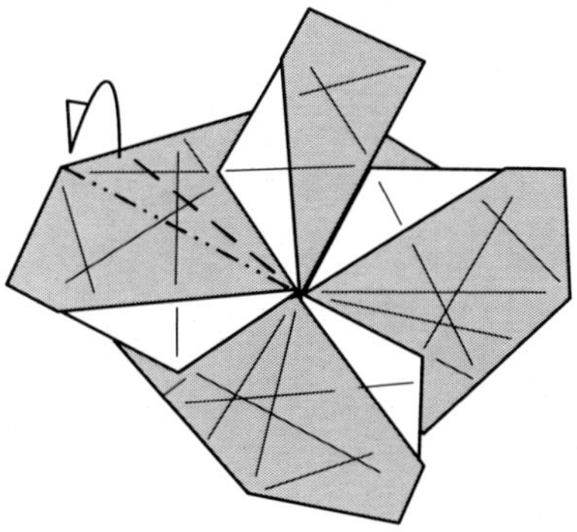

Repeat steps 20–21 three times.

23

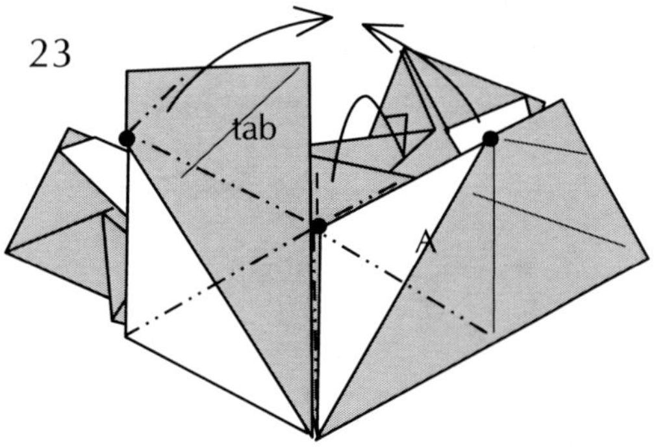

The model closes with a four-way twist lock. One of our sides is shown. Puff out at the lower dot, the upper dots will meet and A will cover the tab.

24

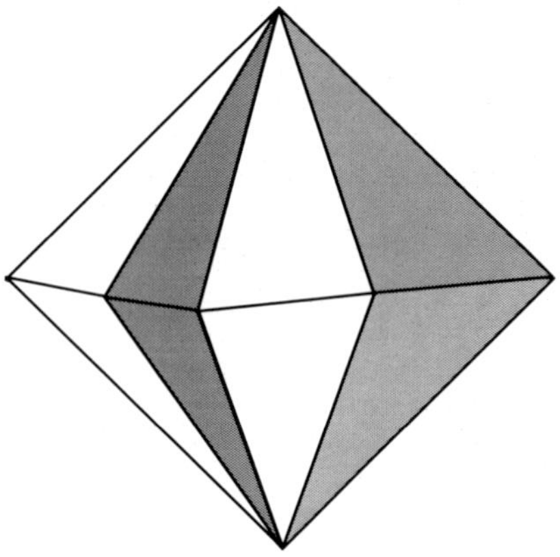

Marble Octahedron

Kaleidoscopic Octahedron

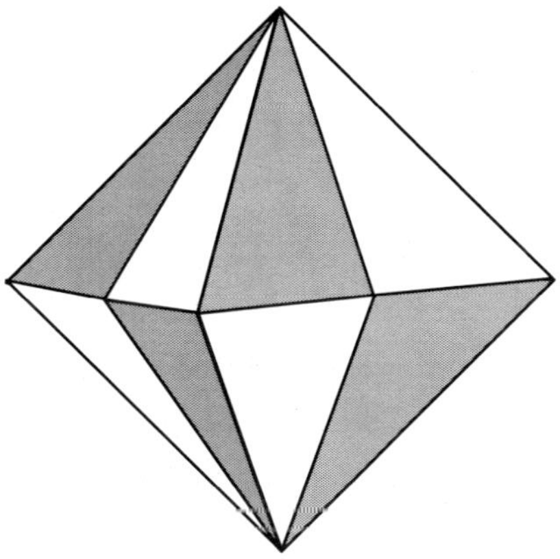

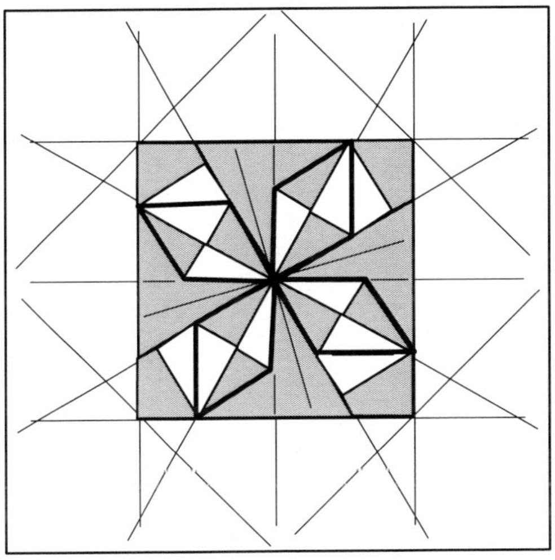

Each of the eight faces of this octahedron has two color-change triangles that alternate on all sides. This model uses square symmetry and closes with a four-way twist lock.

1

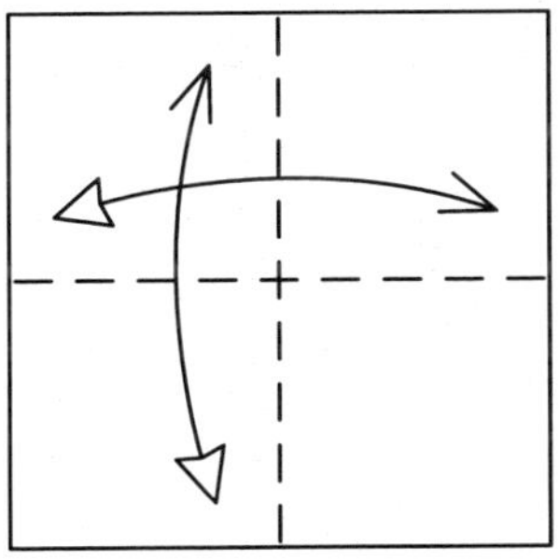

Fold and unfold.

2

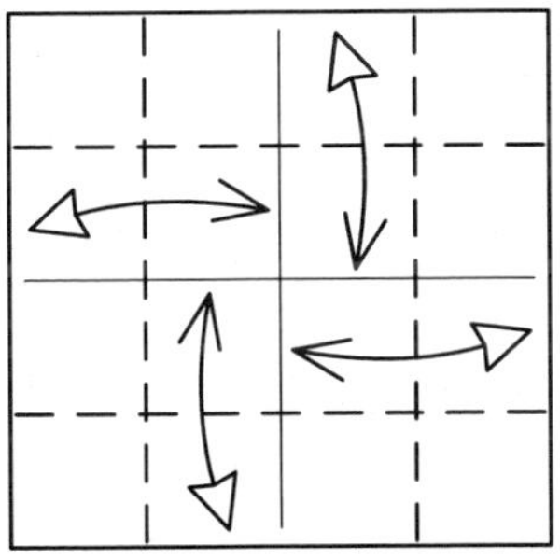

Fold to the center and unfold.

3

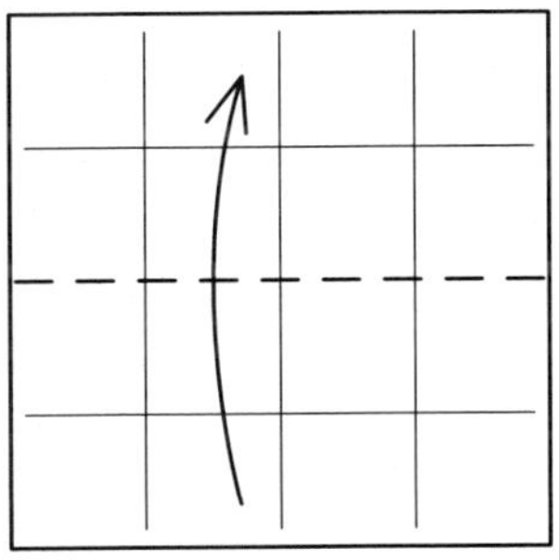

4

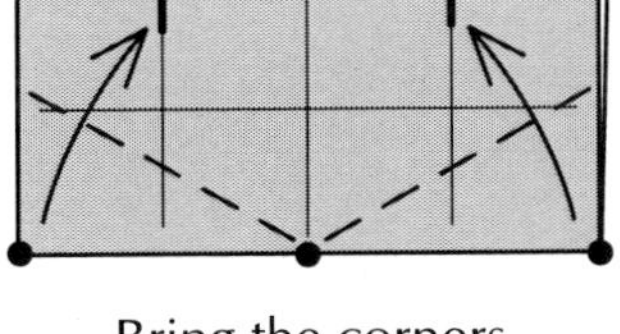

Bring the corners to the lines.

5

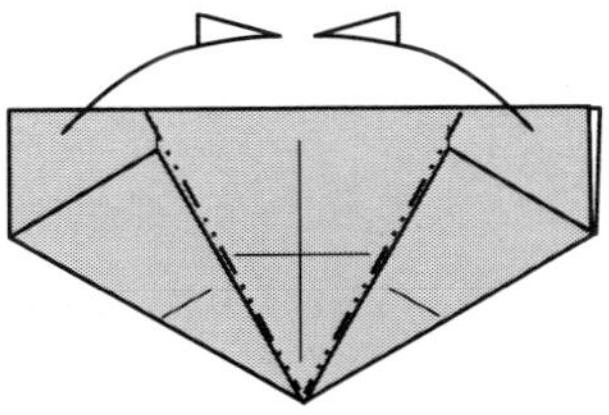

Fold to the center.

6

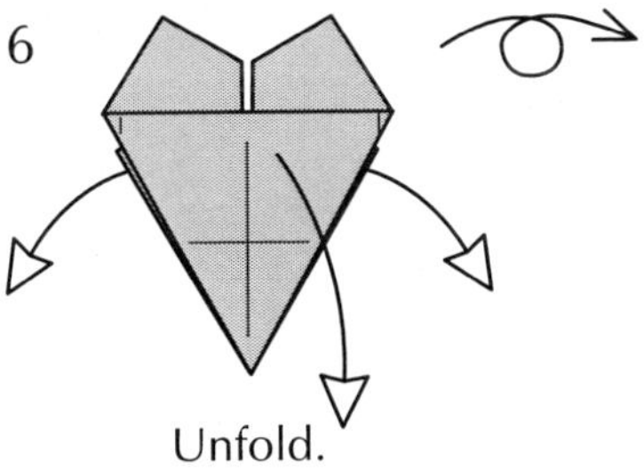

Unfold.

7

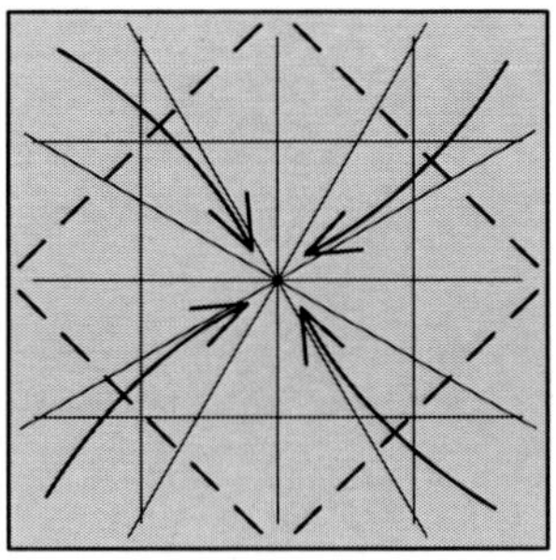

Fold to the center.

8

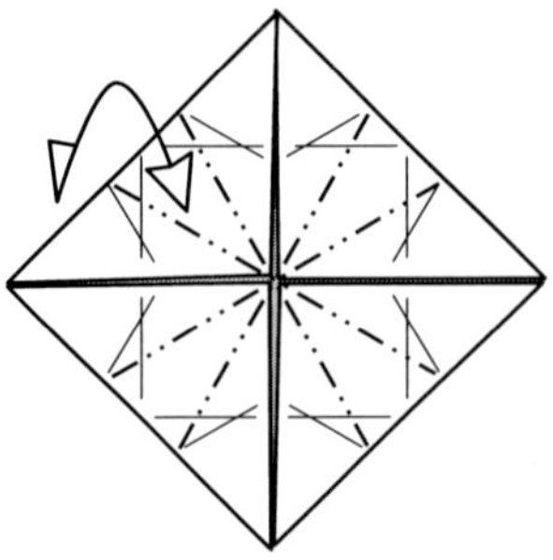

Fold and unfold along
the hidden creases.

9

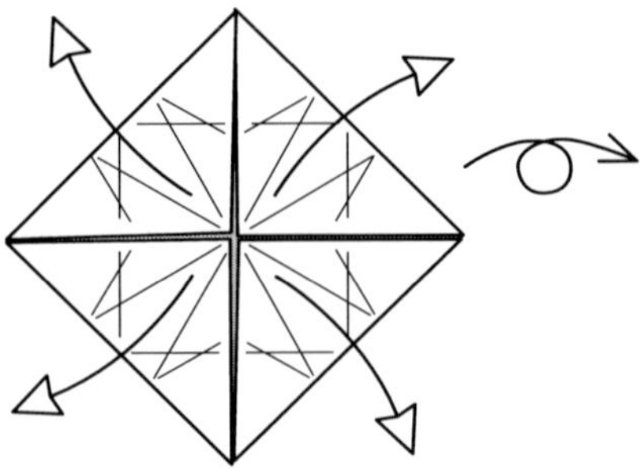

Unfold.

10

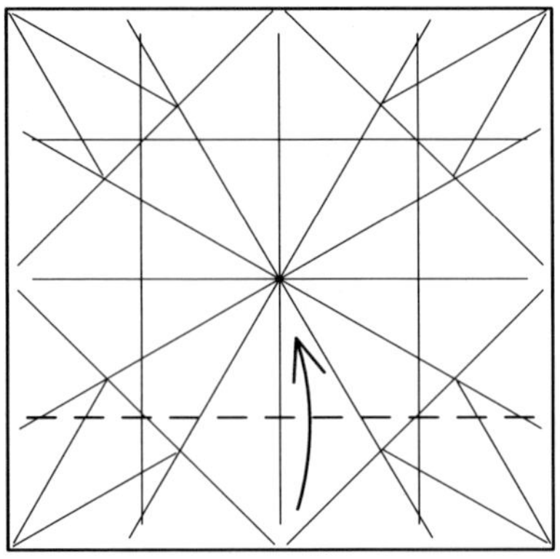

Fold along the crease.

11

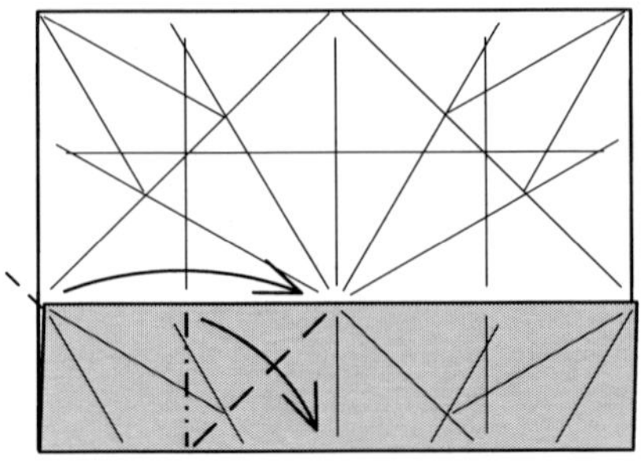

Squash-fold along the creases.

12

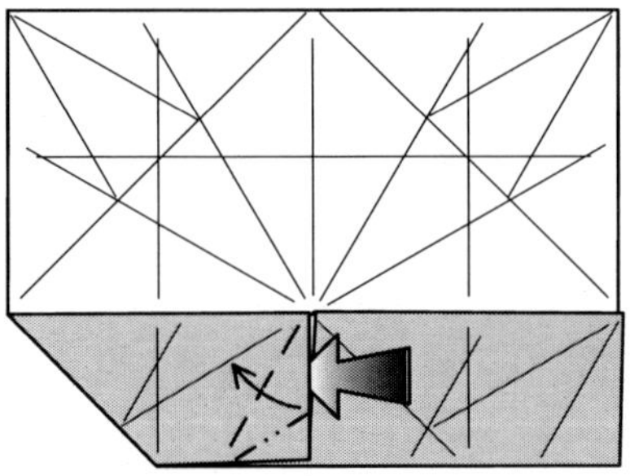

Valley-fold along the crease
for this squash fold.

13

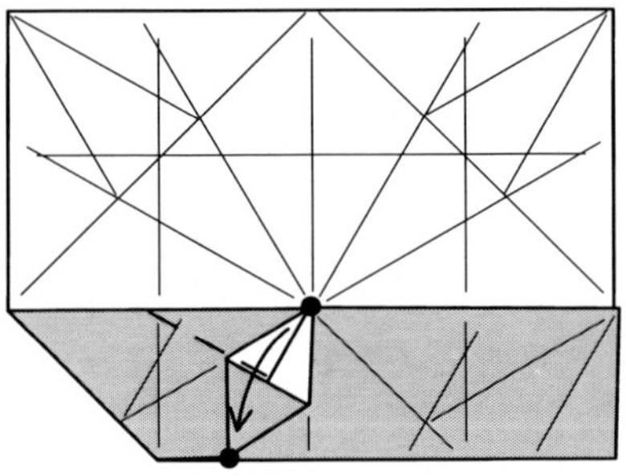

14

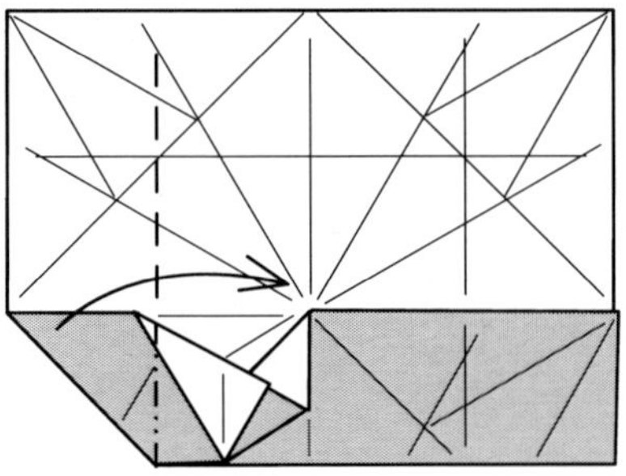

Reverse-fold.

15

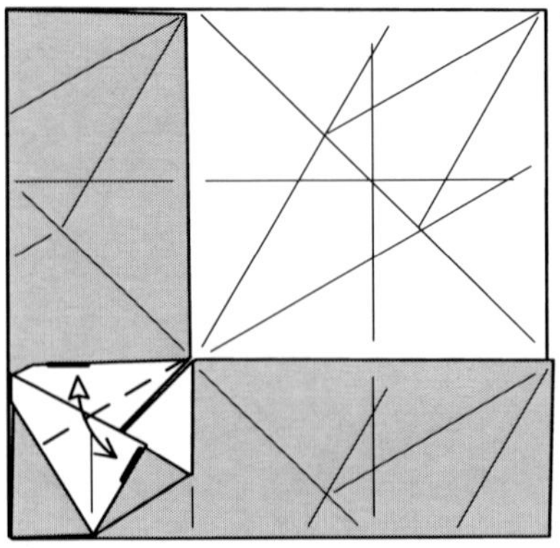

Fold and unfold.

16

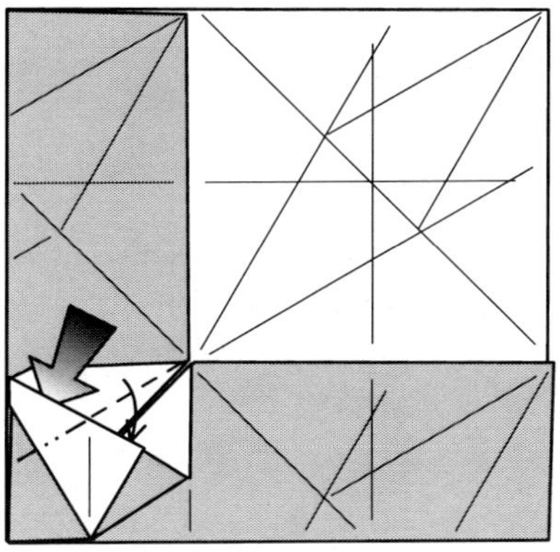

Reverse-fold and rotate 90°.

17

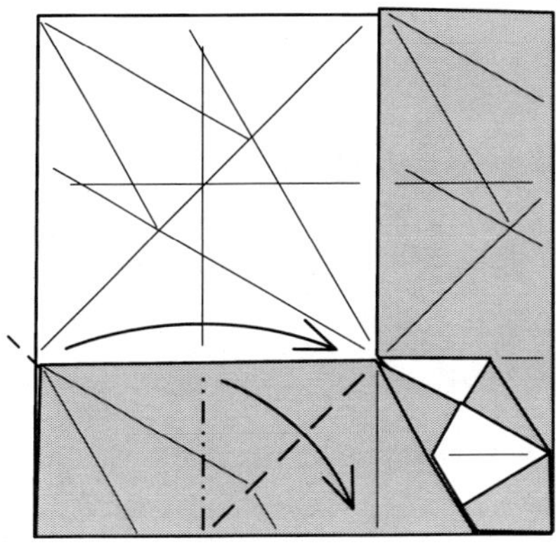

Repeat steps 11–16 three times.

18

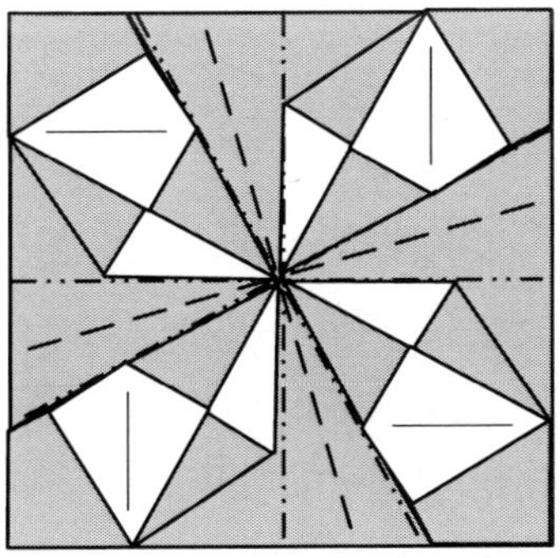

Fold and unfold to bisect the angle.

19

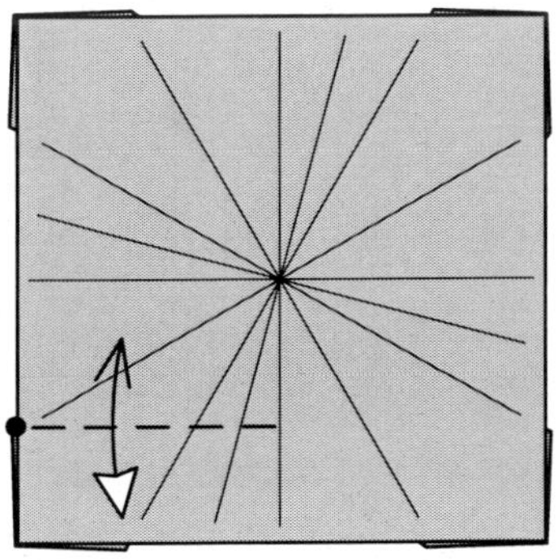

Fold and unfold.

20

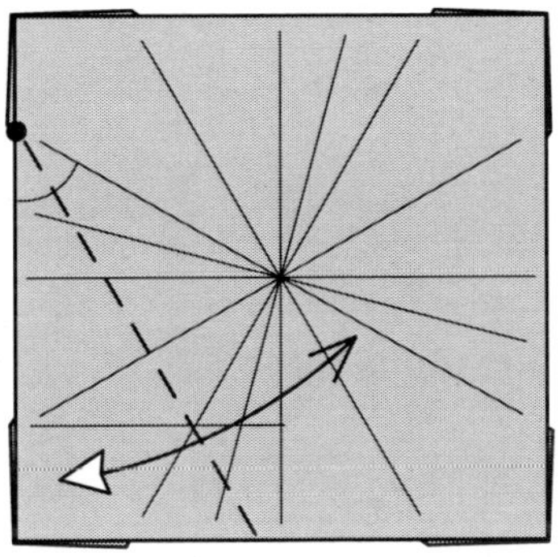

Fold and unfold.

21

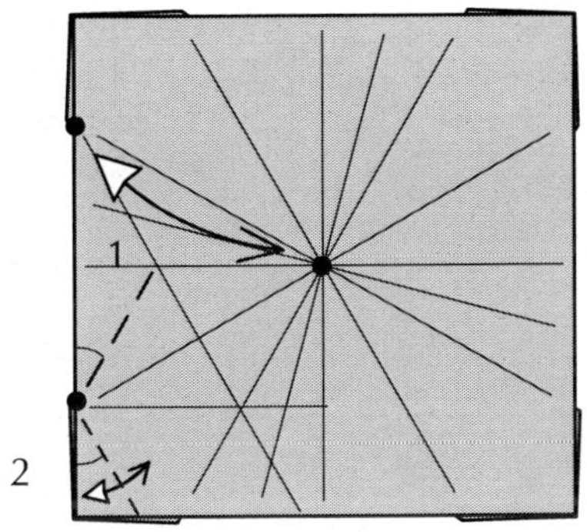

Fold and unfold at 1 and 2. The two angles by the lower dot are about the same. Rotate 90°.

22

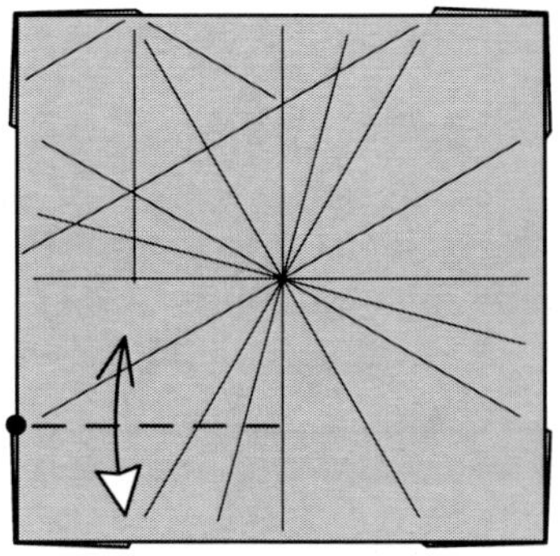

Repeat steps 19–21 three times.

23

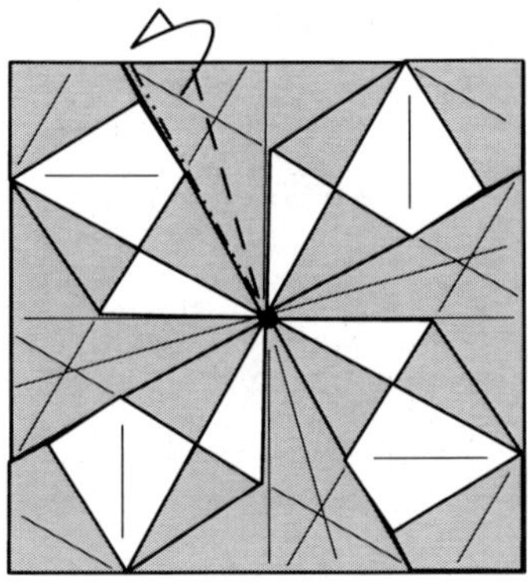

Puff out at the dot.

24

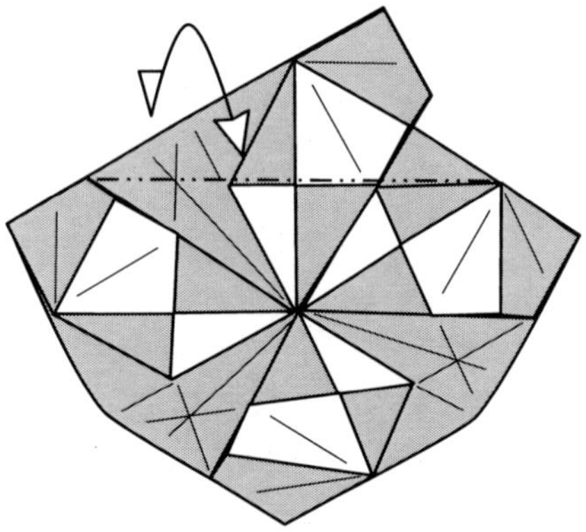

Fold and unfold along the creases. Rotate 90°.

25

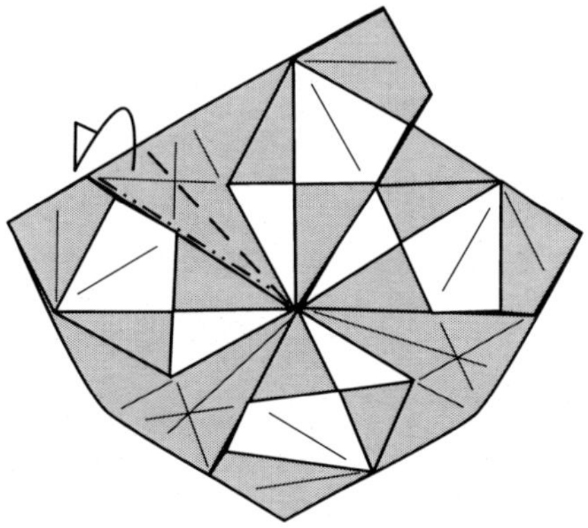

Repeat steps 23–24 three times.

26

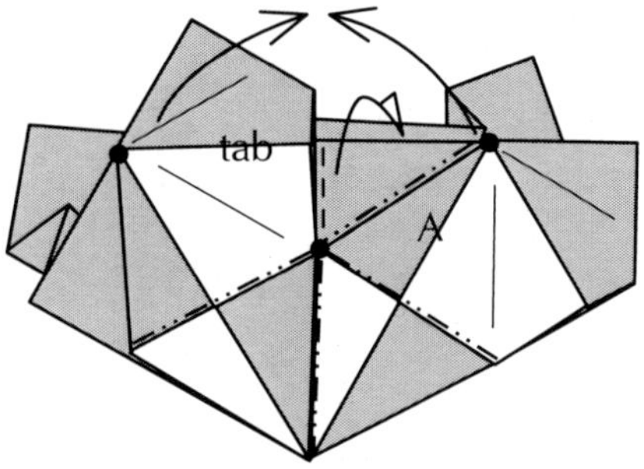

The model closes with a four-way twist lock. One of four sides is shown. Puff out at the lower dot, the upper dots will meet and A will cover the tab.

27

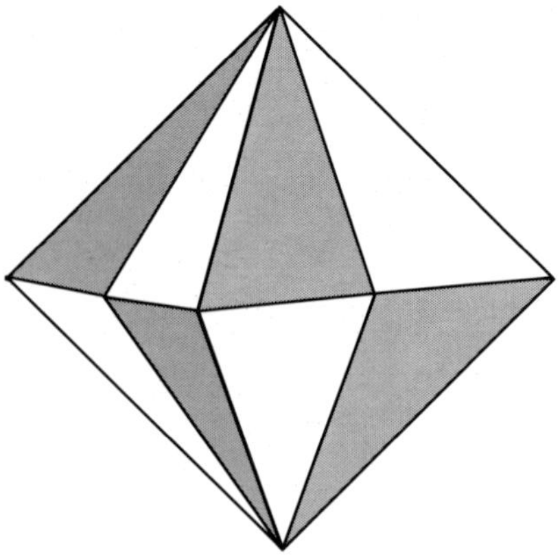

Kaleidoscopic Octahedron

Deltahedra

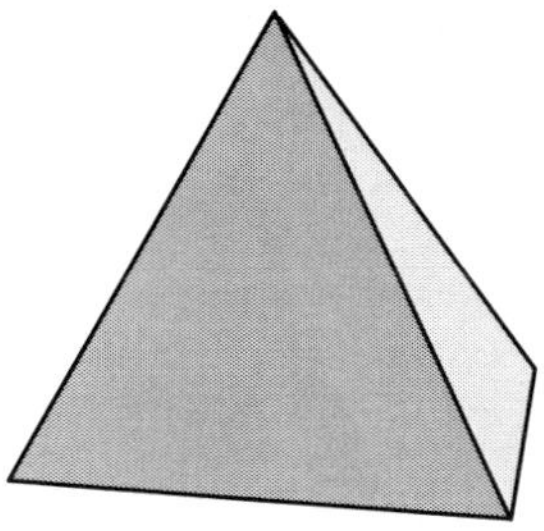

Tetrahedron

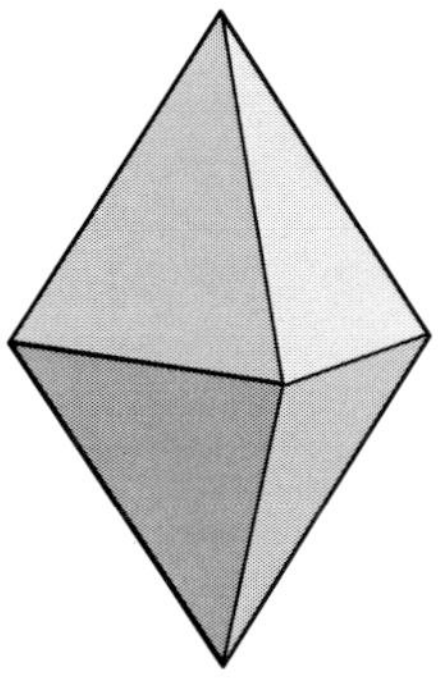

Triangular Dipyramid

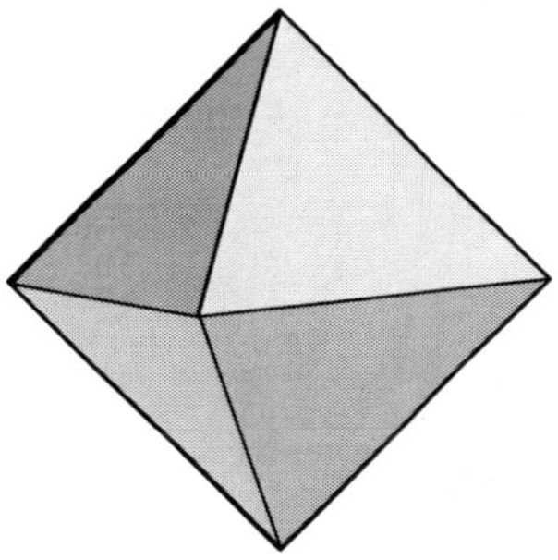

Octahedron

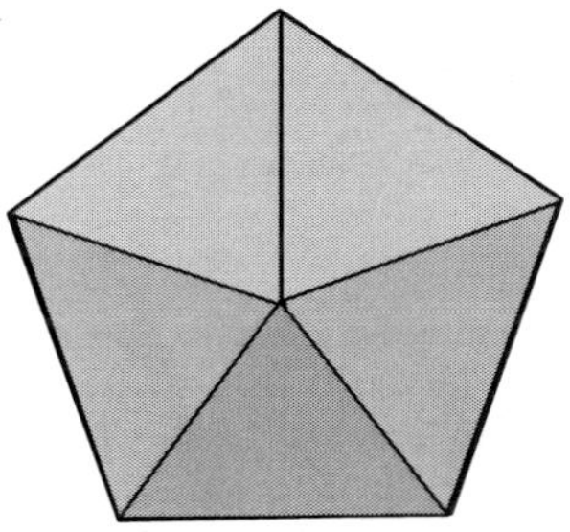

Pentagonal Dipyramid

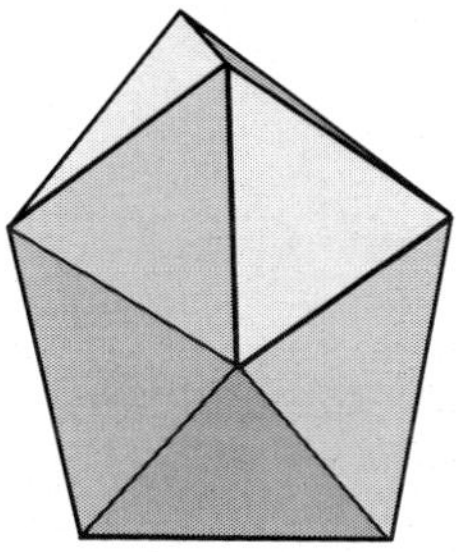

Dodecadeltahedron

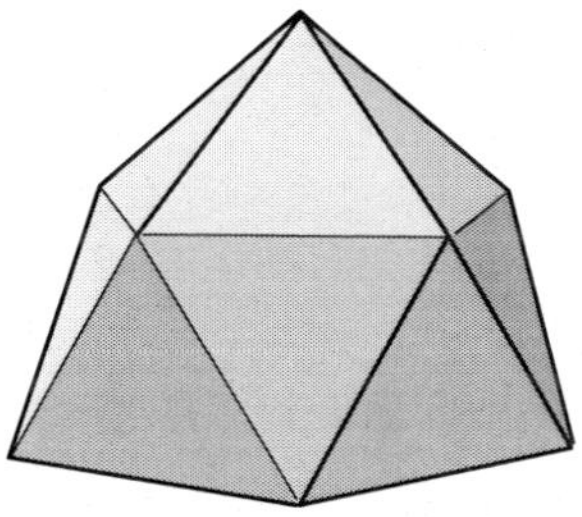

Triaugmented Triangular Prism

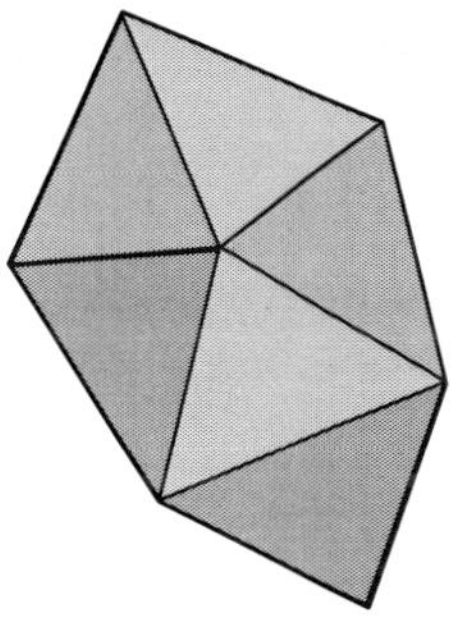

Gyroelongated Square Dipyramid

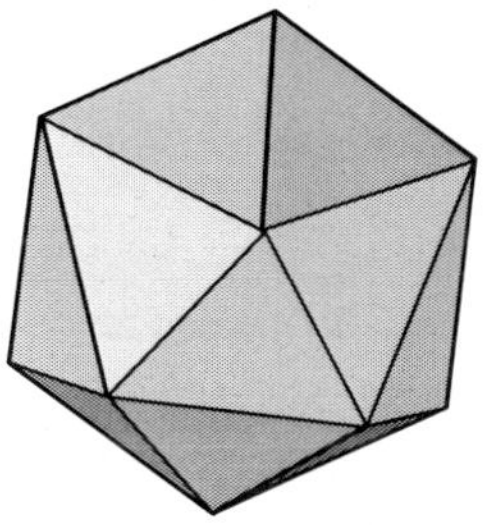

Icosahedron

Deltahedra are polyhedra whose sides are equilateral triangles. There are eight convex deltahedra (convex means any line connecting the vertices are on or inside the solid). Three—the tetrahedron, octahedron, and icosahedron—are Platonic solids. The remaining five are irregular shapes that each have interesting symmetries and other properties, and each has several names.

Here are the diagrams for all but the tetrahedron, which was shown earlier.

Deltahedra	Number of Sides
Tetrahedron	4
Triangular Dipyramid	6
Octahedron	8
Pentagonal Dipyramid	10
Dodecadeltahedron	12
Triaugmented Triangular Prism	14
Gyroelongated Square Dipyramid	16
Icosahedron	20

Triangular Dipyramid

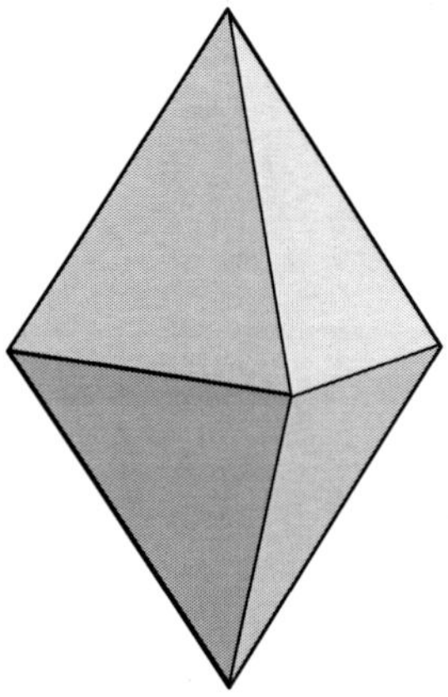

The six faces of this triangular dipyramid are all equilateral triangles.

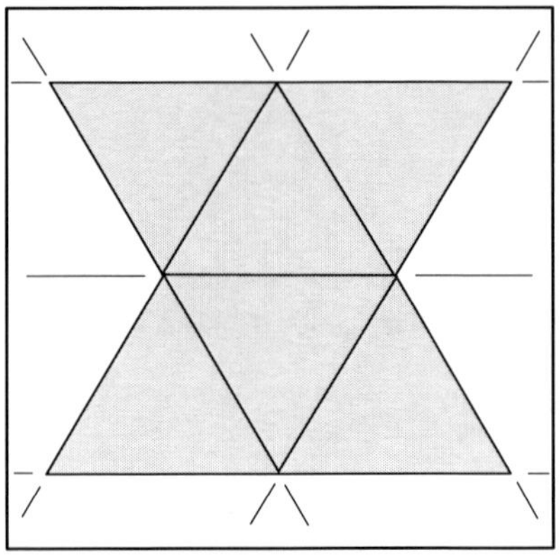

1

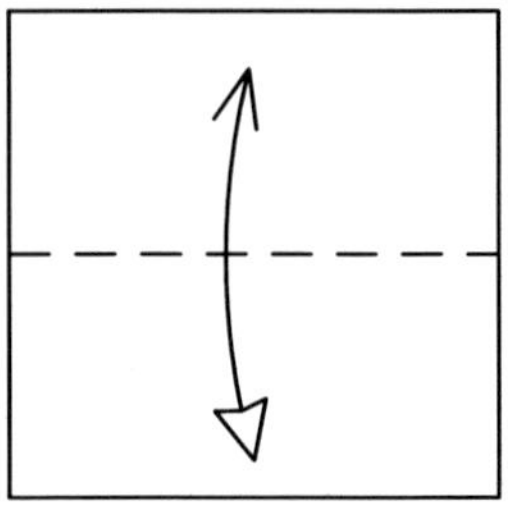

Fold and unfold.

2

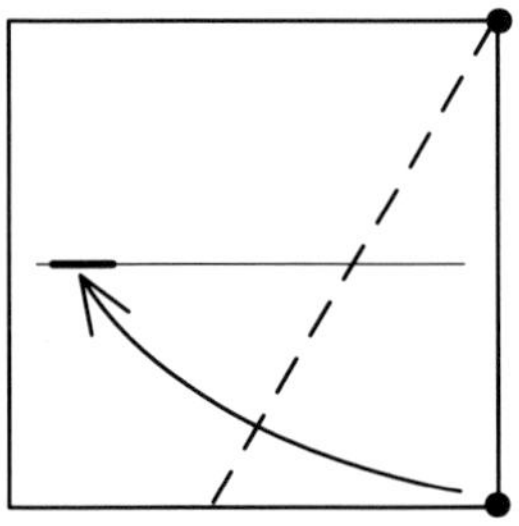

3

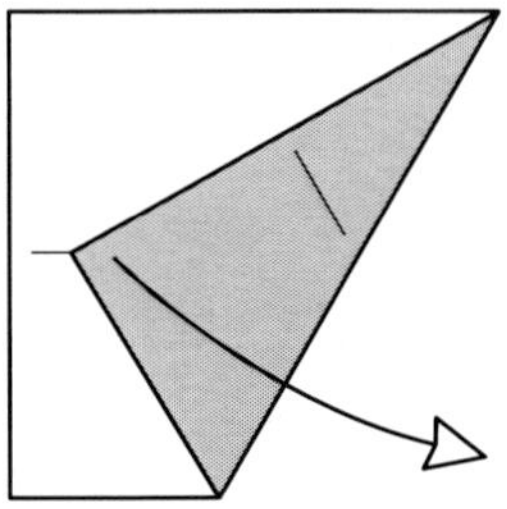

Unfold.

4

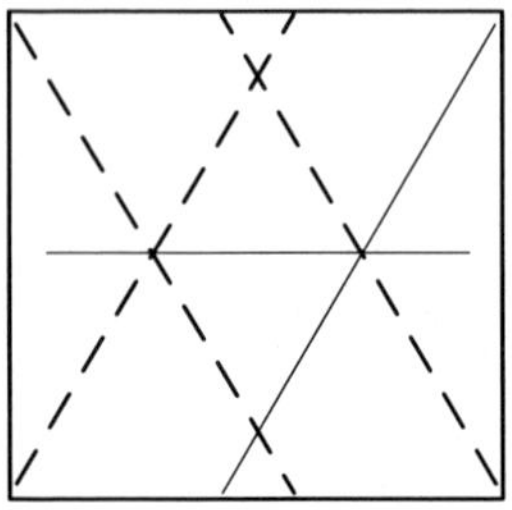

Fold and unfold three times.

5

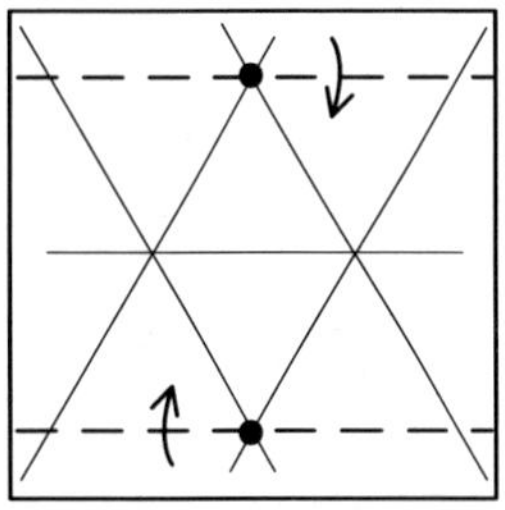

Rotate 90°.

6

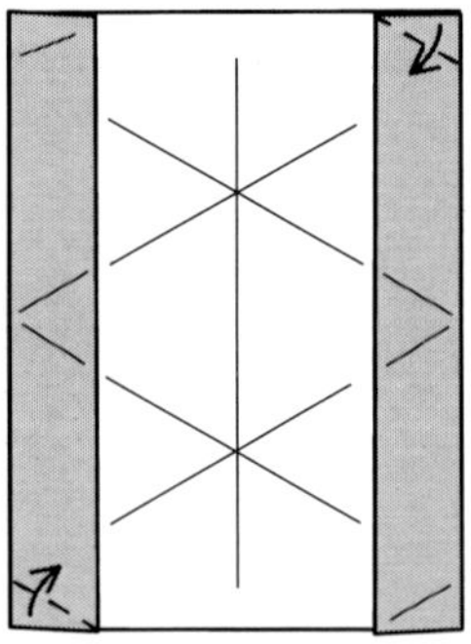

Fold along the creases.

7

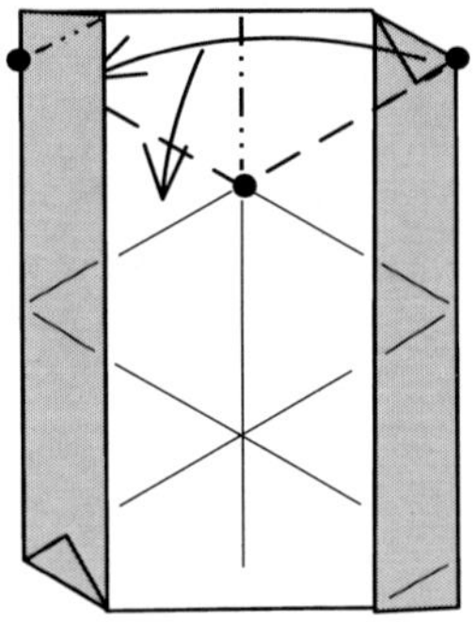

Push in at the center dot. The other dots will meet. Rotate 180°.

8

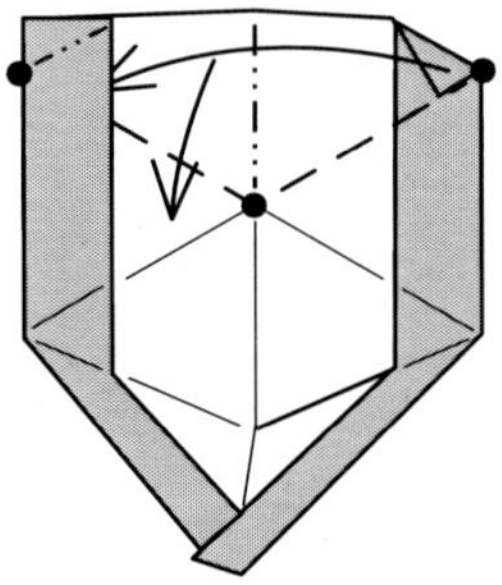

Repeat step 7.

9

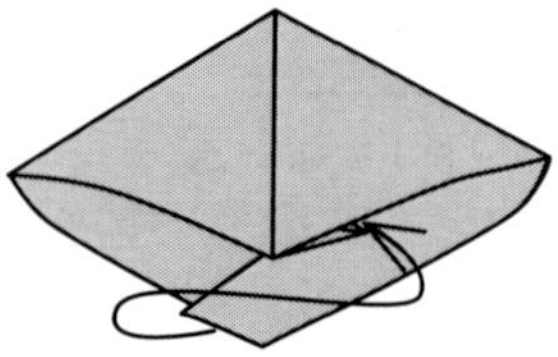

Interlock the tabs and rotate.

10

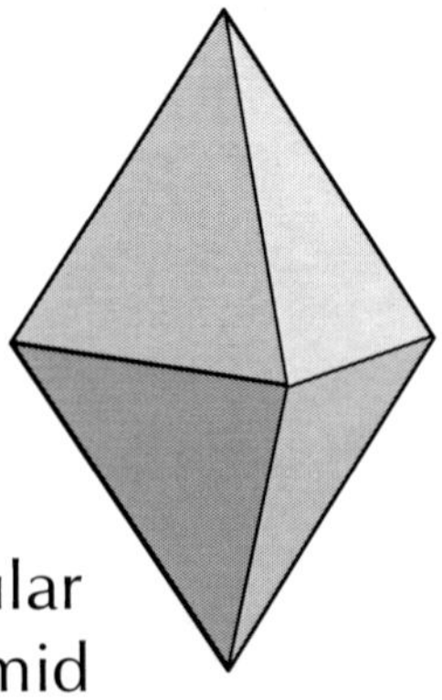

Triangular Dipyramid

Octahedron

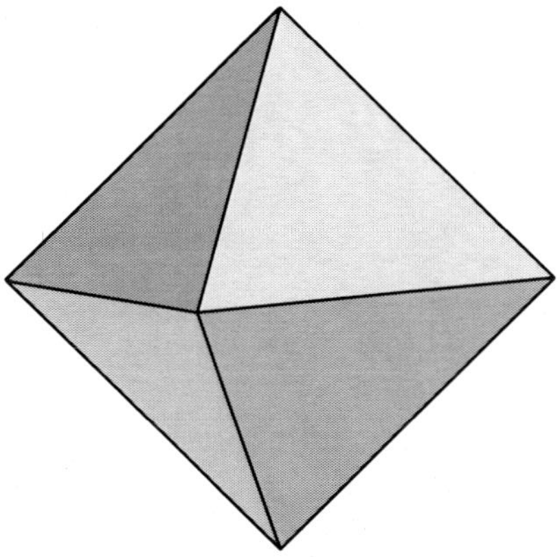

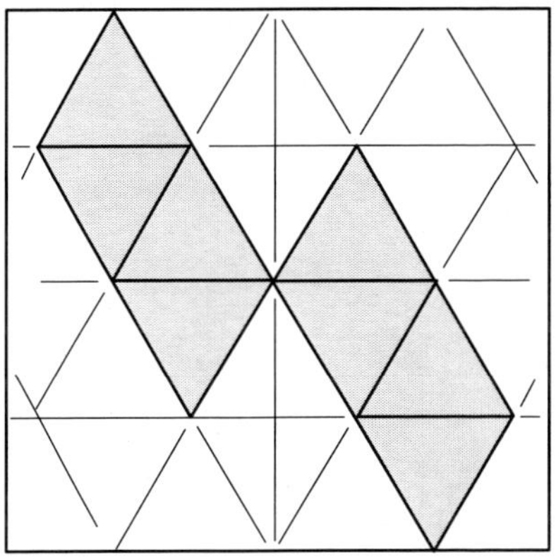

Another design for an origami octahedron was shown earlier. This model was designed to be as easy as possible. It uses odd symmetry and finshes by inflating the model.

In designing origami polyhedra, or any origami subject, there are many ways it can be accomplished. Each version highlights different aspects that are important to the designer.

1

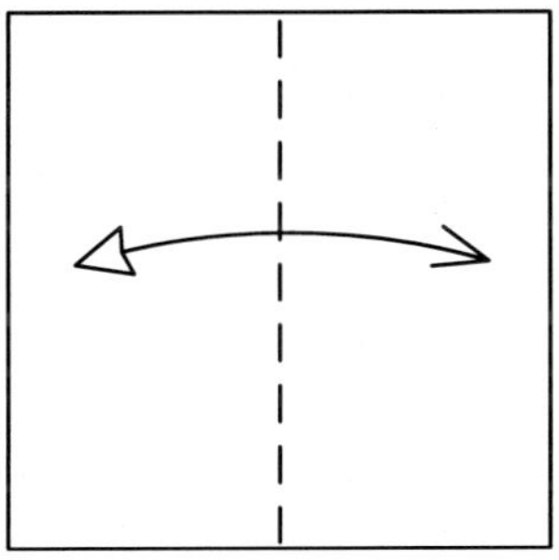

Fold and unfold.

2

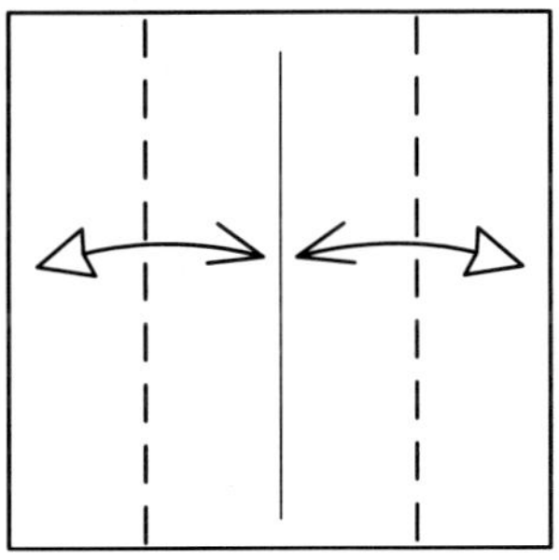

Fold and unfold.

3

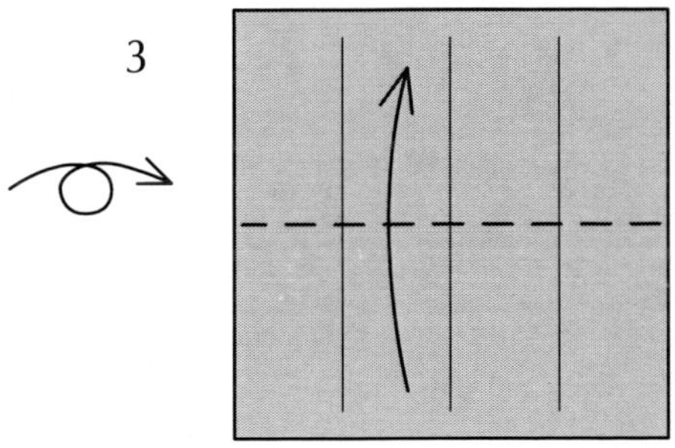

4

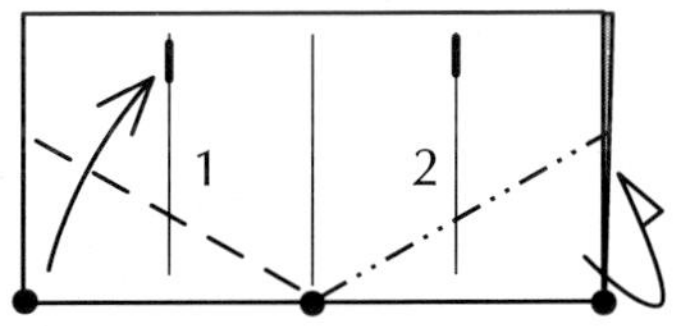

1. Bring the lower left corner to the line.
2. Turn over and repeat.

5

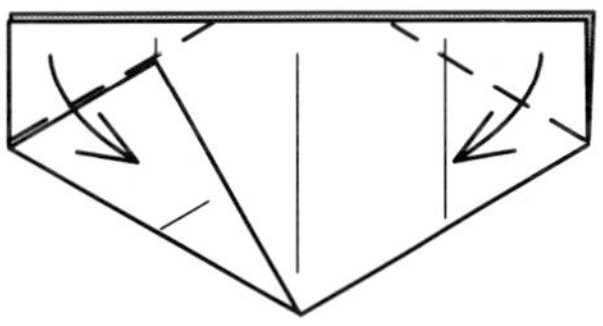

Repeat behind.

6

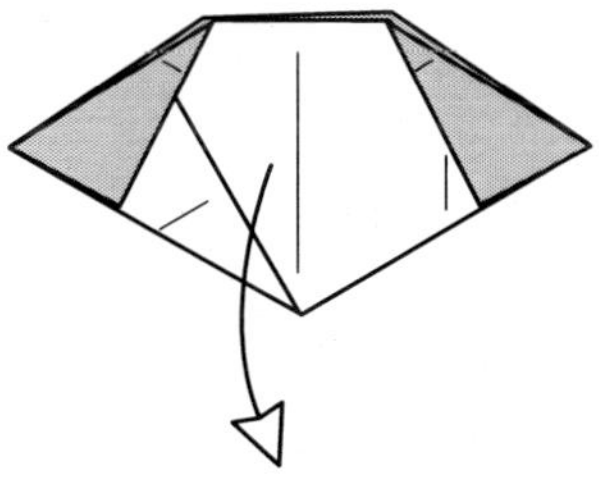

Unfold alomost everything.

7

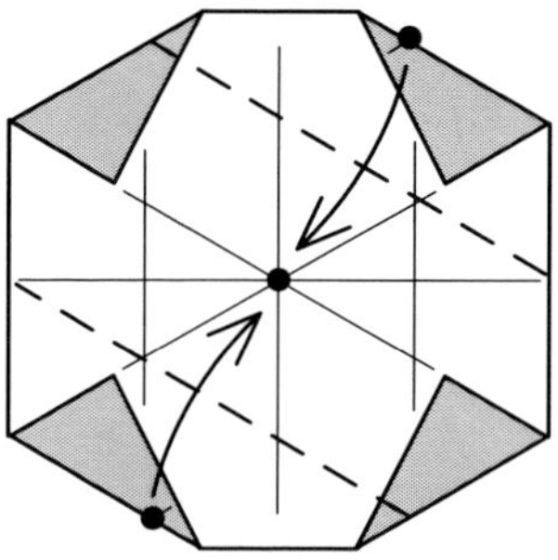

Fold to the center.

8

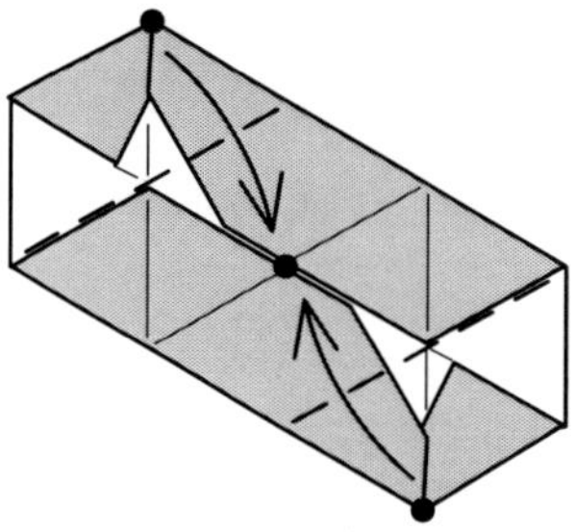

Fold to the center.

9

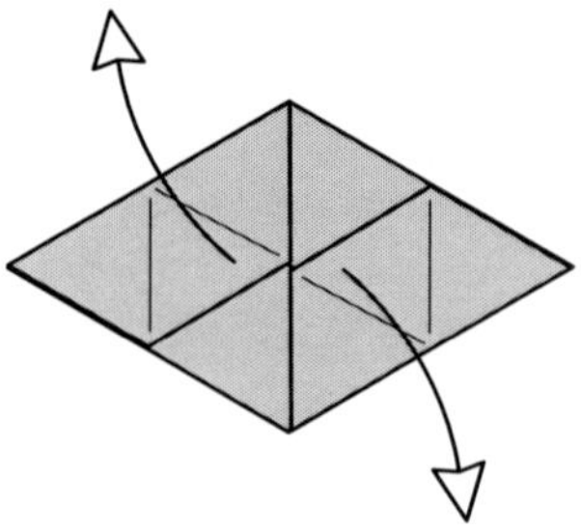

Unfold everything and rotate 90°.

10

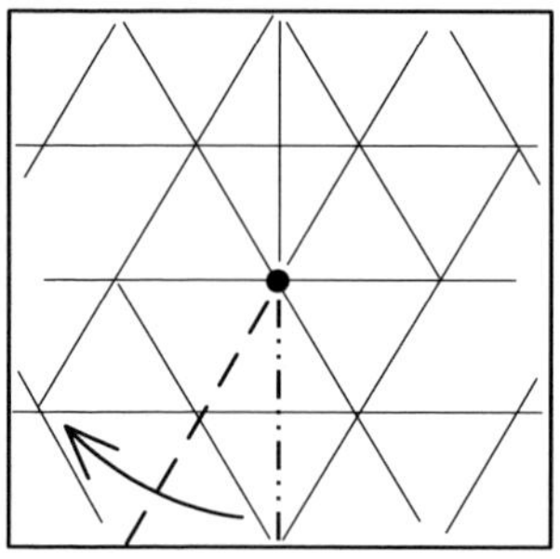

Push in at the dot and fold along the creases.

11

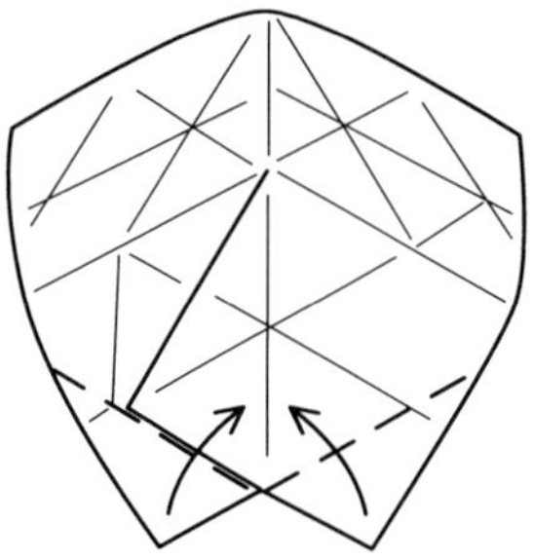

Fold along the creases and rotate 180°.

12

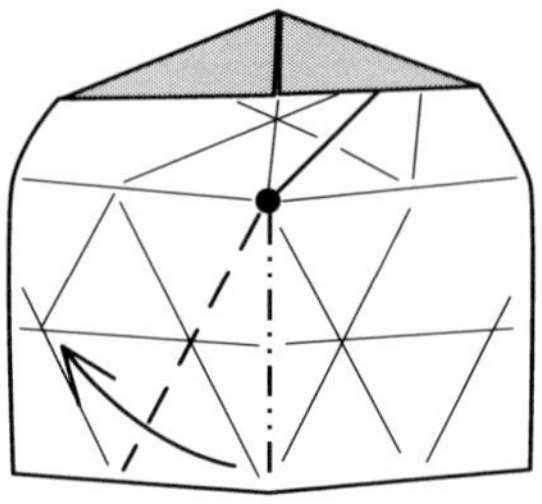

Repeat steps 10–11. Flatten.

13

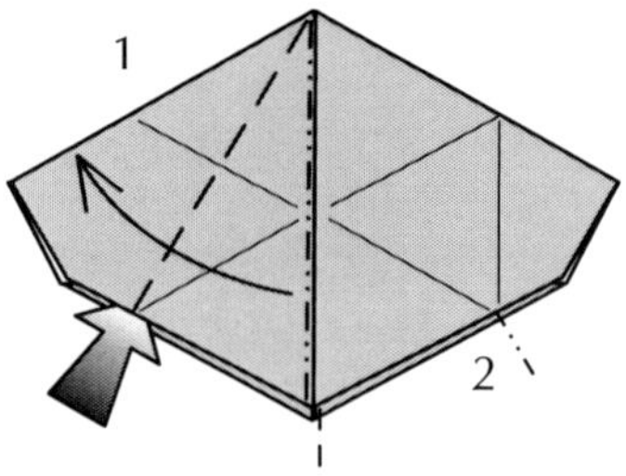

1. Fold the top layer.
2. Turn over and repeat.

14

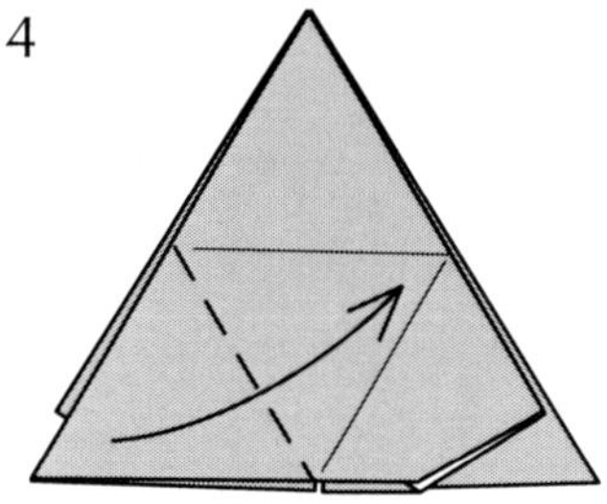

Fold along the crease. Turn over and repeat.

15

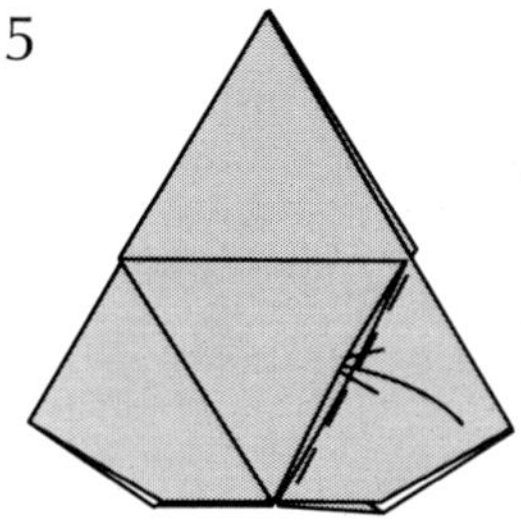

Tuck inside the pocket. Turn over and repeat.

16

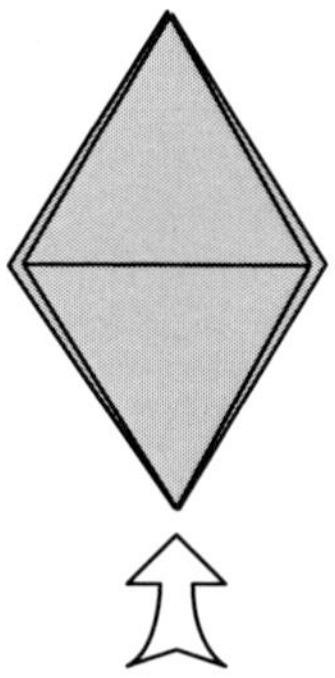

Inflate.

17

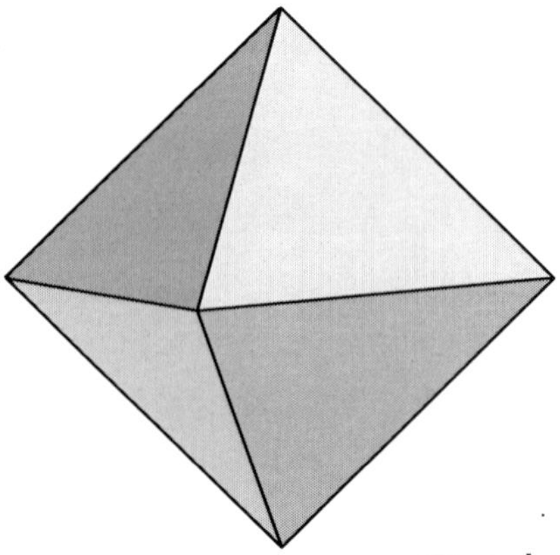

Octahedron

Pentagonal Dipyramid 60°

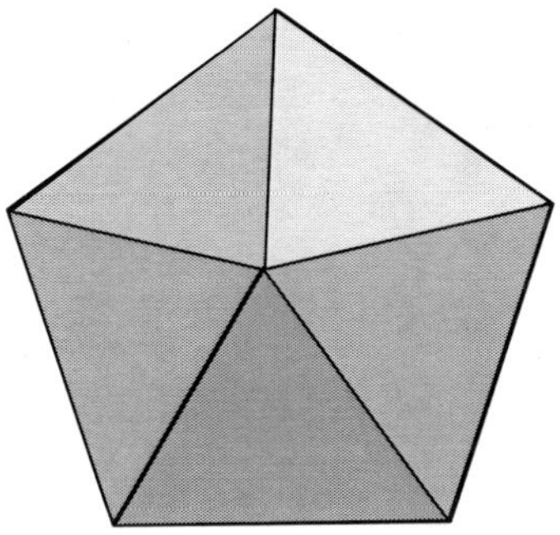

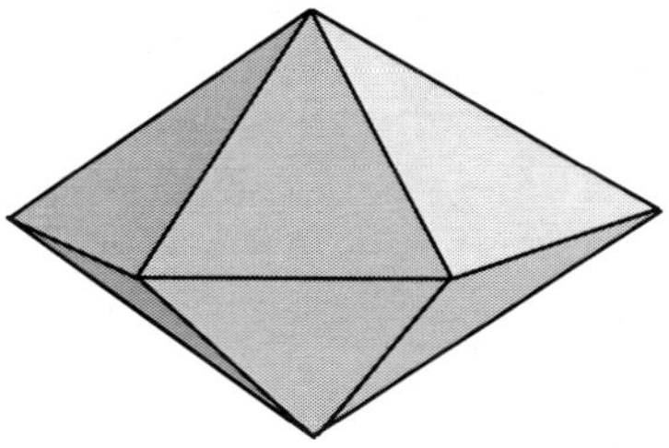

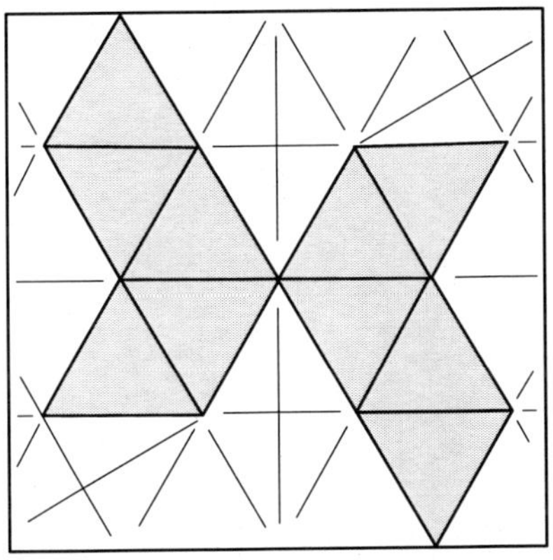

All the sides of this pentagonal diamond are equilateral triangles. The ratio of the height to the diameter is .61803 to 1.

1

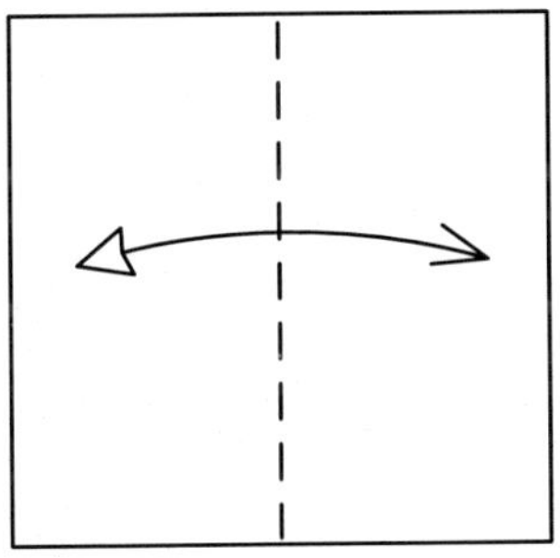

Fold and unfold.

2

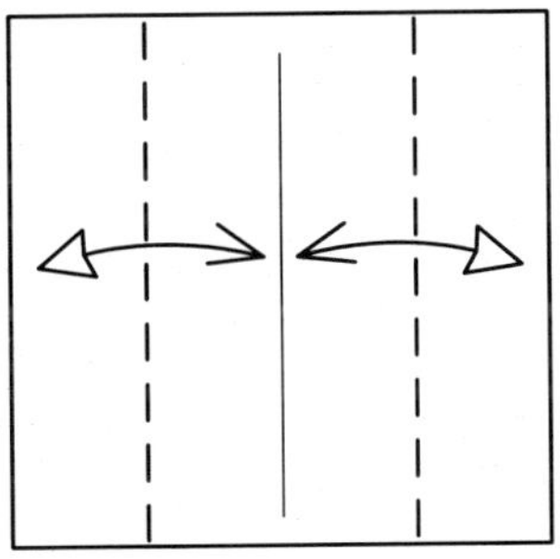

Fold and unfold.

3

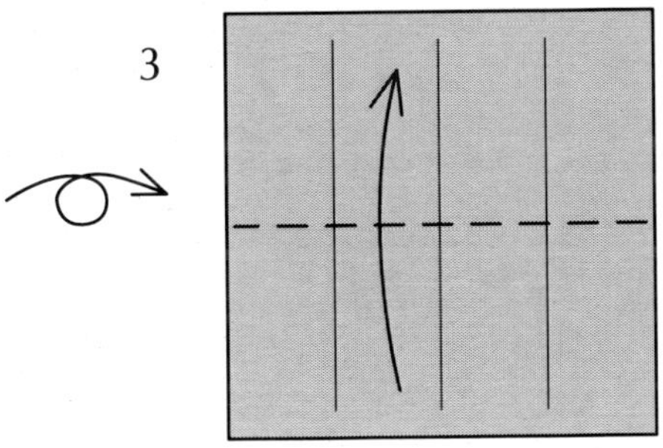

4

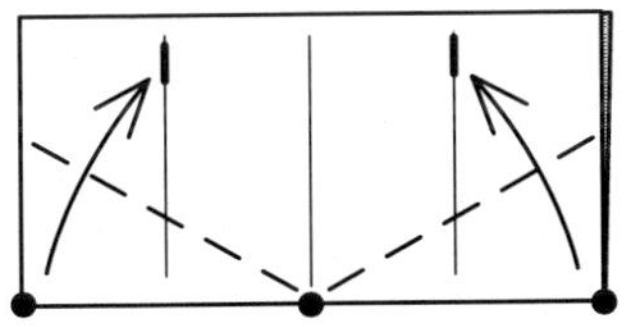

5

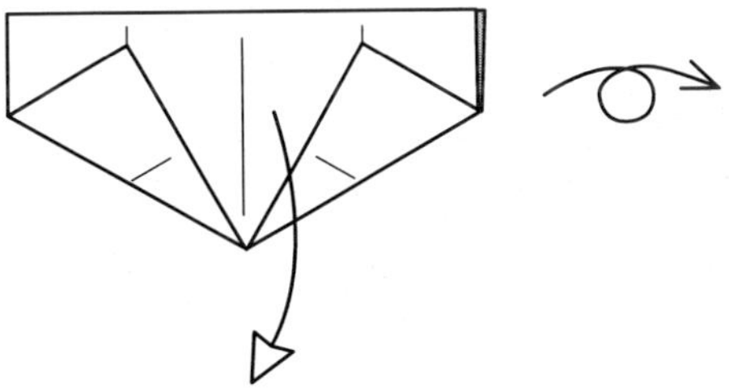

Unfold and rotate 90°.

6

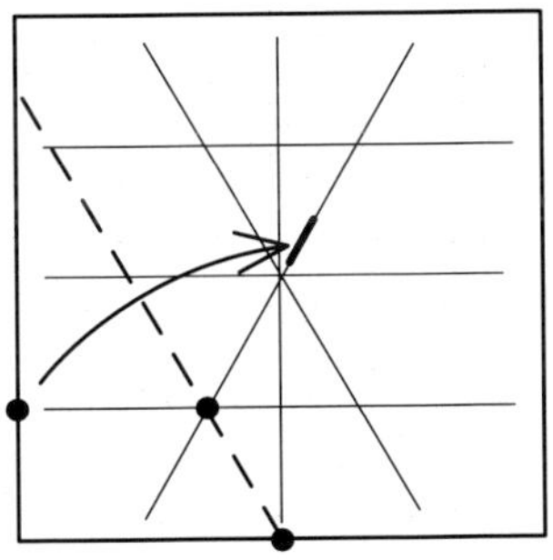

Fold and unfold.

7

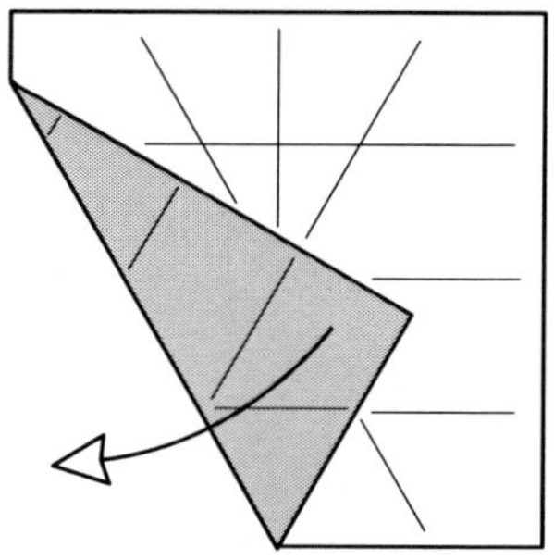

Unfold.

8

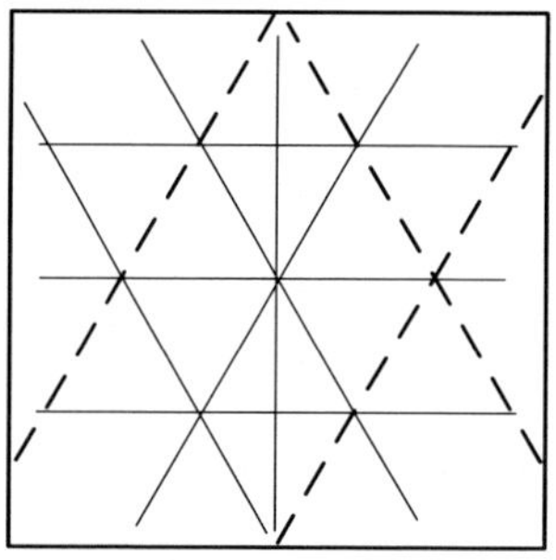

Fold and unfold.

9

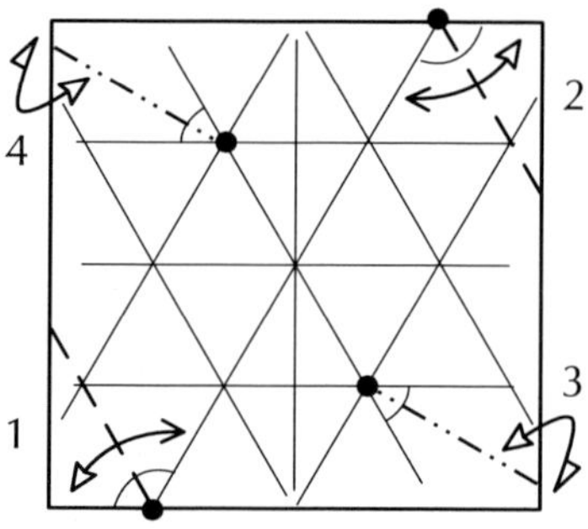

Fold and unfold.

10

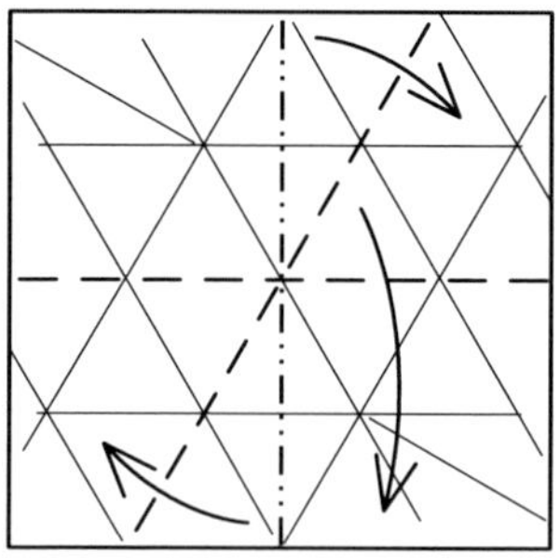

11

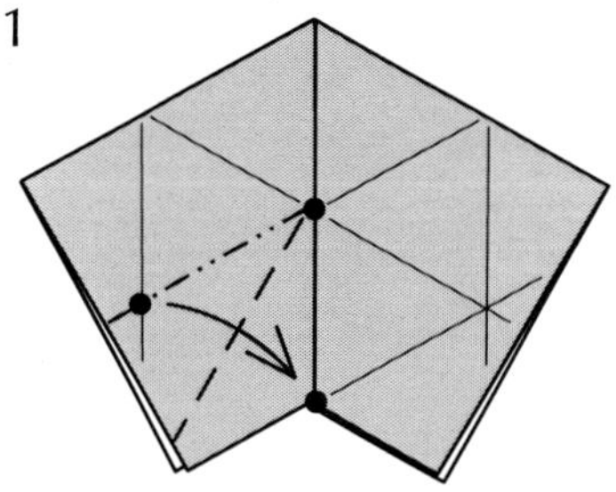

Puff out at the upper dot.
Turn over and repeat.

12

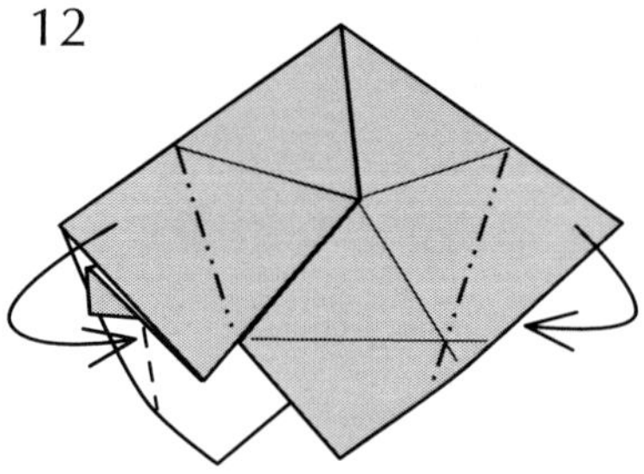

Fold along the creases
for these reverse folds.
Rotate 180°.

13

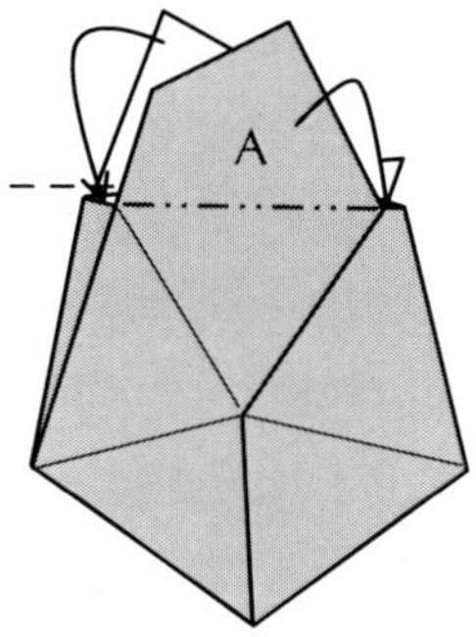

Tuck A into the small
pocket on the right.
Turn over and repeat.

14

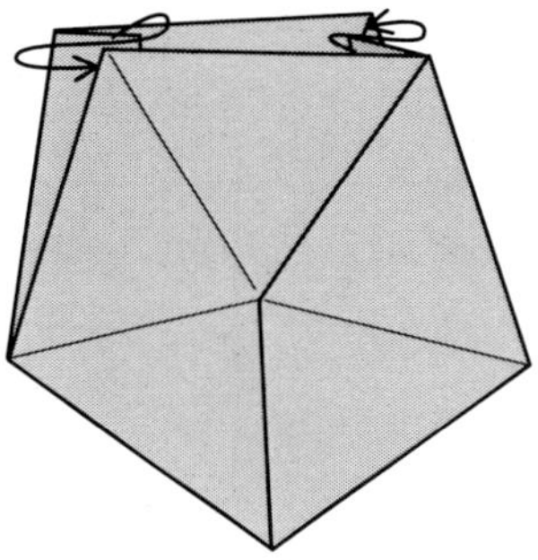

Tuck the small tabs
into the pockets.

15

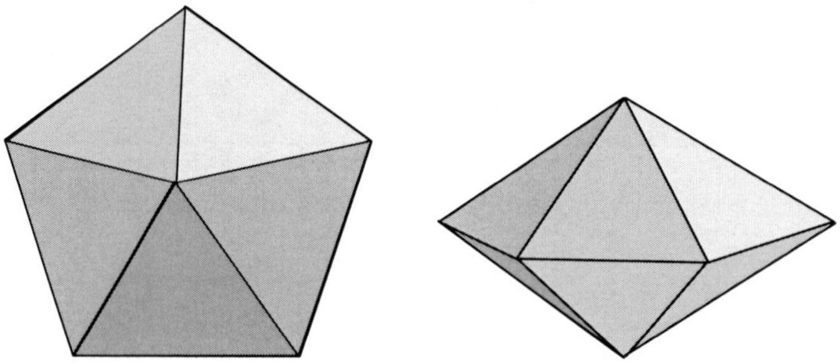

Pentagonal
Dipyramid 60°

Dodecadeltahedron

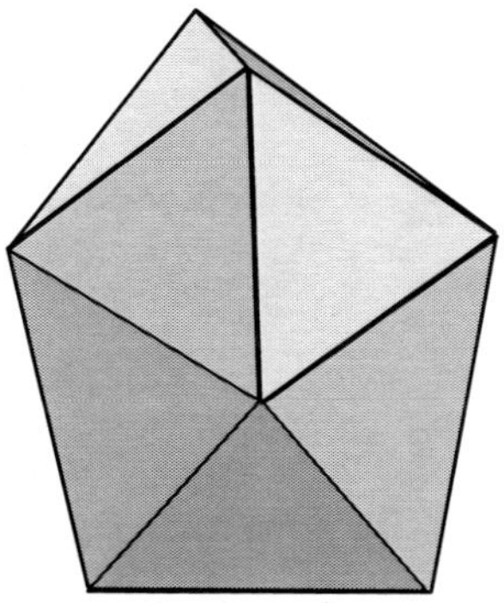

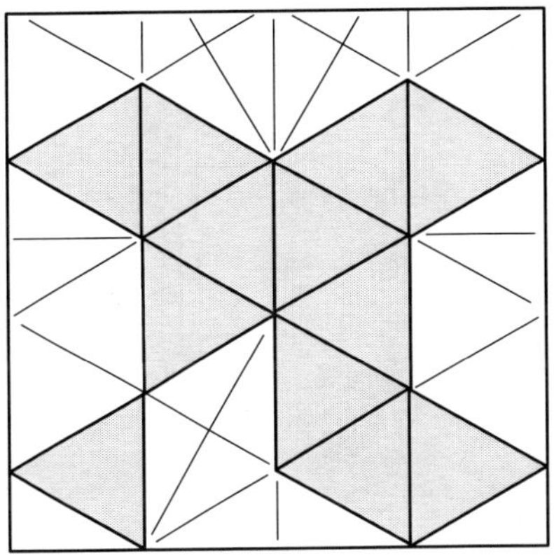

The twelve-sided dodecadeltahedron can also be called the Siamese Dodecahedron or Snub Disphenoid.

1

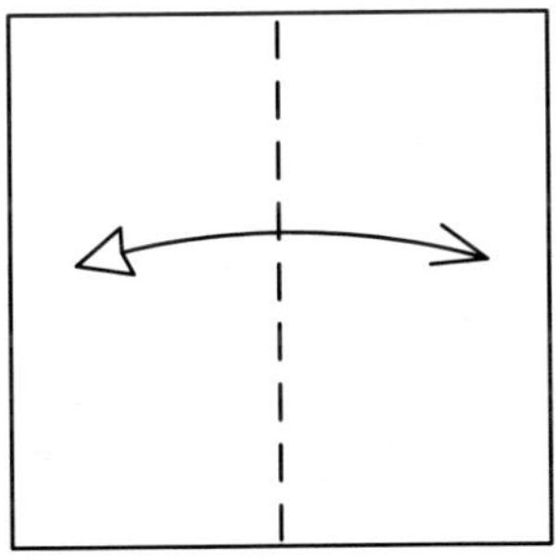

Fold and unfold.

2

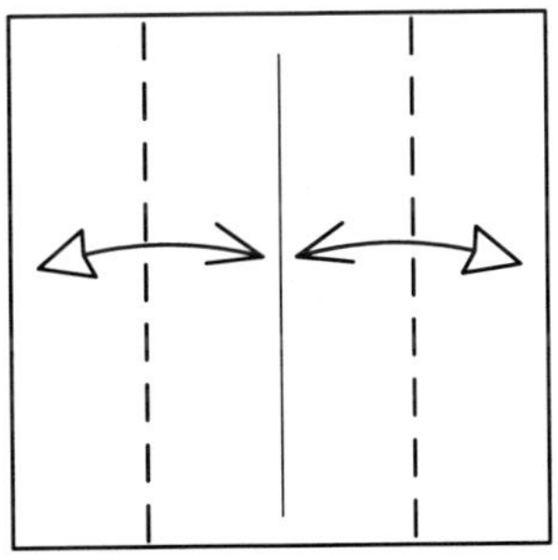

Fold and unfold.

3

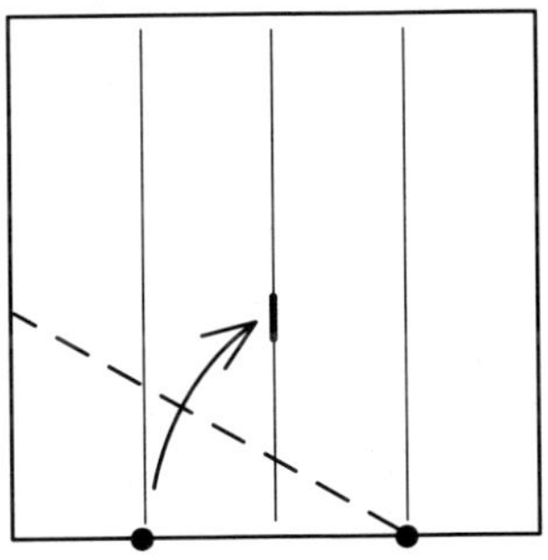

Bring the left dot to the line.

4

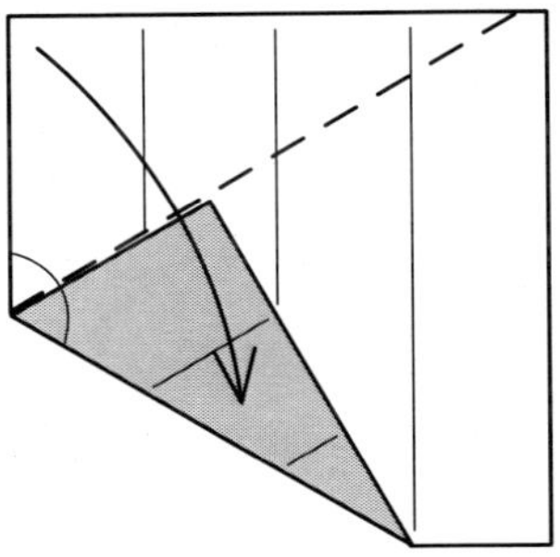

The valley line does not meet the top corner.

5

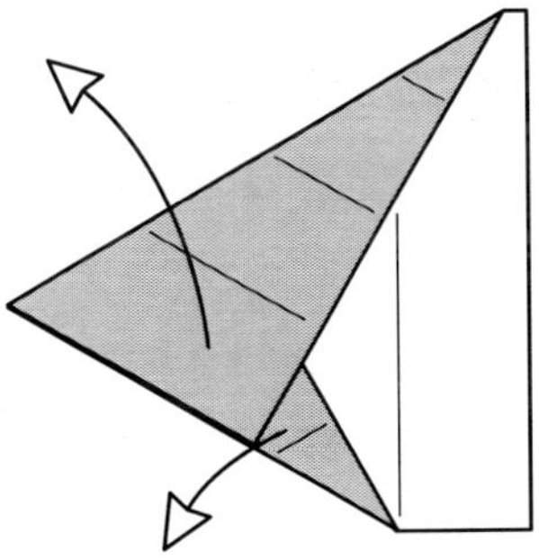

Unfold.

6

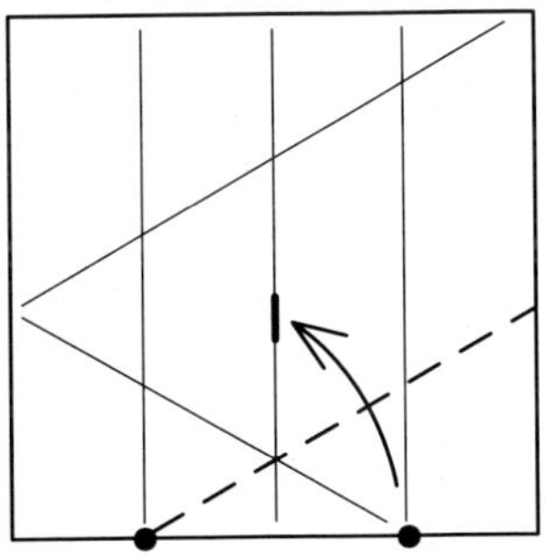

Repeat steps 3–5 on the right.

7

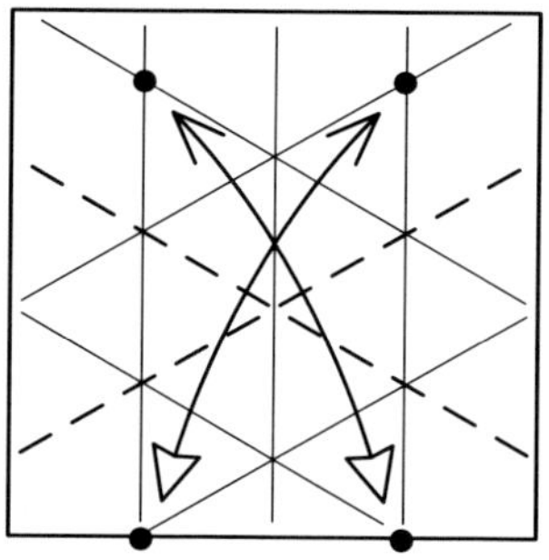

Fold and unfold.

8

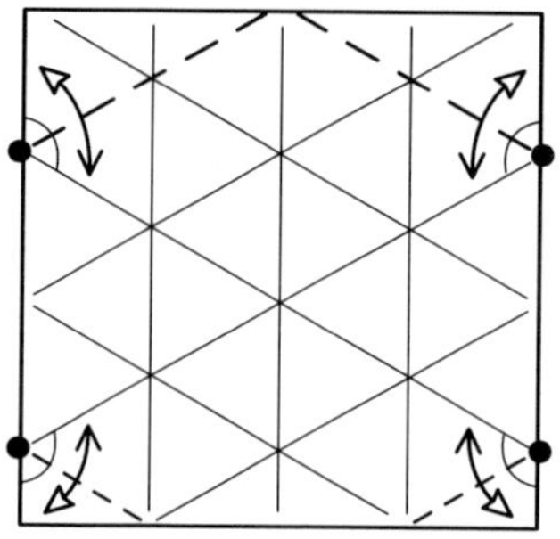

Fold and unfold.

9

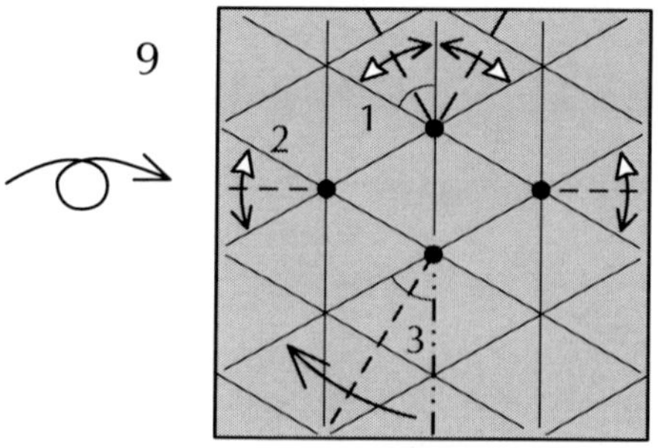

1. Fold and unfold at the top.
2. Fold and unfold on the left and right.
3. Puff out at the dot.

10

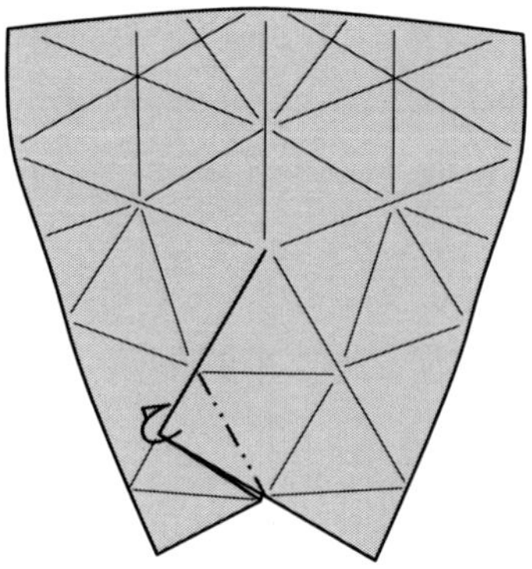

11

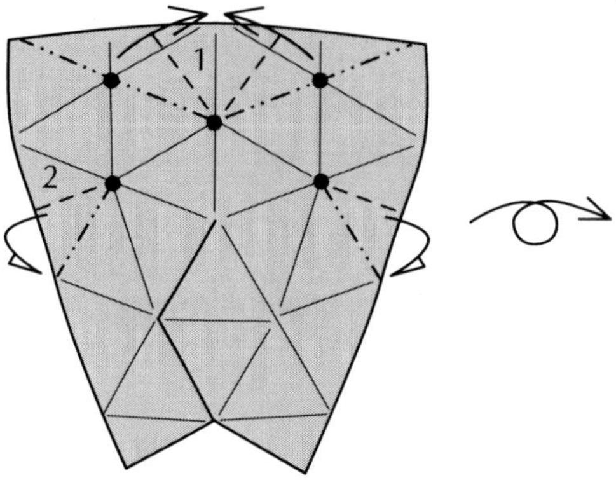

1. Puff out at the lower dot, the other dots will meet.
2. Puff out a the dots on the left and right.

12

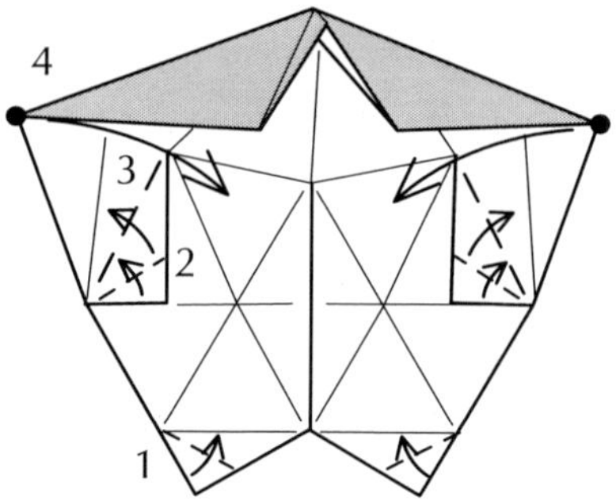

1. Fold to the line.
2. Fold to the line.
3. Continue folding the flap along the crease.
4. The corner will come down and cover the fold in part 3.

Repeat on the right.

13

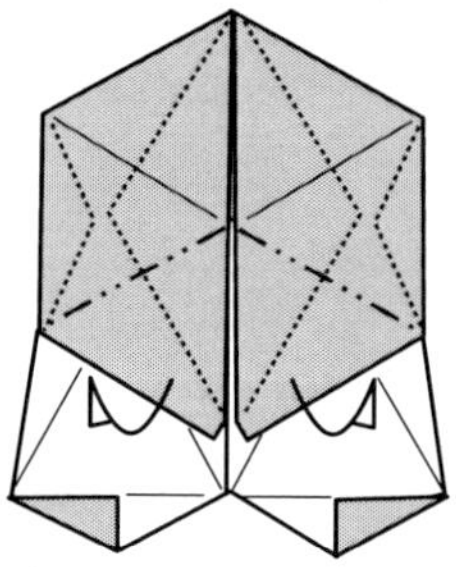

Fold behind and cover the hidden flaps.

14

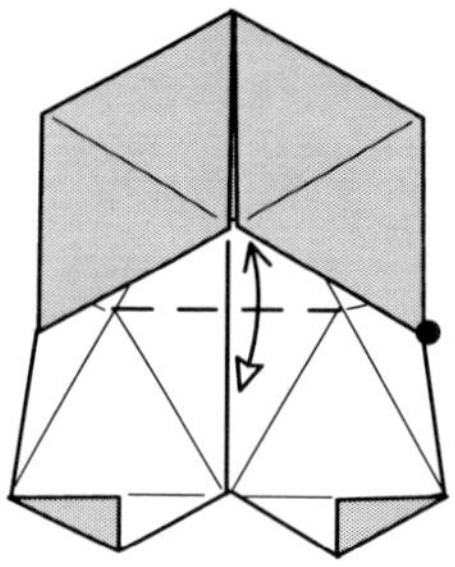

Fold and unfold along the crease. Rotate the dot to the top and center.

15

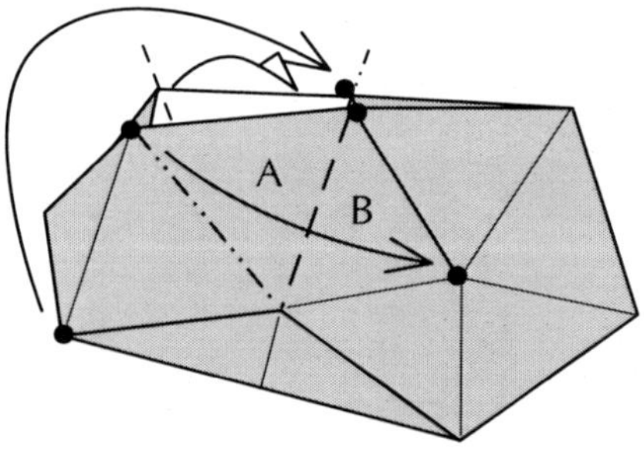

Triangle A will cover B so two dots will meet. Repeat behind at the same time so three dots (on the left and two at the top) meet.

16

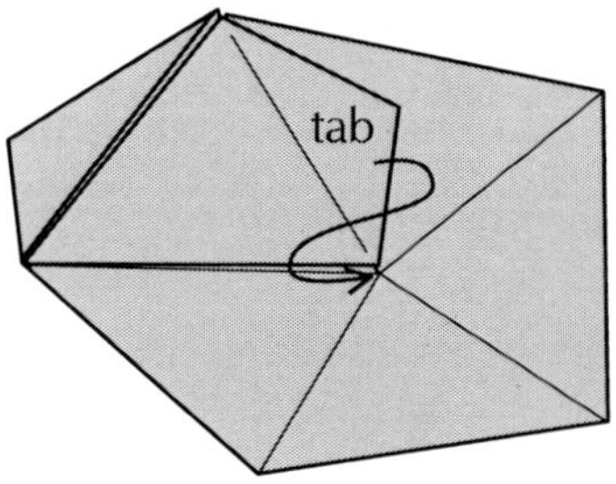

Tuck the tab into the pocket. Repeat behind. Rotate.

17

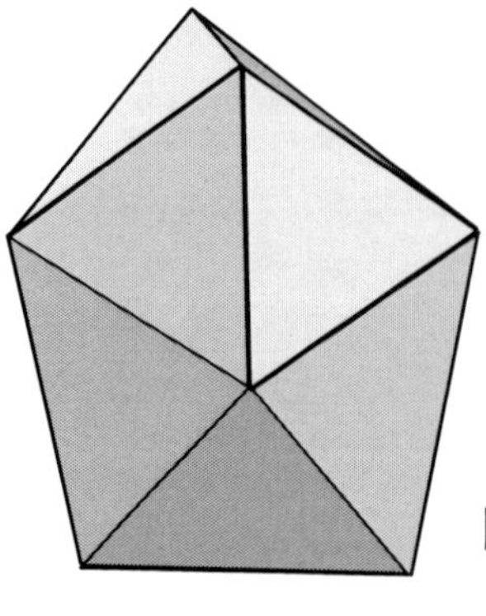

Dodecadeltahedron

Triaugmented Triangular Prism

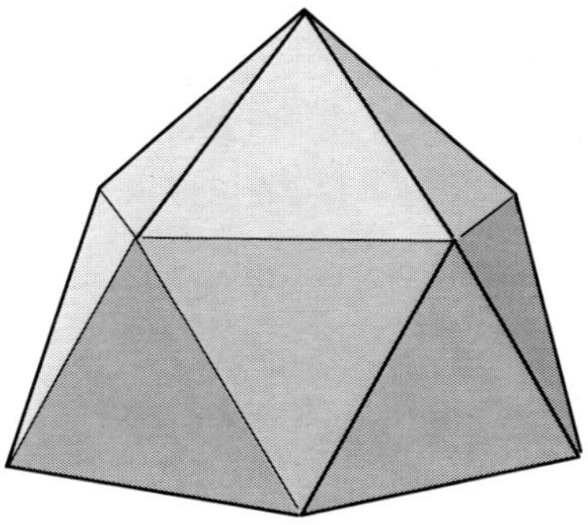

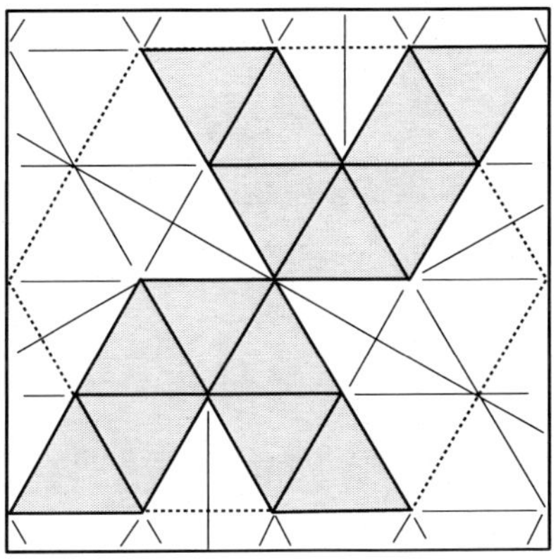

This deltahedron is formed from fourteen equilateral triangles. The crease pattern shows odd symmetry. A hexagon is folded, as shown by the dotted lines, and is used to generate the crease pattern.

1

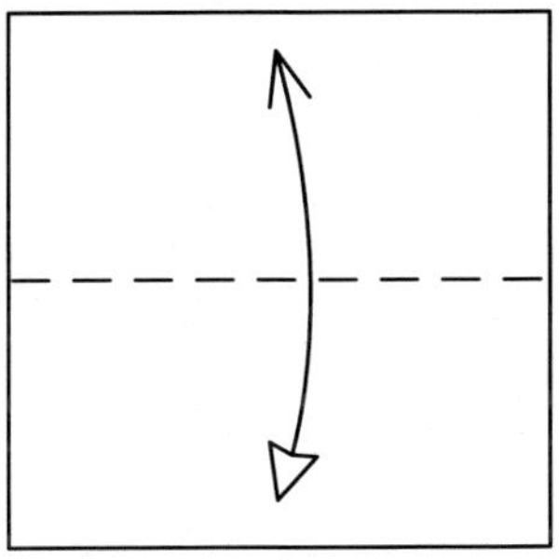

Fold and unfold.

2

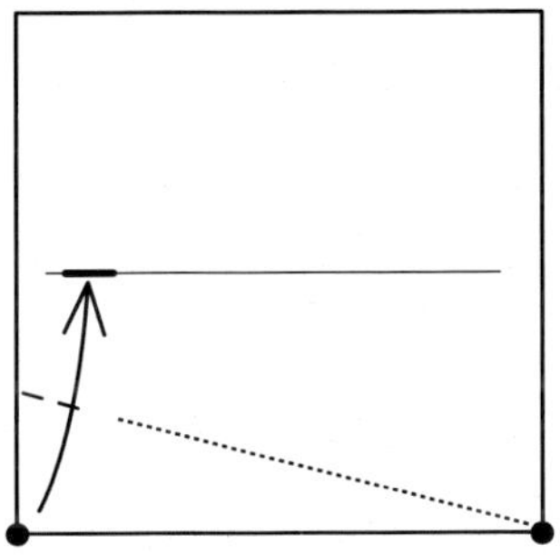

Bring the corner to the line and crease on the left.

3

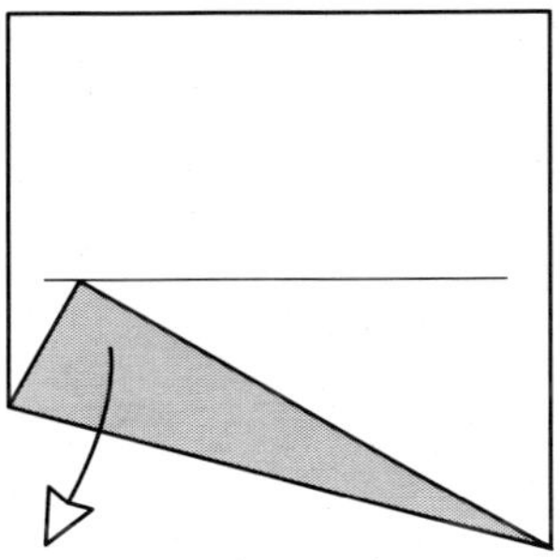

Unfold.

4

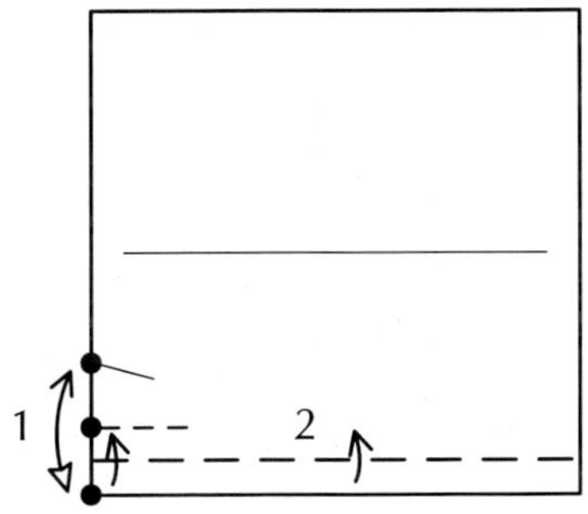

1. Fold and unfold on the left.
2. Fold to the dot.
Rotate 180°.

5

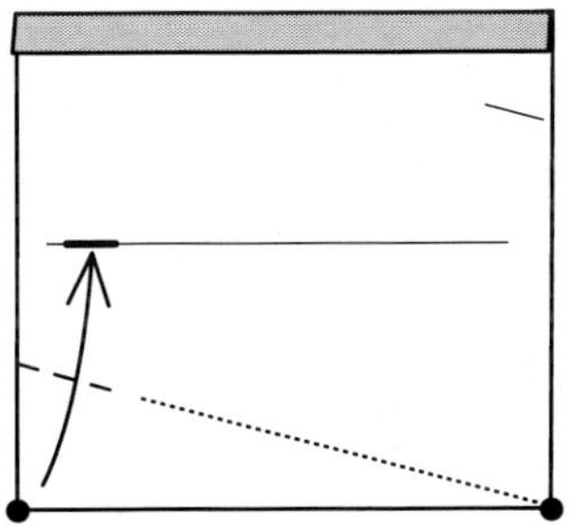

Repeat steps 2–5.

6

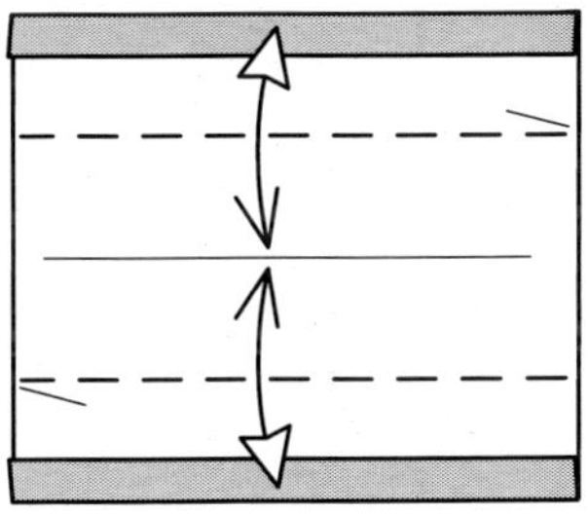

Fold to the center and unfold.

7

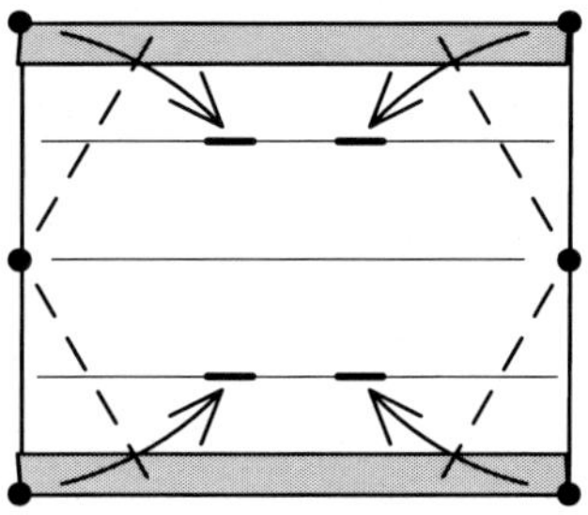

Bring the corners
to the lines.

8

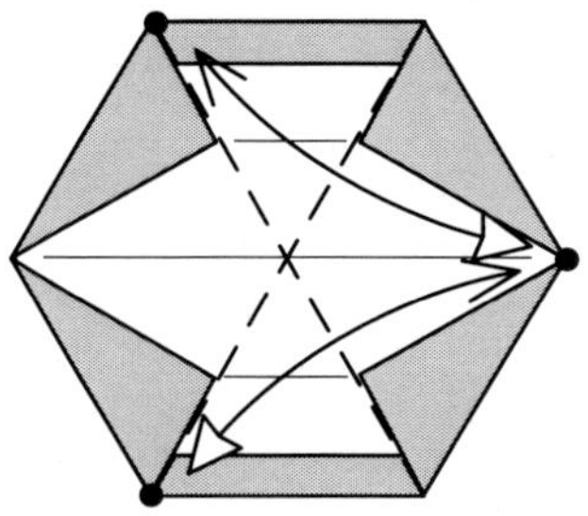

This is a regular hexagon.
Fold and unfold.

9

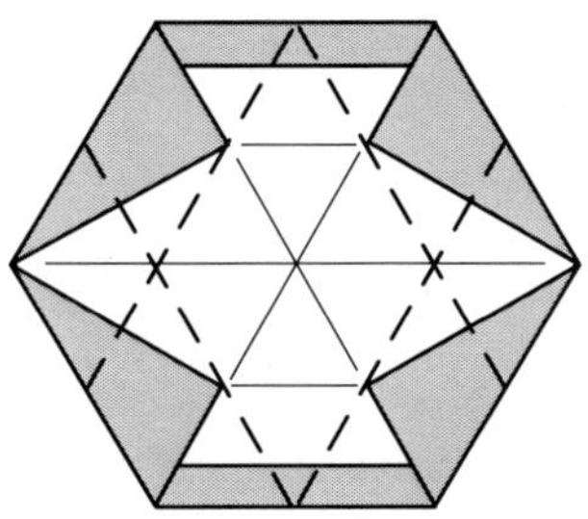

Fold to the center and
unfold, on four sides.

10

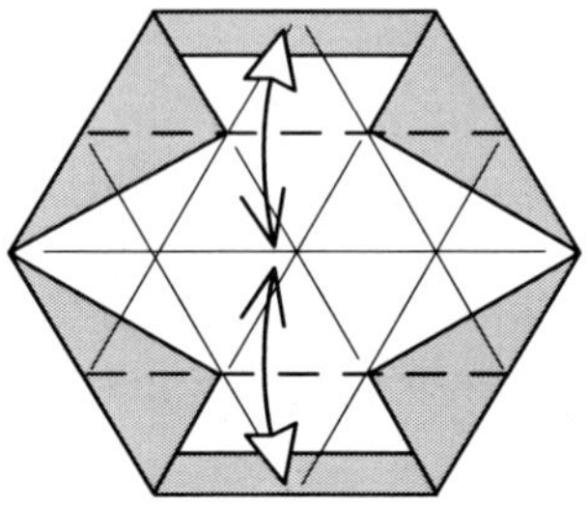

Fold and unfold along
the lower creases.

11

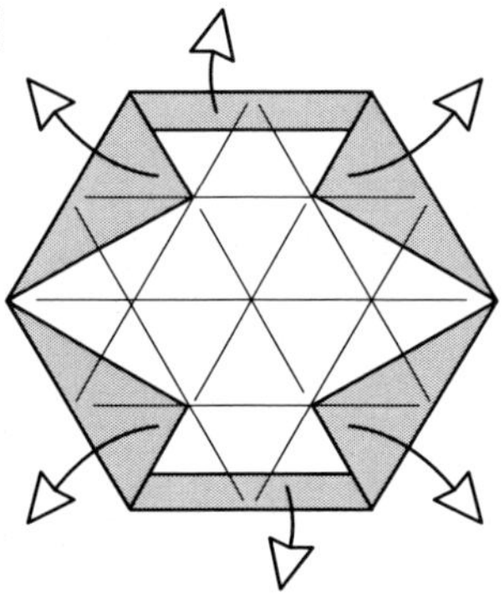

Unfold everything.

12

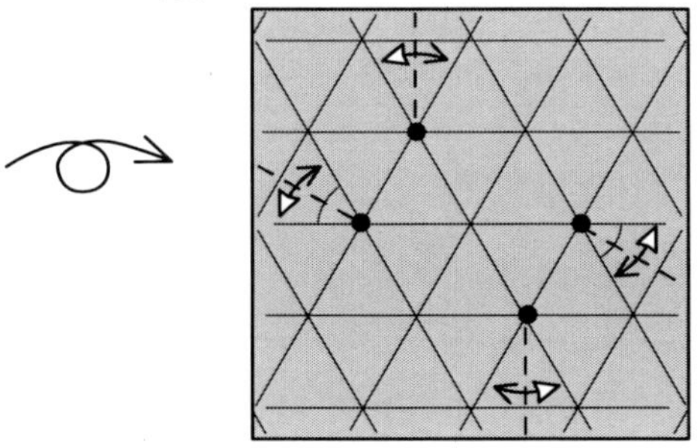

Fold and unfold.

13

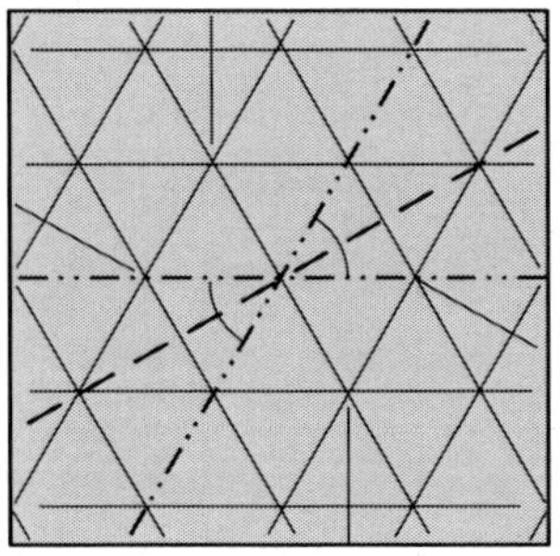

Fold and unfold to bisect
the angles. Mountain-fold
along the creases.

14

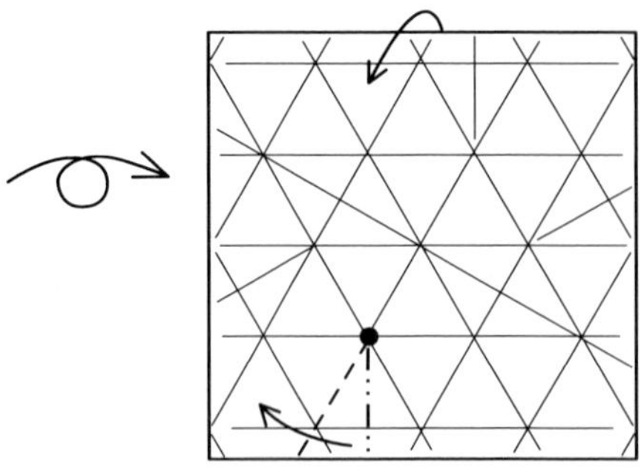

Push in at the dot.

15

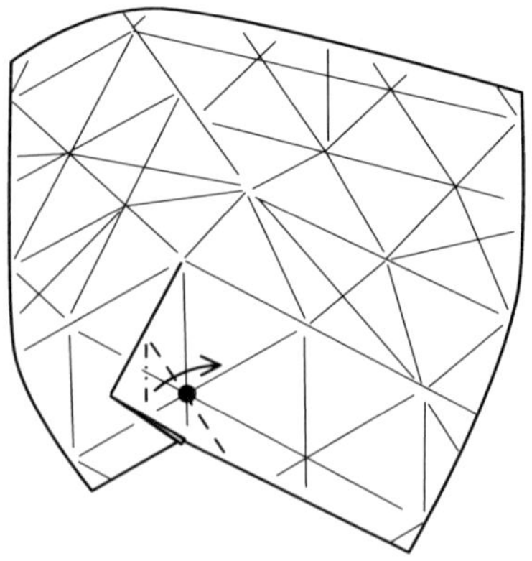

Squash-fold through
the dot.

16

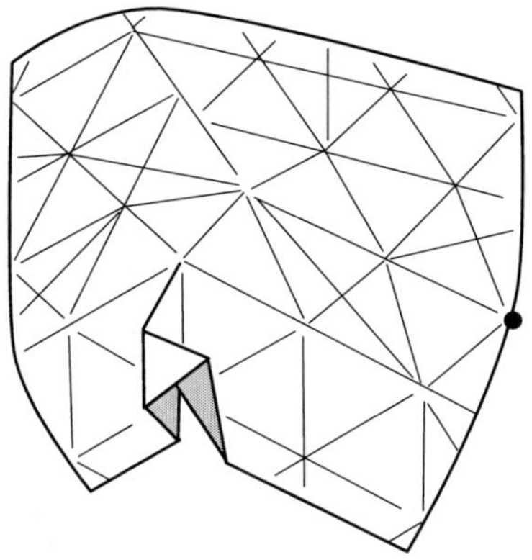

Rotate 180° and repeat steps 14–15. Rotate to view the outside so the dot is at the center and top.

17

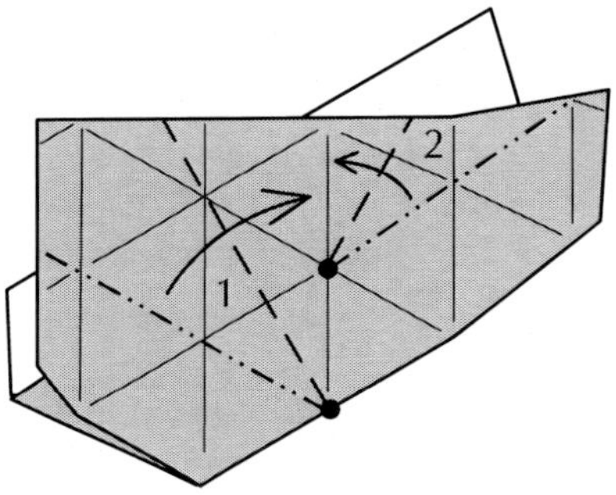

Fold along the creases and puff out at the dots. Turn over and repeat.

18

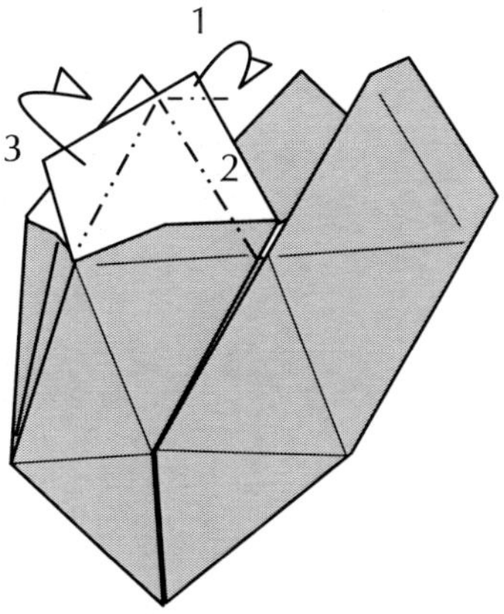

Fold behind at 1, 2, and 3. Turn over and repeat.

19

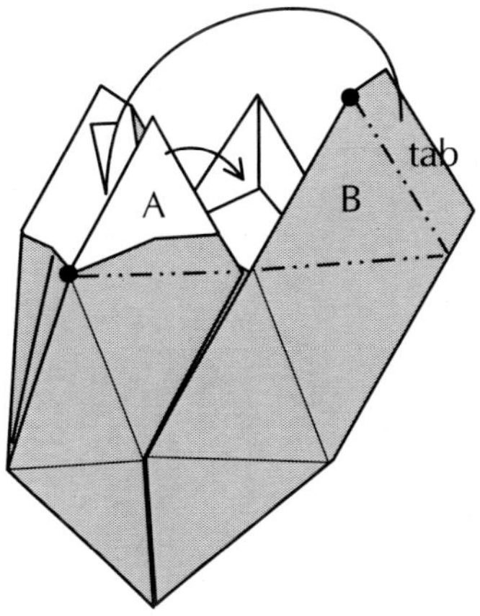

Region B will cover A and the tab will wrap around A. The dots will meet. Turn over and repeat.

20

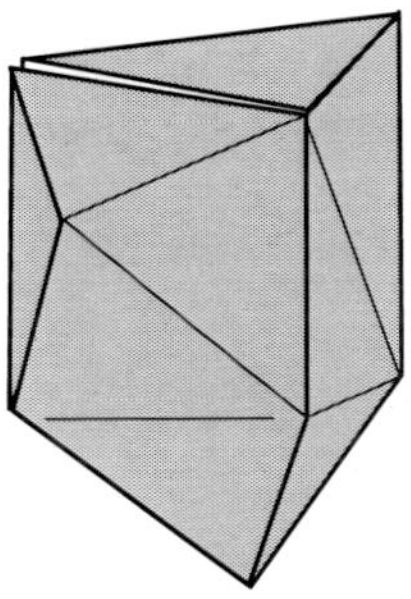

Rotate the top to the bottom.

21

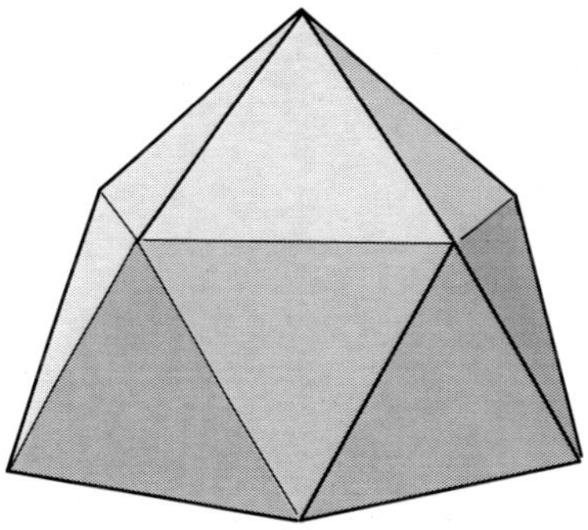

Triaugmented Triangular Prism

Gyroelongated Square Dipyramid

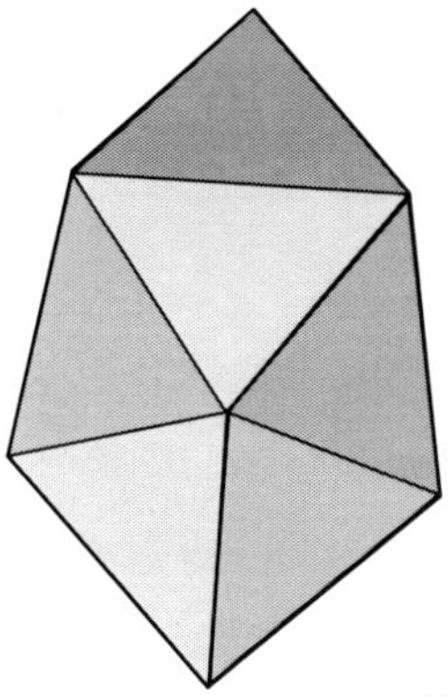

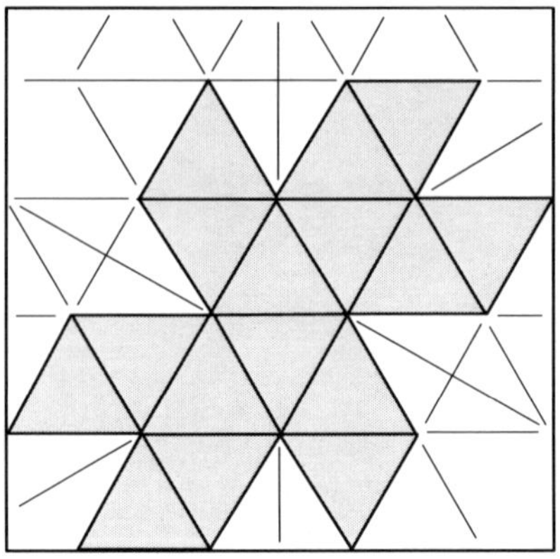

This Gyroelongated Square Dipyramid is a deltahedrn with 16 sides. It can also be called a hexadecahedron— hexa for 6 and deca for 10. This forms an egg shaped polyhedron. The crease pattern shows a tab on top, and the layout has odd symmetry.

1

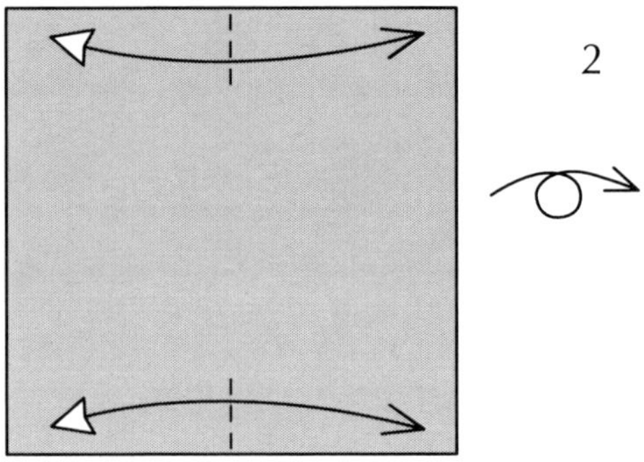

Fold and unfold on the top and bottom.

2

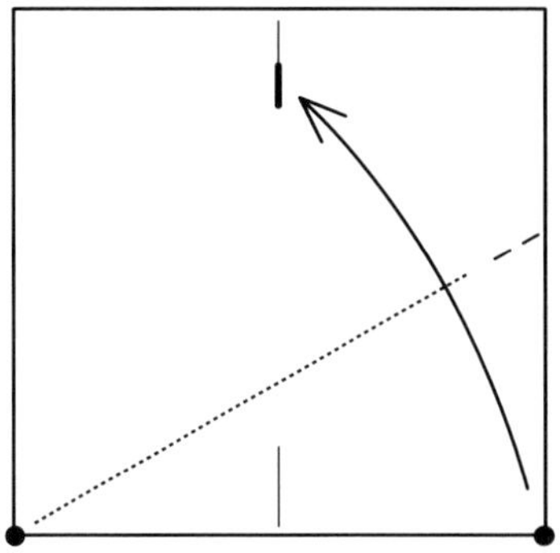

Bring the corner to the line and crease on the right.

3

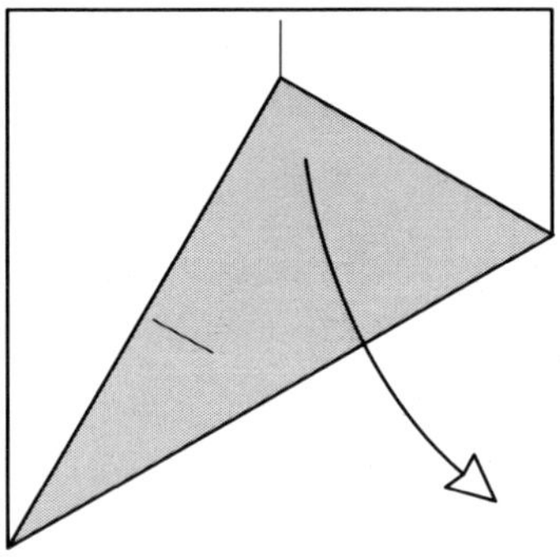

Unfold.

4

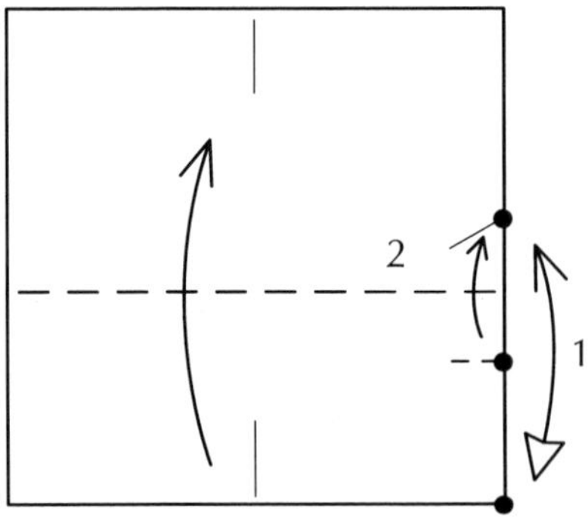

1. Fold and unfold on the right.
2. Fold up.

5

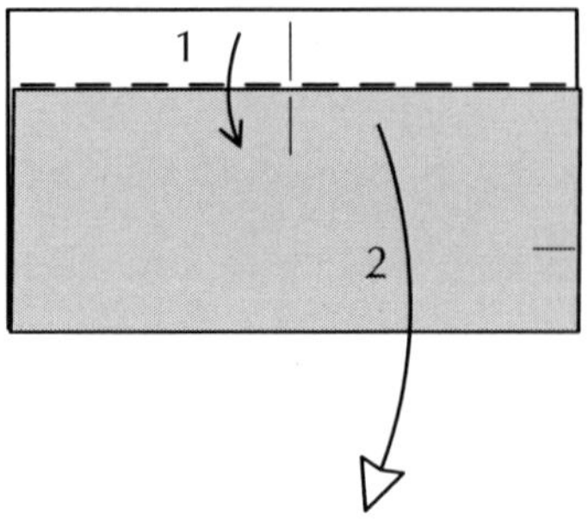

1. Fold along the edge.
2. Unfold.

6

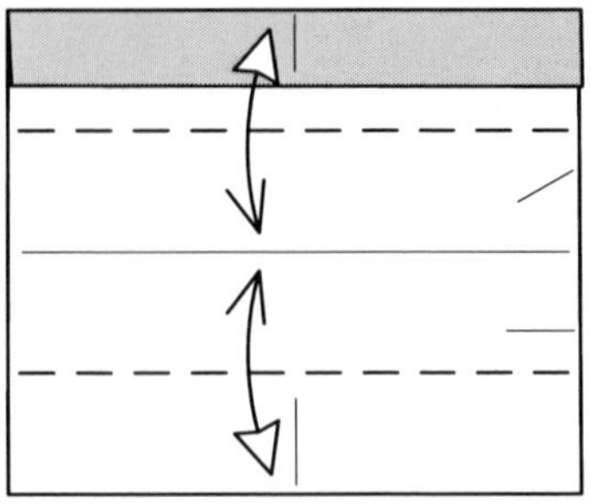

Fold to the center and unfold.

7

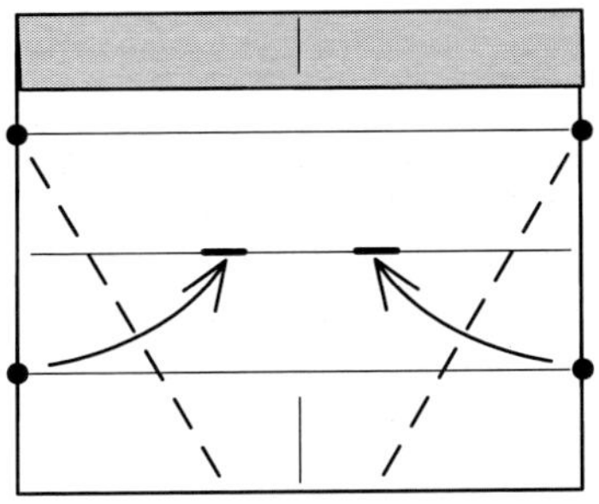

8

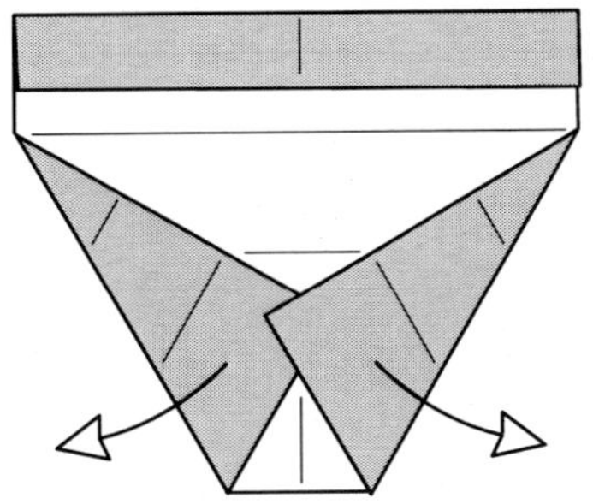

Unfold.

9

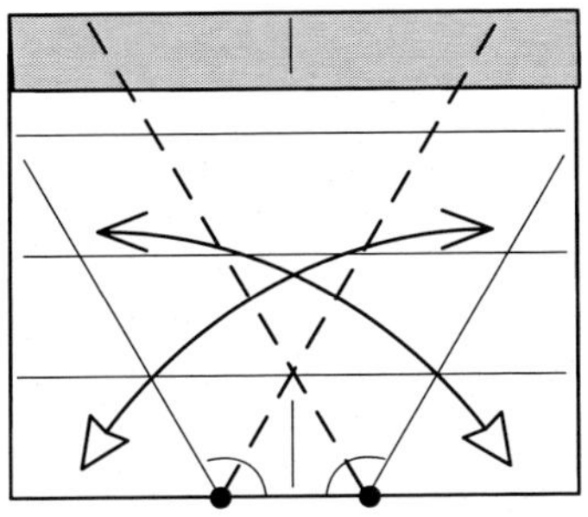

Fold and unfold.
Rotate 180°.

10

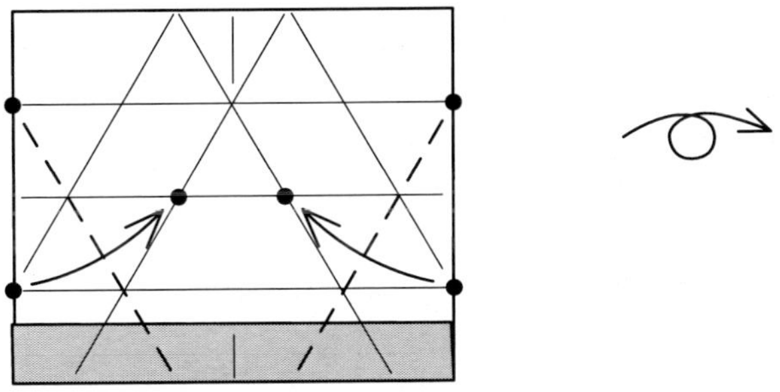

Repeat steps 7–9.

11

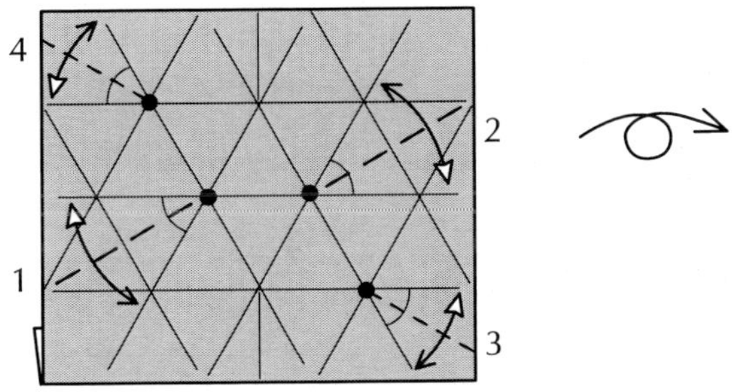

Fold and unfold.

12

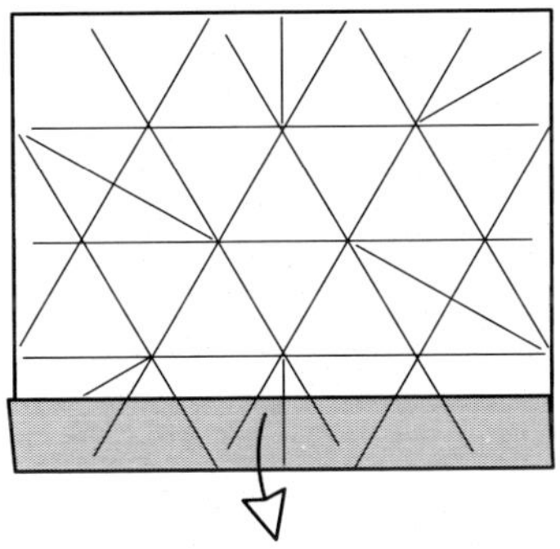

Unfold.

13

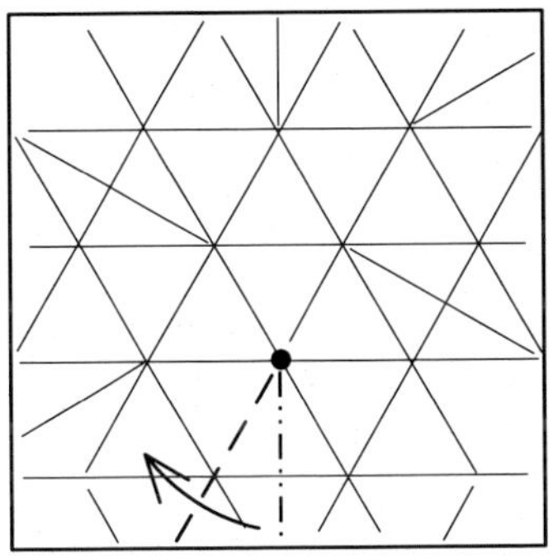

Push in at the dot.

14

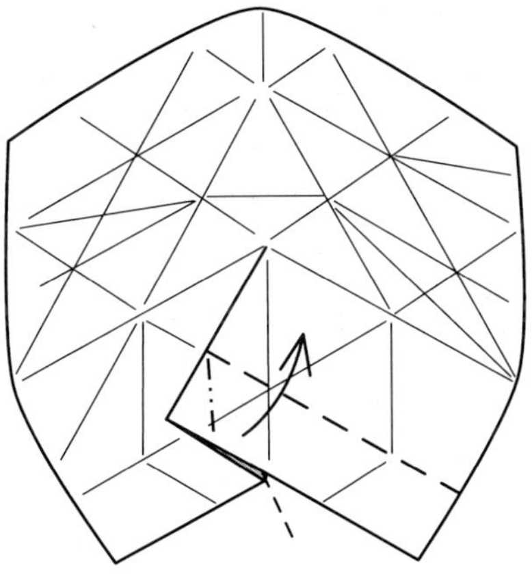

Squash-fold.

15

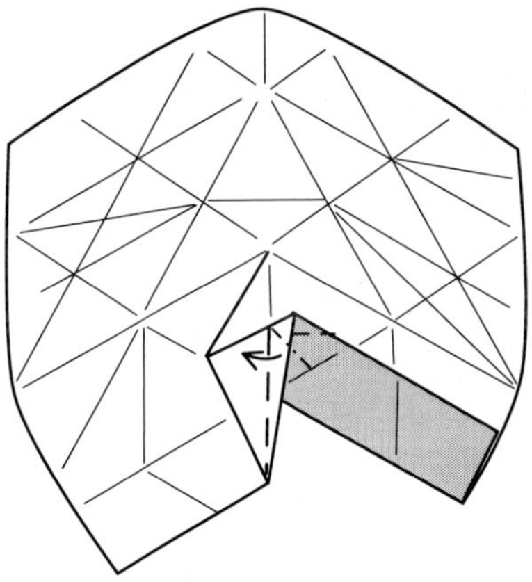

Make a thin squash fold.

16

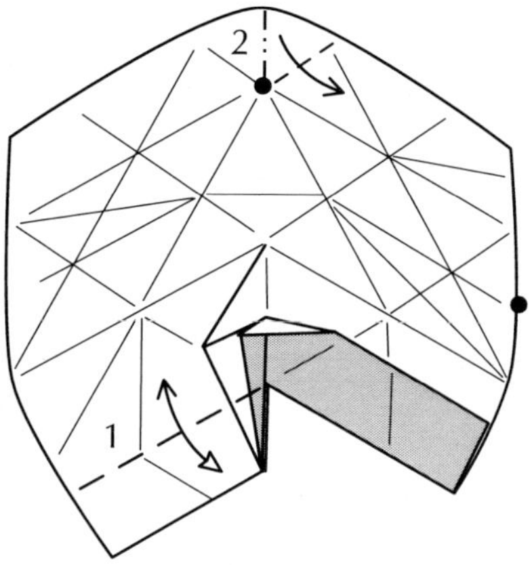

1. Fold and unfold along the crease.
2. Push in at the dot.
3. Rotate the dot to the front and top.

17

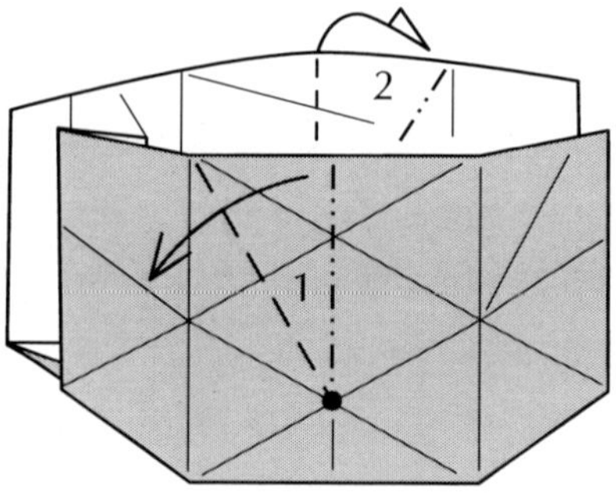

1. Puff out at the dot.
2. Turn over and repeat.

18

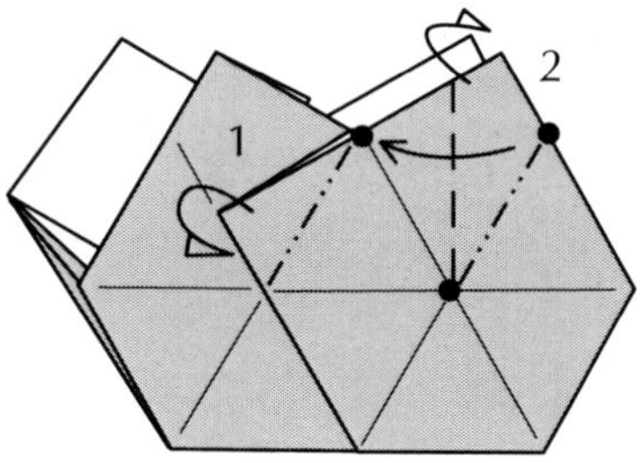

1. Fold behind and repeat behind.
2. Puff out at the lower dot.
   Repeat behind, but note
   the flap is longer.

19

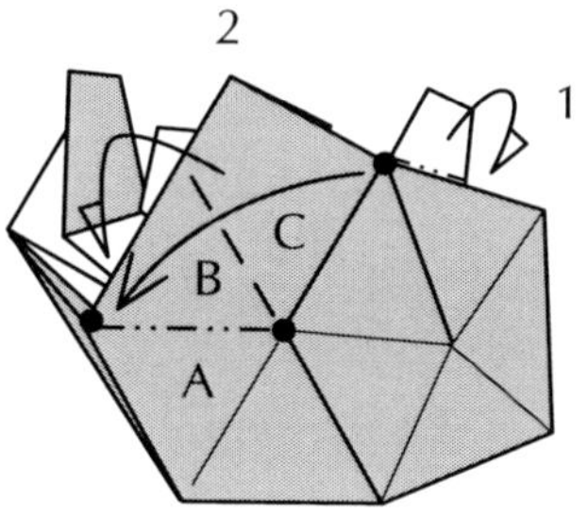

1. Fold behind.
   (Do not repeat behind.)
2. Puff out at the center dot.
   Flatten B and C behind A.
   Turn over and repeat.

Rotate the center dot to the right.

20

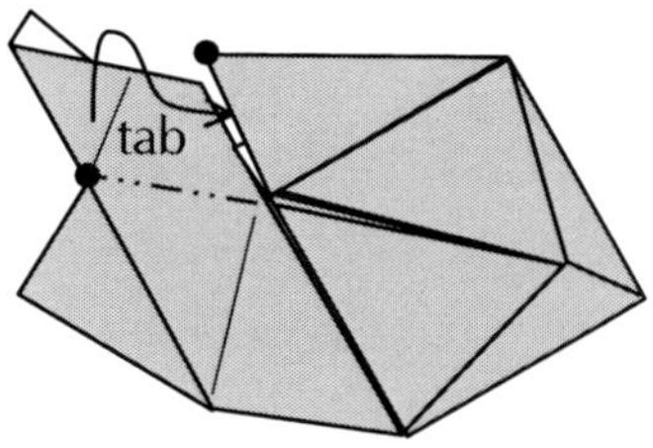

Tuck the tab into the pocket so the dots meet.

21

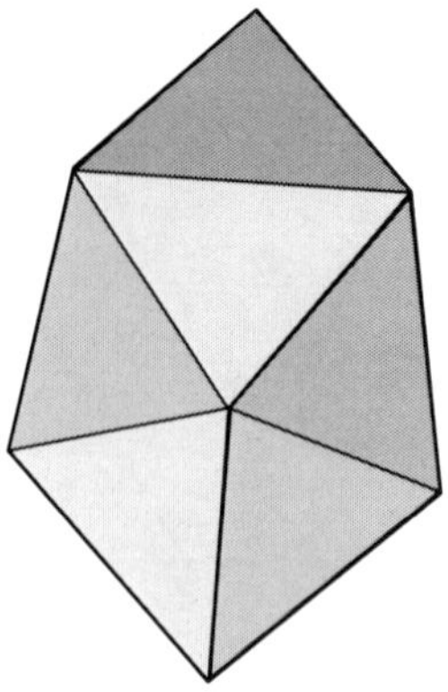

Gyroelongated
Square Dipyramid

Icosahedron

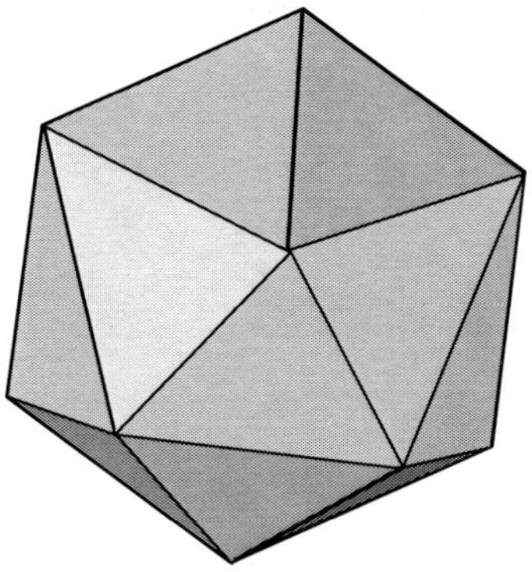

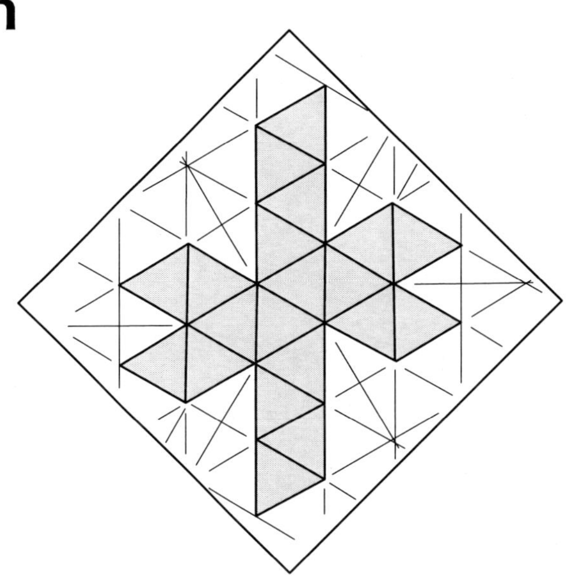

The icosahedron is composed of 20 equilateral triangles. Plato attributed this model to water because if its ability to roll.

The layout shows a band of ten triangles down the diagonal with five triangles on each side. Odd symmetry is used.

1

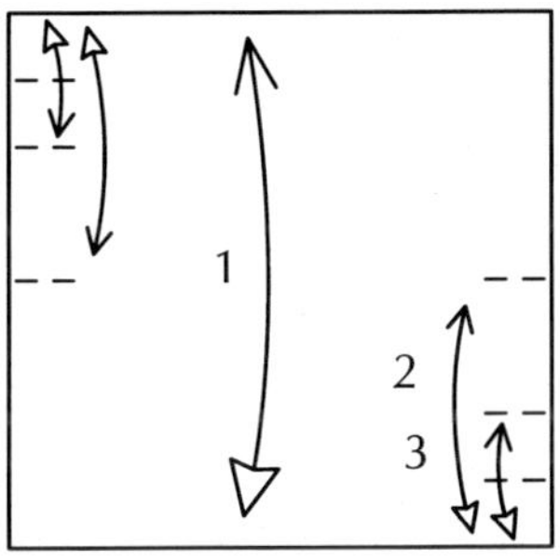

Make small marks by folding and unfolding in half three times, on the left and right.

2

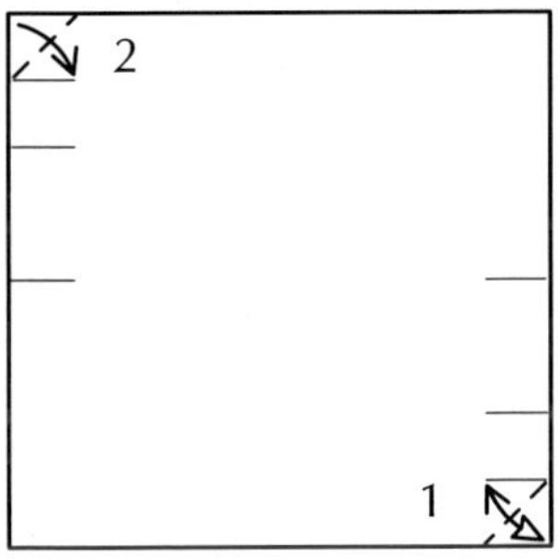

1. Fold and unfold.
2. Fold to the line.
Rotate 45°.

3

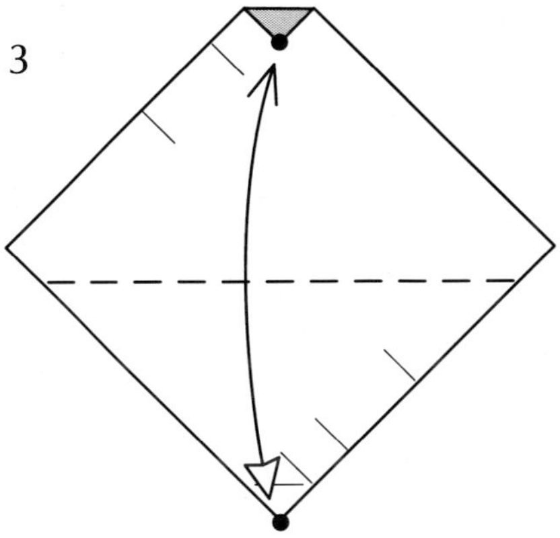

Fold and unfold.
Rotate 180°.

4

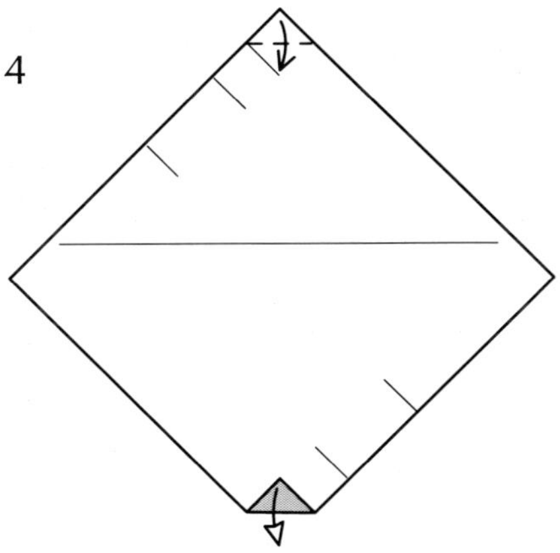

Repeat steps 2–3.

5

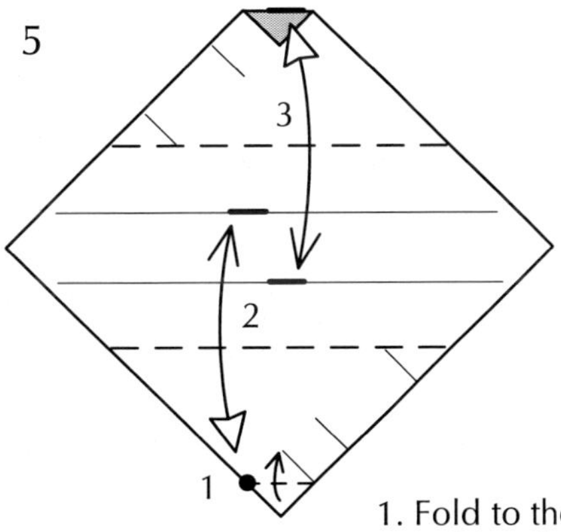

1. Fold to the line.
2. Fold and unfold.
3. Fold and unfold.

6

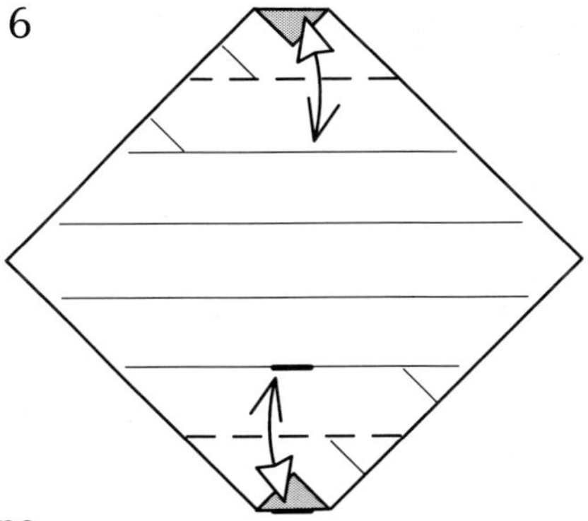

Fold and unfold.

7

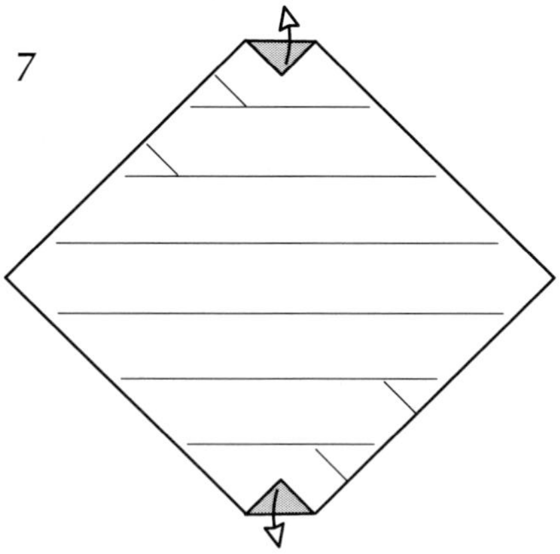

Unfold and rotate 45°.

8

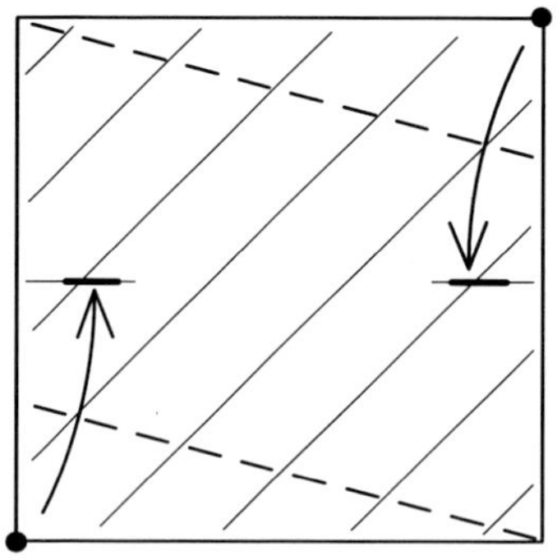

Bring the corners
to the lines.

9

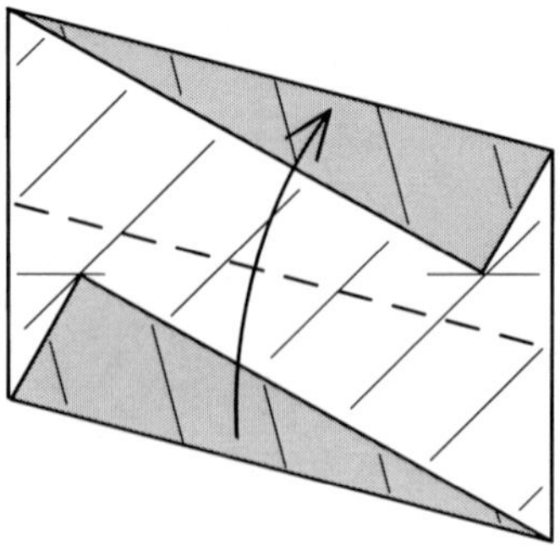

Fold in half.

10

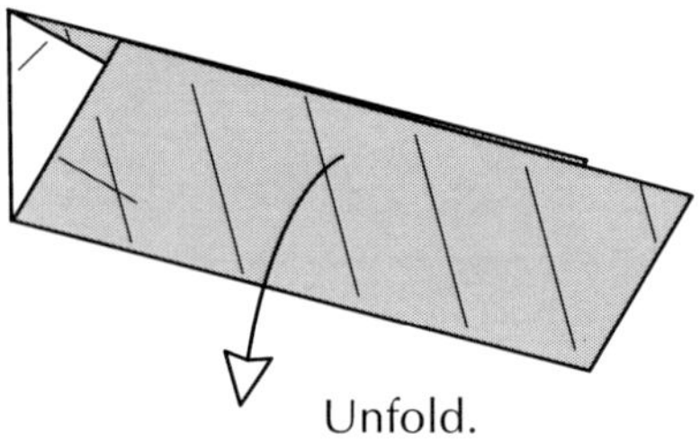

Unfold.

11

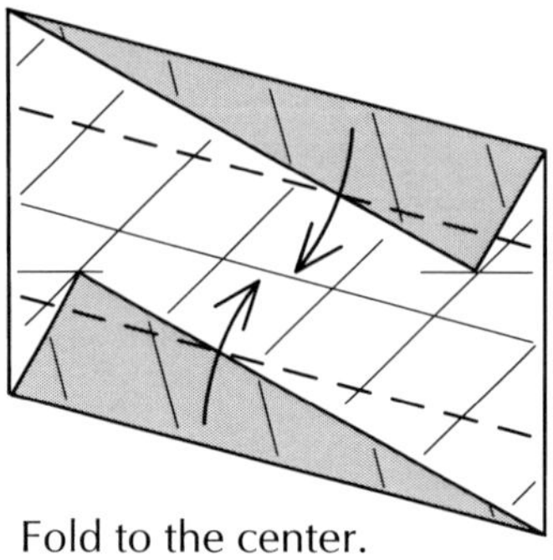

Fold to the center.

12

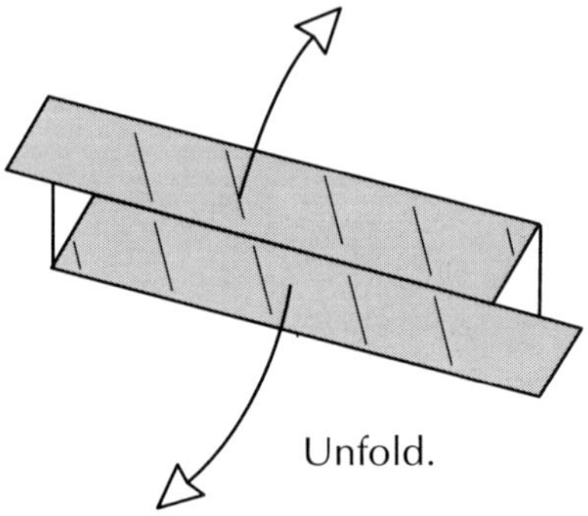

Unfold.

13

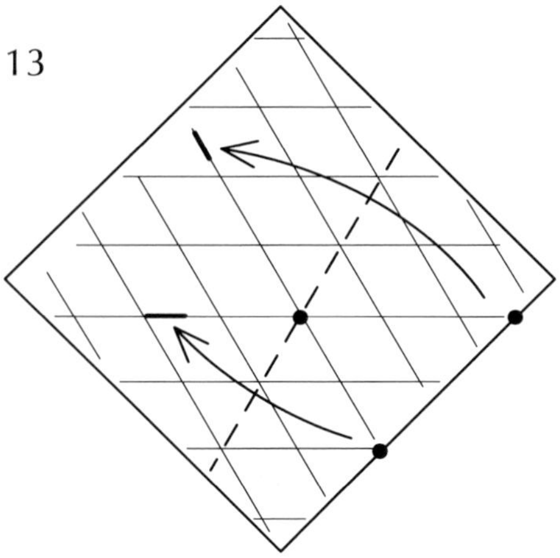

14

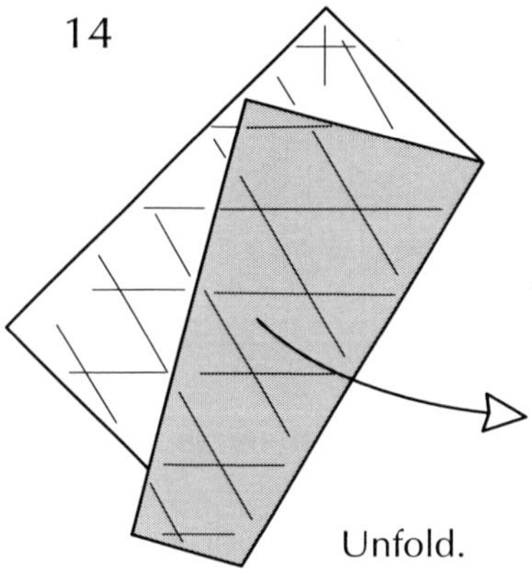

Unfold.

15

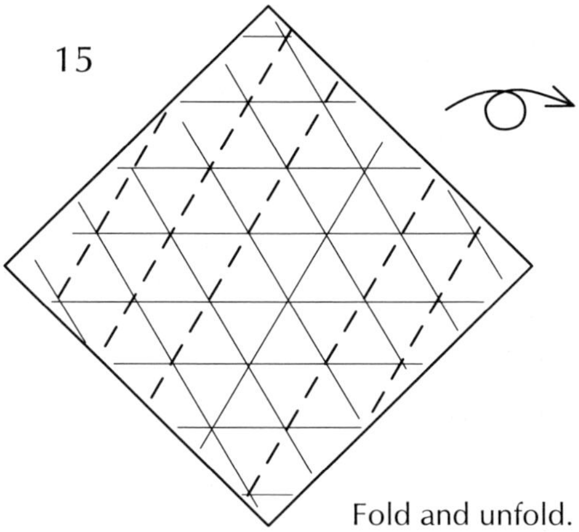

Fold and unfold.
Rotate 90°.

16

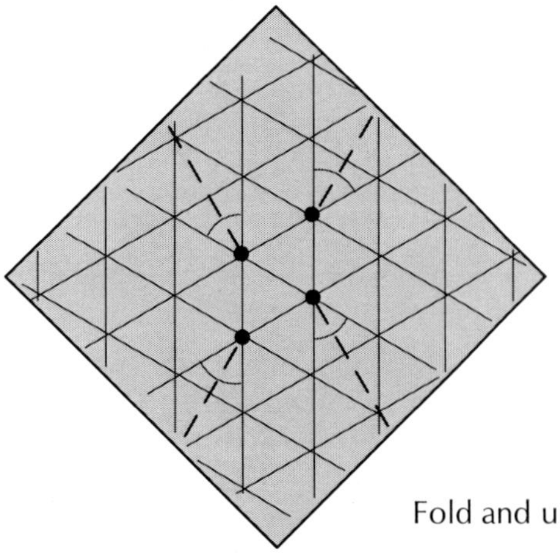

Fold and unfold.

17

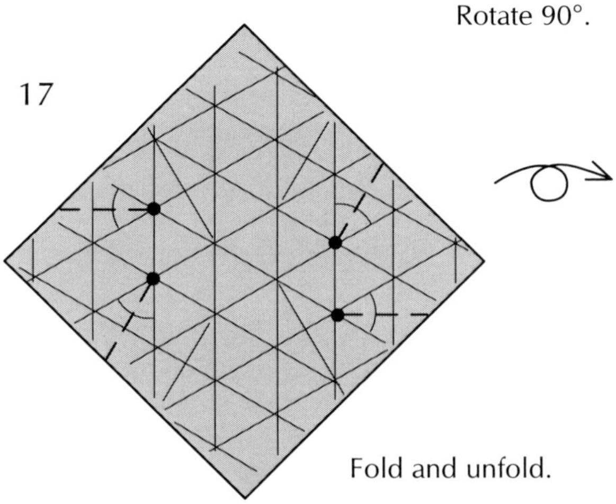

Fold and unfold.

18

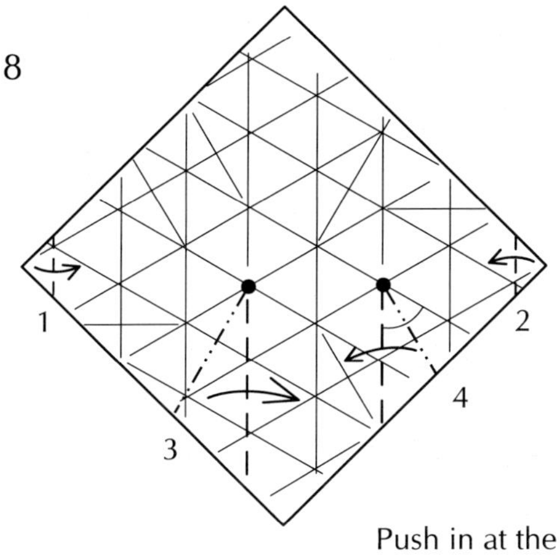

Push in at the dots.

19

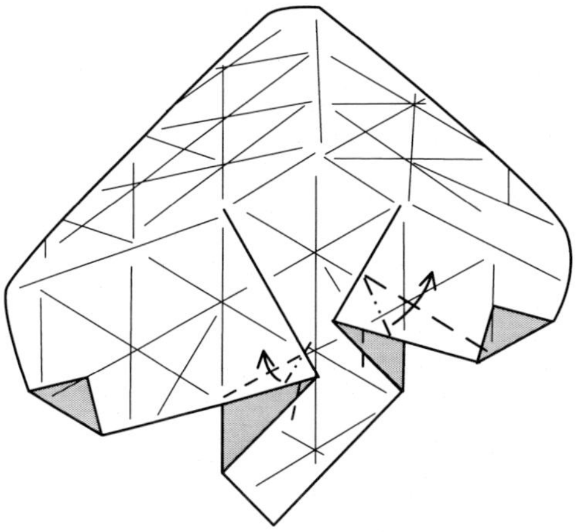

Make squash folds.

20

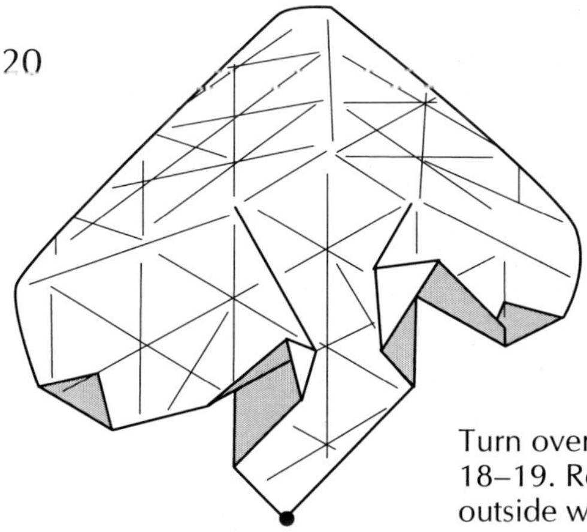

Turn over and repeat steps 18–19. Rotate to view the outside with the dot at the top.

21

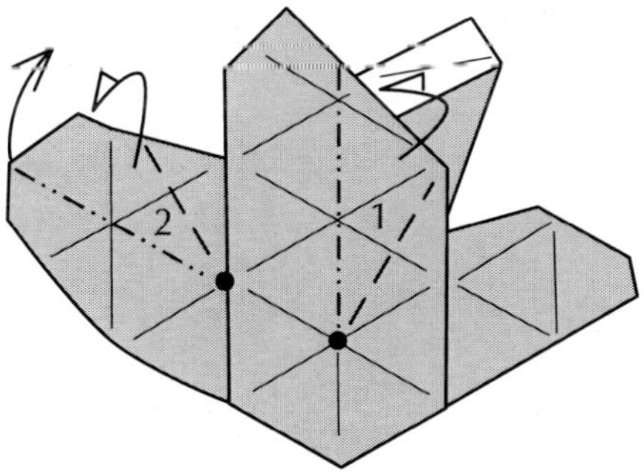

Puff out at the dots at 1 and 2. Turn over and repeat.

22

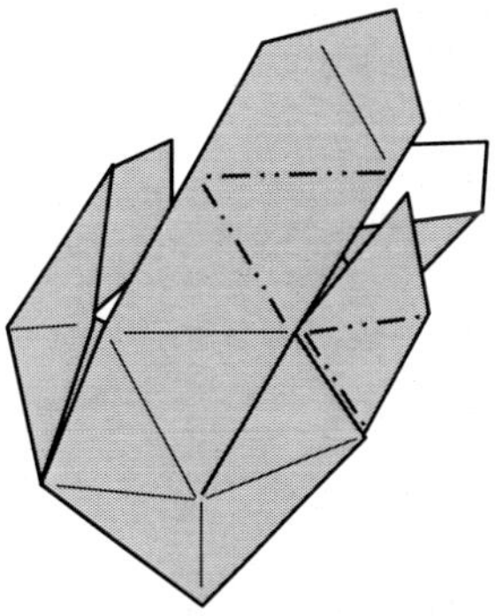

Fold and unfold along the creases. Turn over and repeat.

23

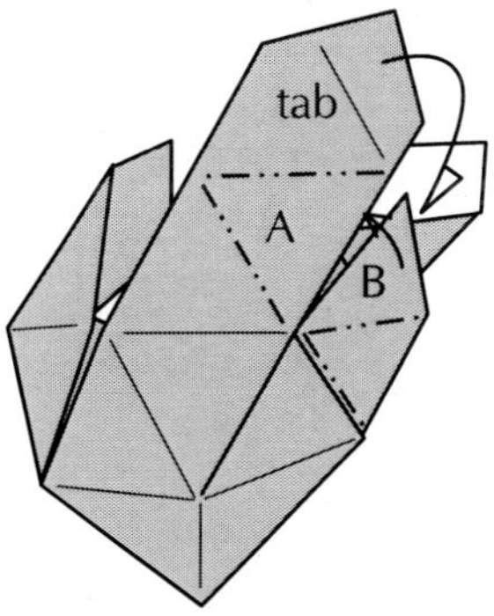

Triangle A will cover B and the tab will wrap around B. Turn over and repeat.

A typical method for these polyhedra is to do steps 22–23 on one side, unfold back to step 22, turn over and repeat on the other side, then refold the previous side to lock the model.

24

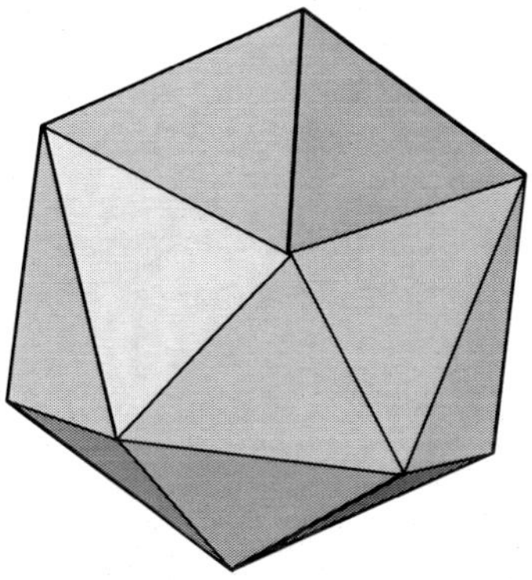

Icosahedron

Dimpled Polyhedra

Here is a quartet of dimpled polyhedra. Two of them, the cubehemioctahedron and dimpled hexagonal dipyramid, are versions of a cube with sunken corners.

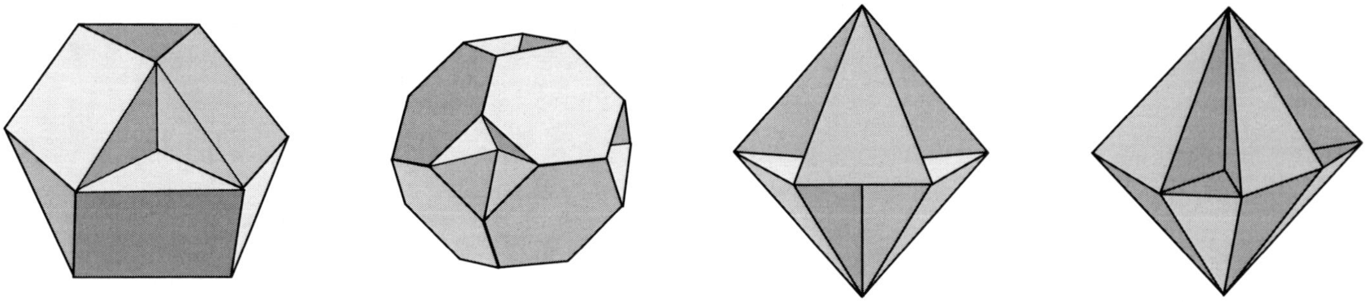

Cubehemioctahedron

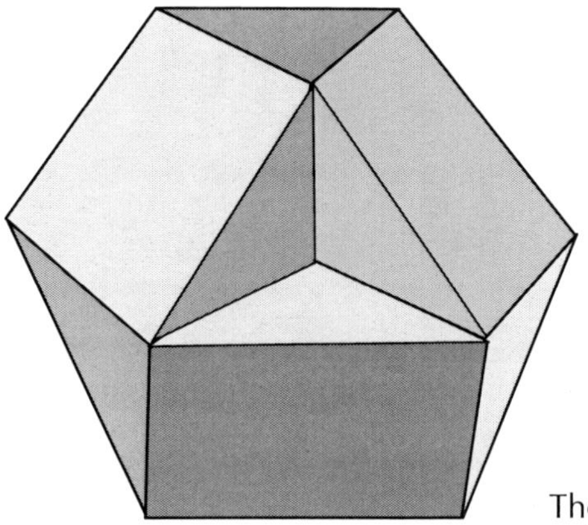

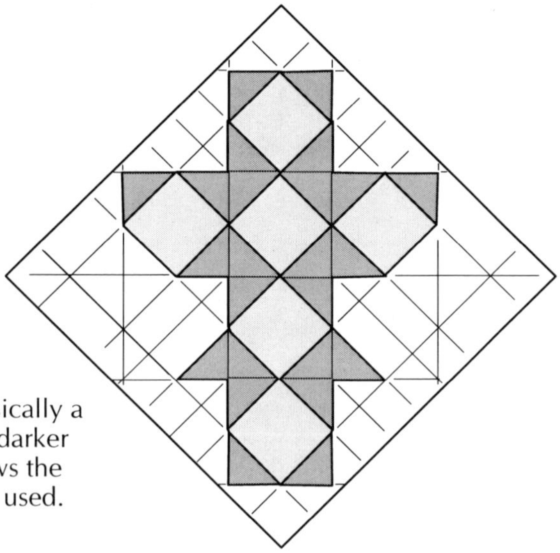

The cubehemioctahedron is basically a cube with sunken corners. The darker paper in the crease pattern shows the sunken sides. Even symmetry is used.

1

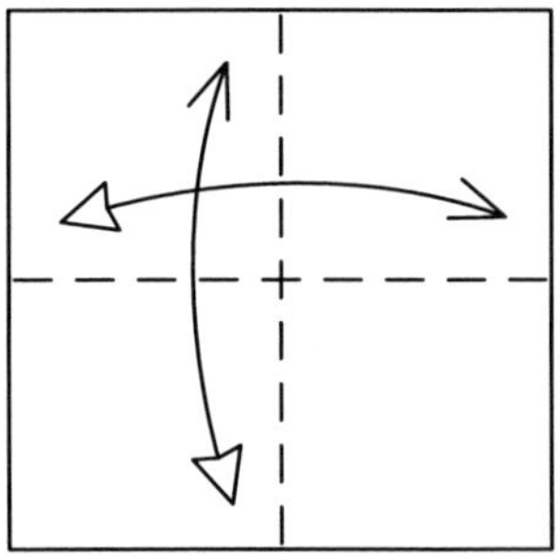

Fold and unfold.
Rotate 45°.

2

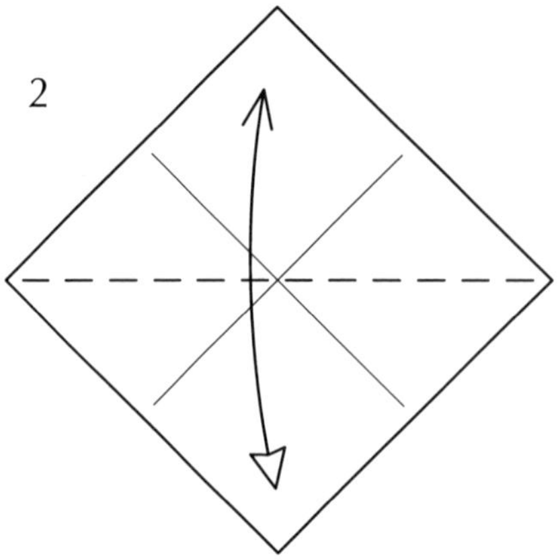

Fold and unfold.

3

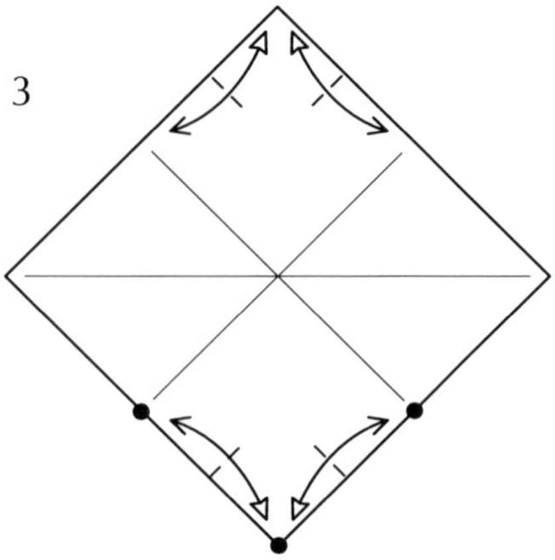

Fold and unfold on the edges.

4

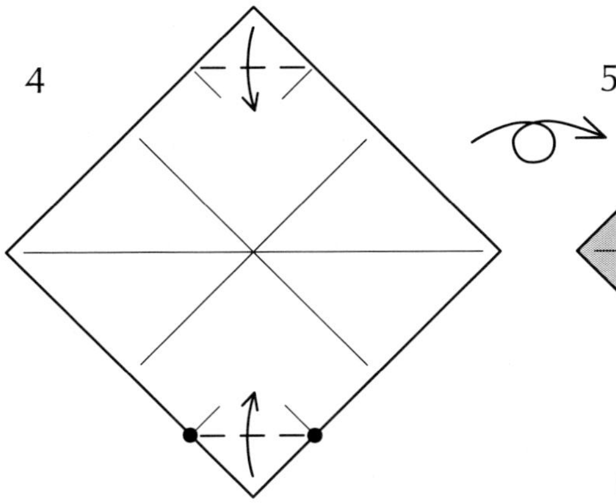

5

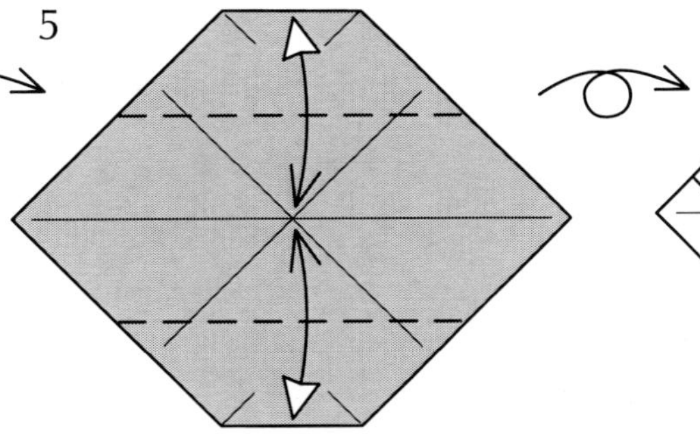

Fold to the center and unfold.

6

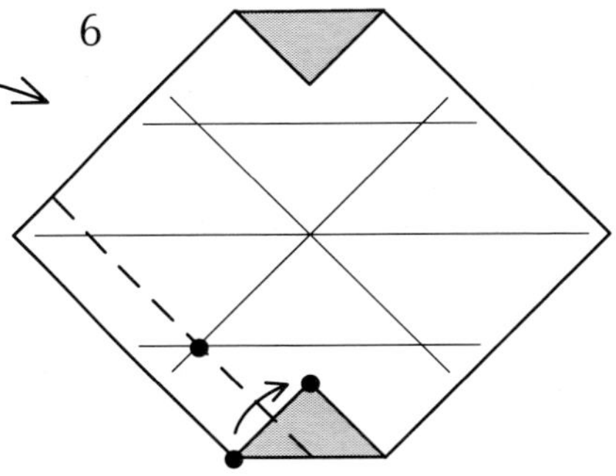

7

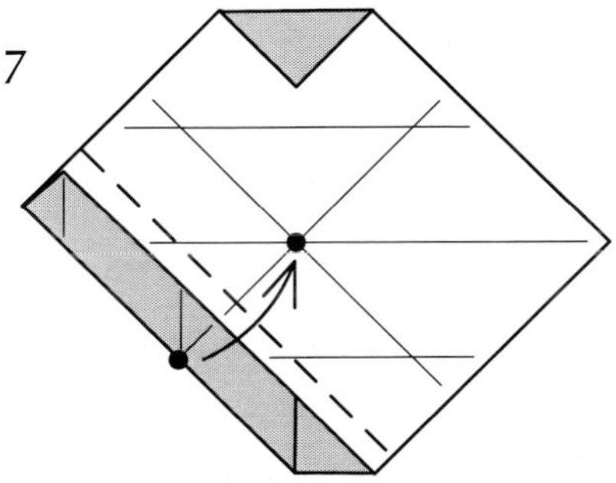

8

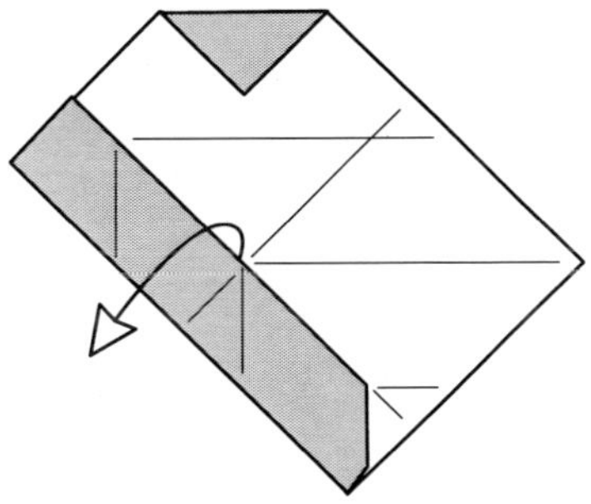

Unfold.

9

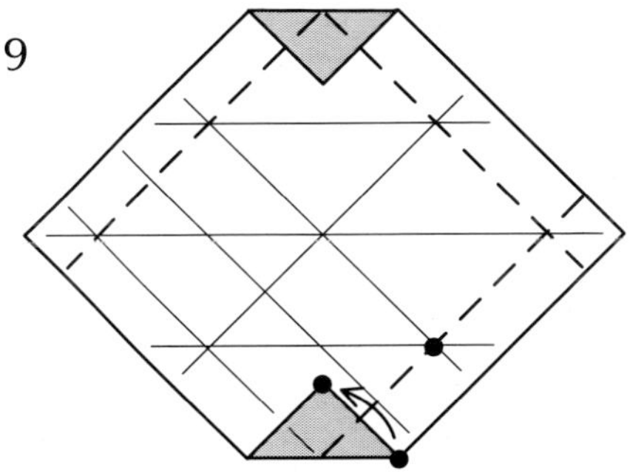

Repeat steps 6–8 three times. Begin each part on the valley lines.

10

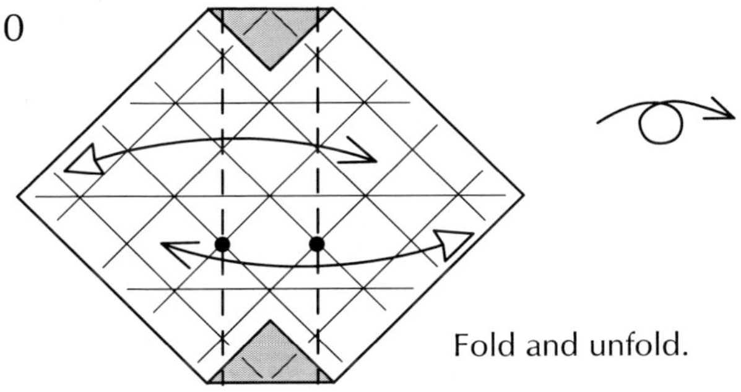

Fold and unfold.

11

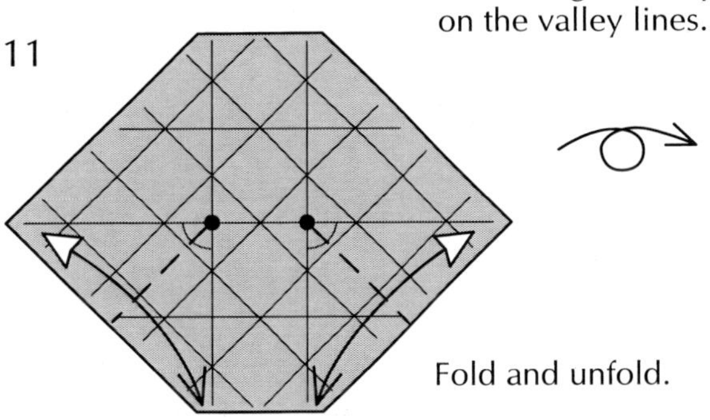

Fold and unfold.

12

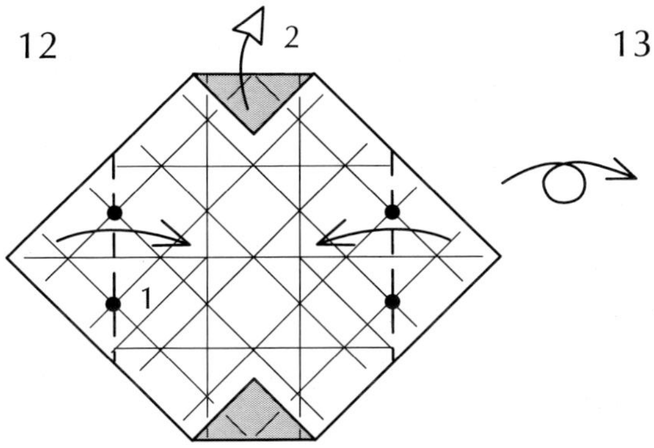

1. Make valley folds.
2. Unfold.

13

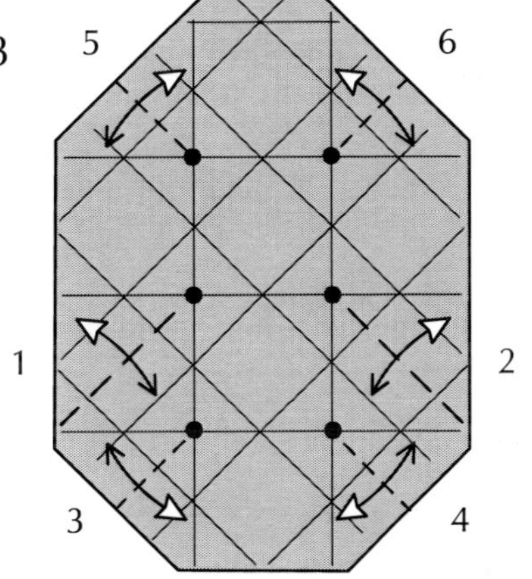

1 and 2. Fold and unfold along the creases.
3 to 6. Fold and unfold.

14

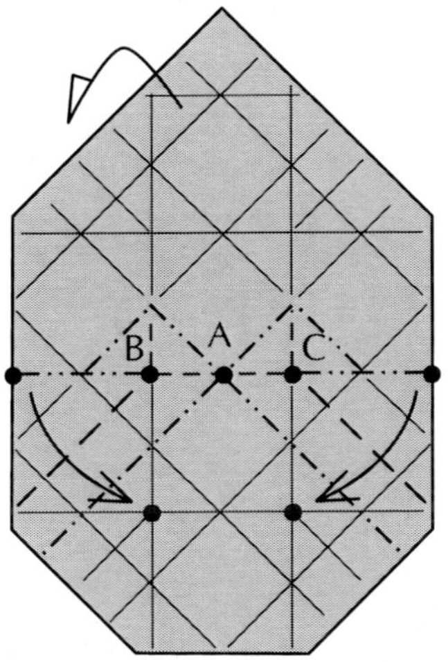

Begin at dot B. Push in at B to make a sunken triangle. Repeat at dot C. Puff out at dot A. The other pairs of dots will meet.

15

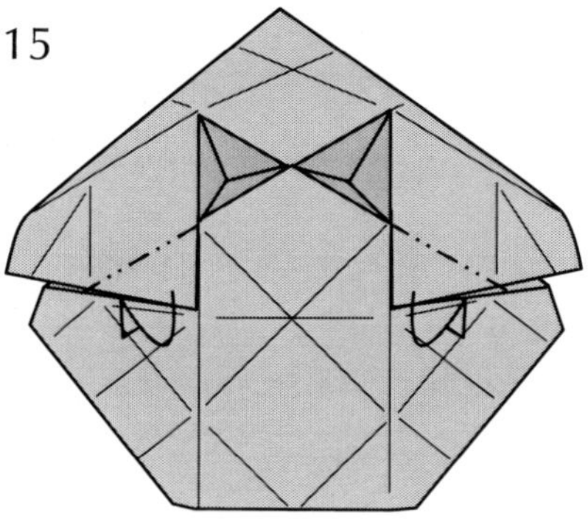

Mountain-fold the layers together.

16

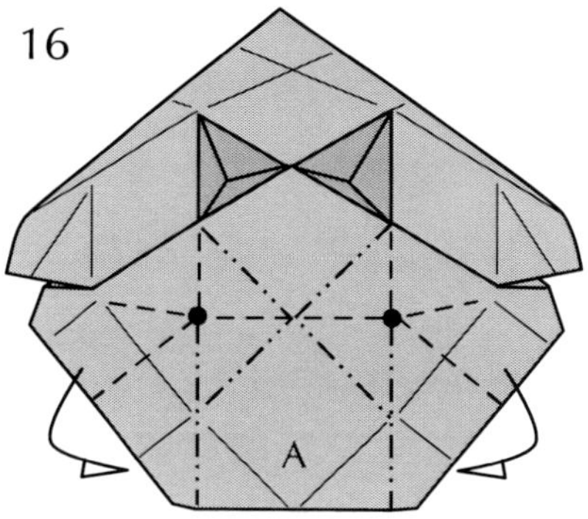

Push in at the dots to form two sunken triangles. Flatten the layers behind region A.

17

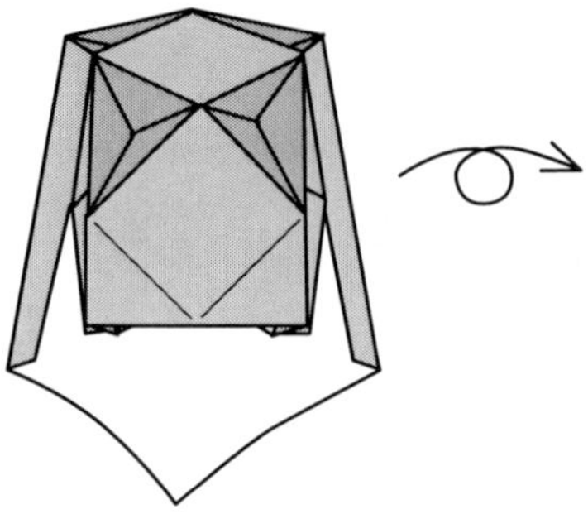

18

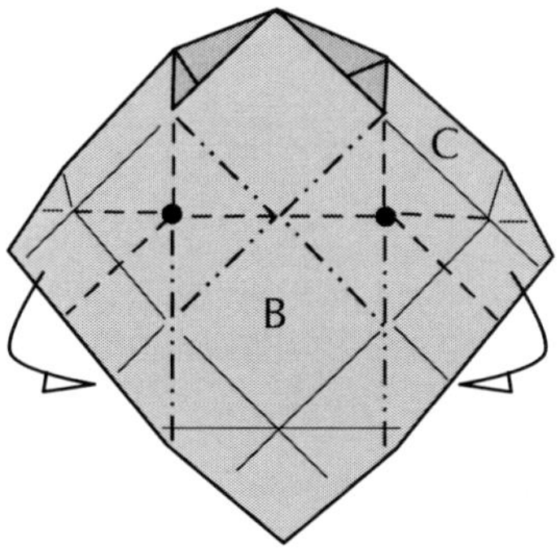

Push in at the dots. Flatten the layers behind region B. Rotate the top to the bottom and view region C, so it becomes front and center.

19

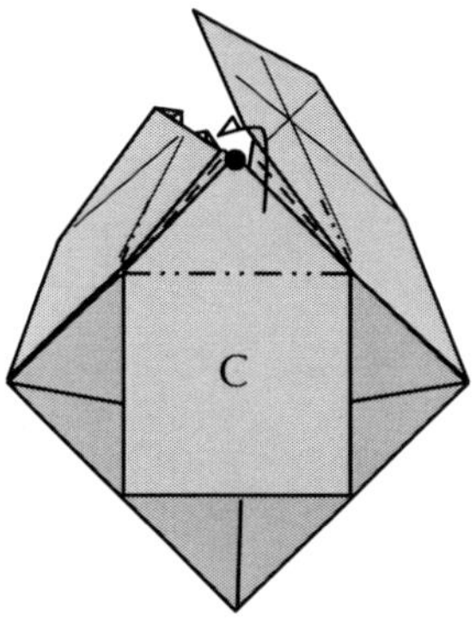

Fold inside towards the center and repeat behind, so the two dots (only one drawn) meet. This begins the formation of two sunken triangles.

20

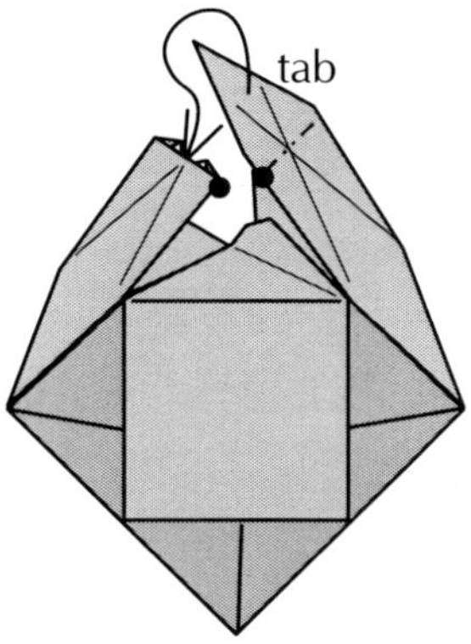

Tuck the tab inside the pocket so the dots meet.

21

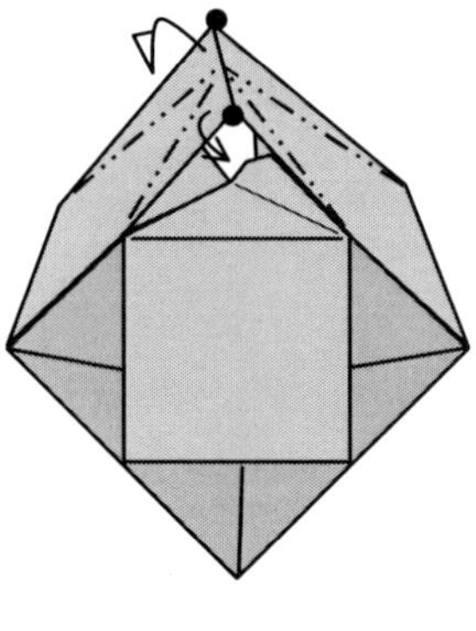

Push in at the dots to complete the sunken triangles. (Repeat behind.) This is my favorite step!

22

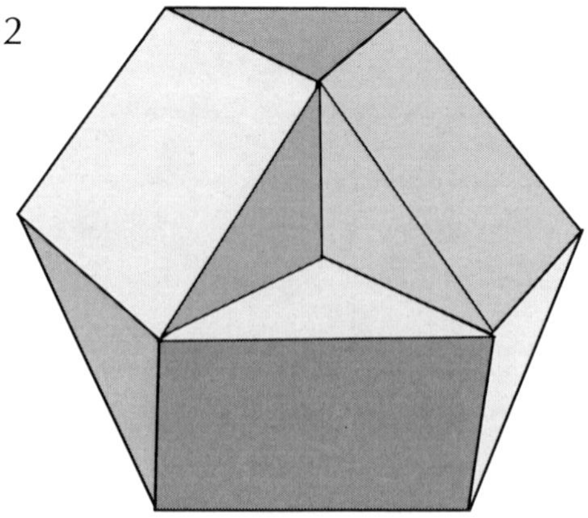

Cubehemioctahedron

Dimpled Truncated Octahedron

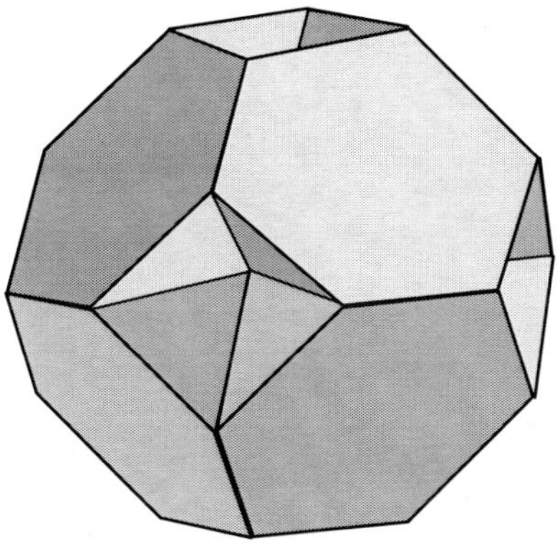

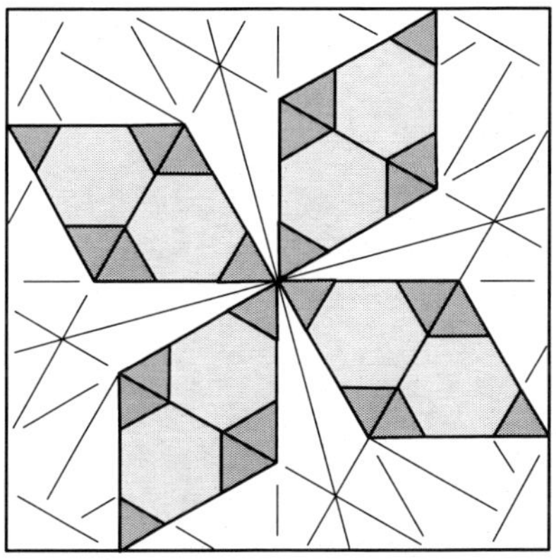

This dimpled truncated octahedron comes from an octahedron design with sunken sides. The crease pattern shows square symmetry.

1

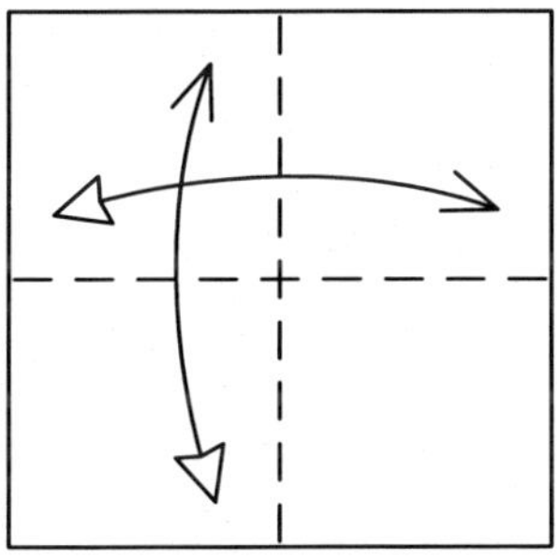

Fold and unfold.

2

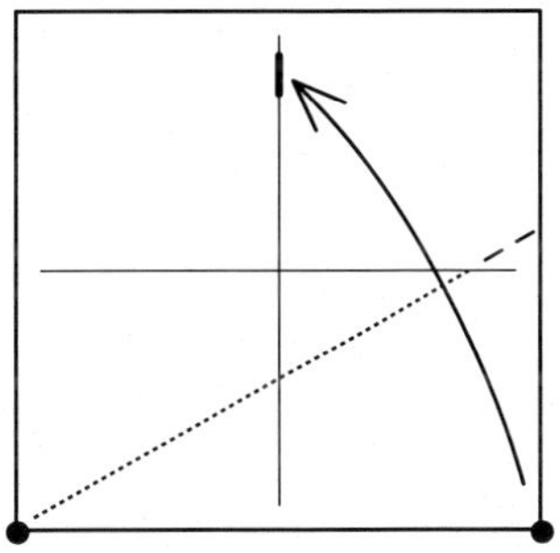

Bring the left dot to the line. Crease on the right.

3

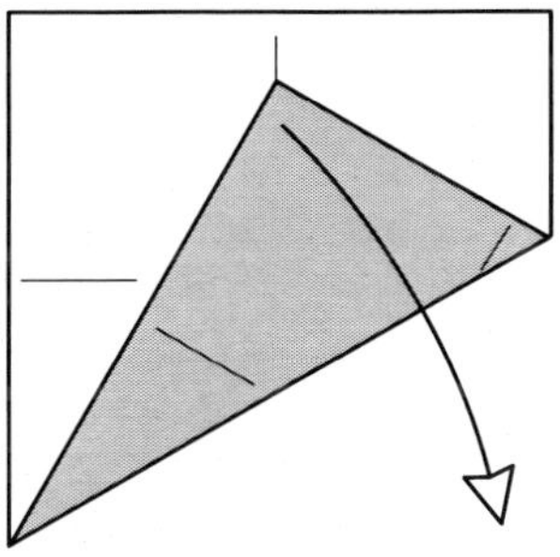

Unfold and rotate 90°.

4

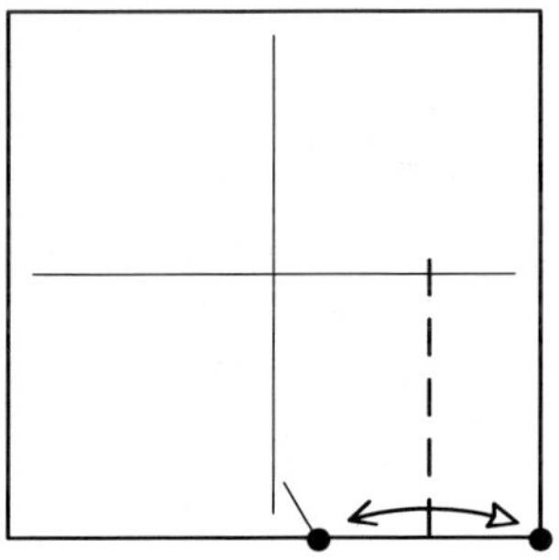

Fold and unfold.

5

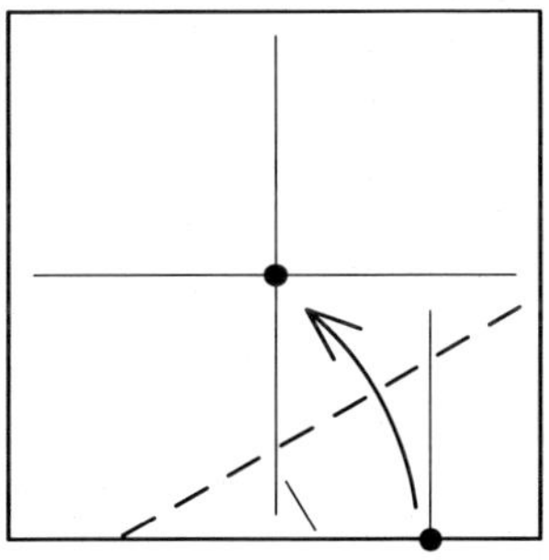

6

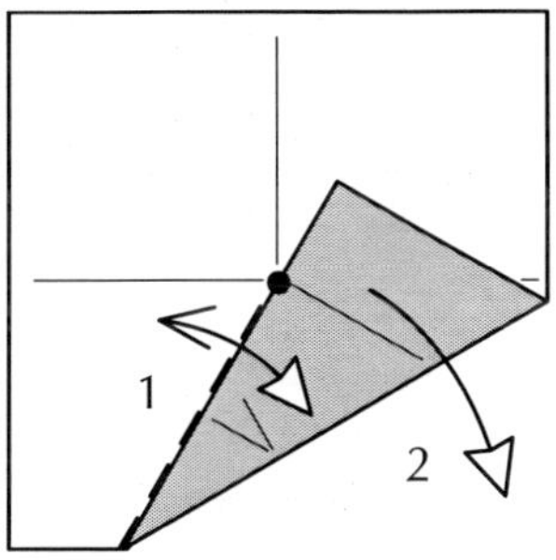

1. Fold and unfold along the edge.
2. Unfold.

7

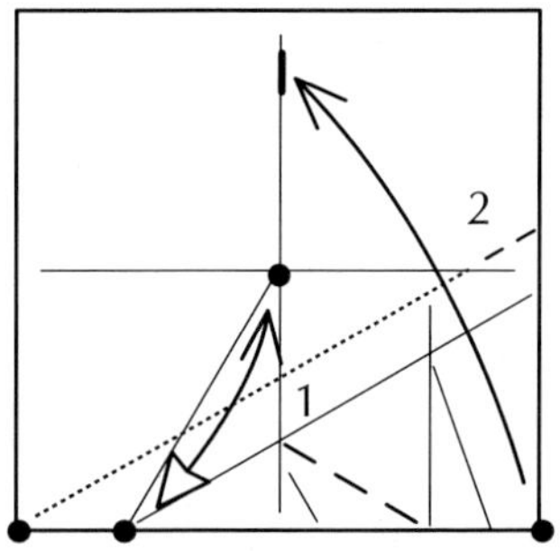

1. Fold and unfold.
2. Repeat steps 2–7 three times.

8

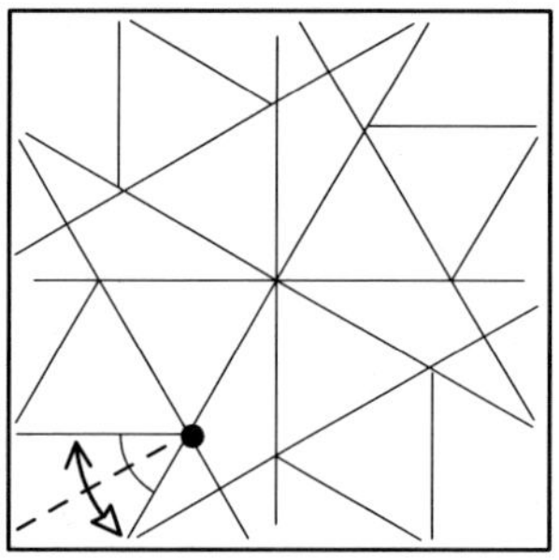

Fold and unfold to bisect the angle.

9

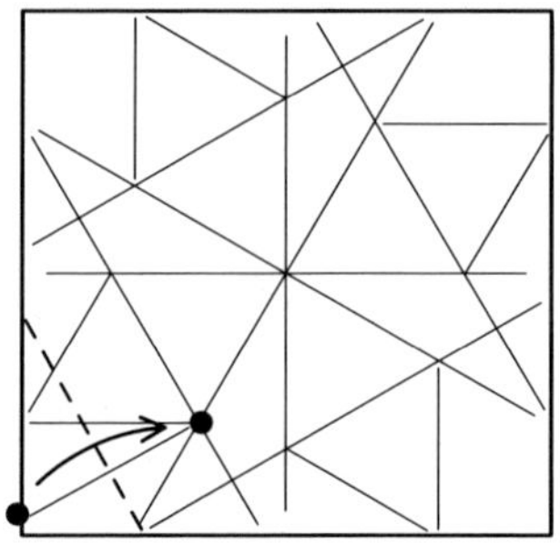

10

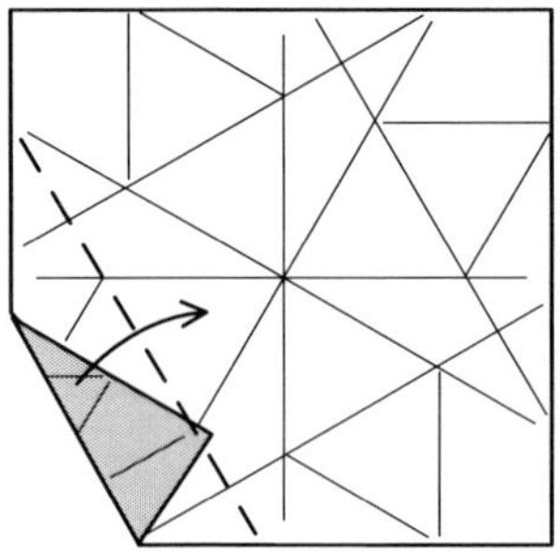

Fold along the crease.

11

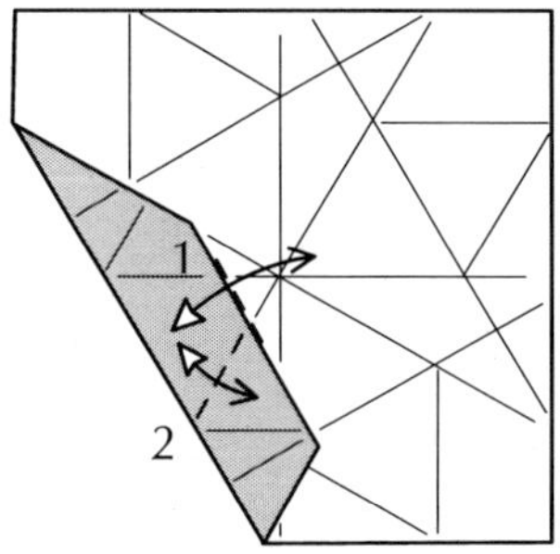

1. Fold and unfold.
2. Fold and unfold along the crease.

12

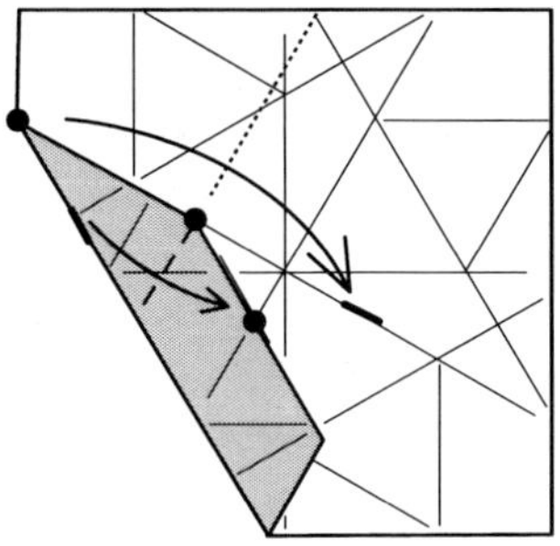

Begin by folding at the center dot. The other dots and lines meet.

13

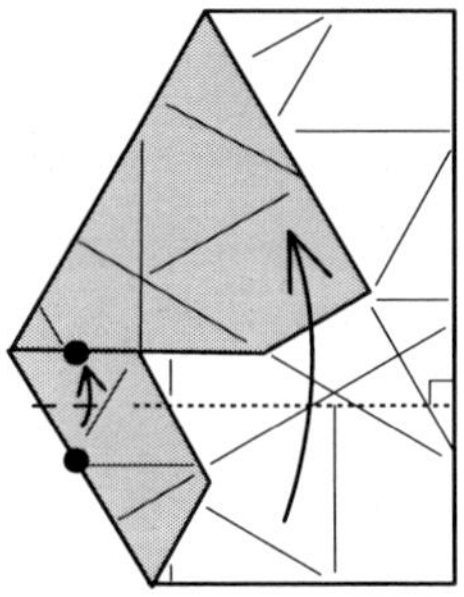

Fold along a horizontal line so the dots meet.

14

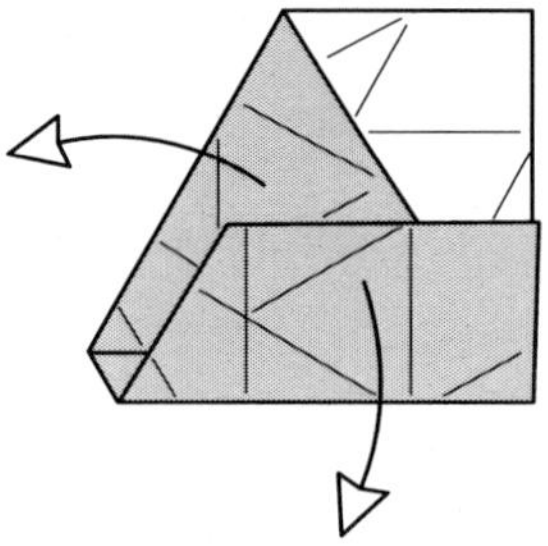

Unfold everything.

15

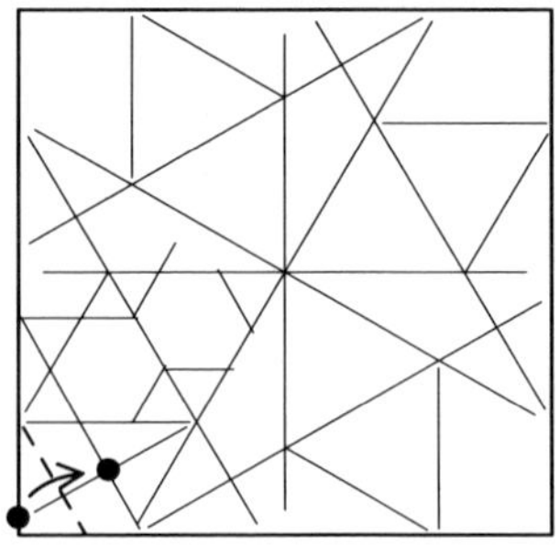

Rotate 90°.

16

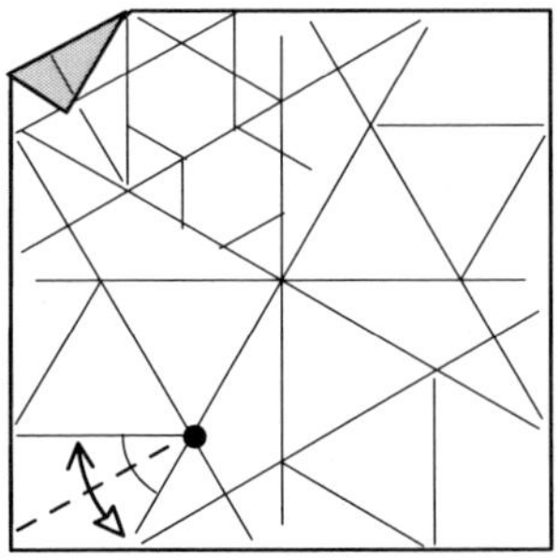

Repeat steps 8–15 three times.

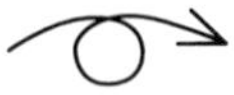

17

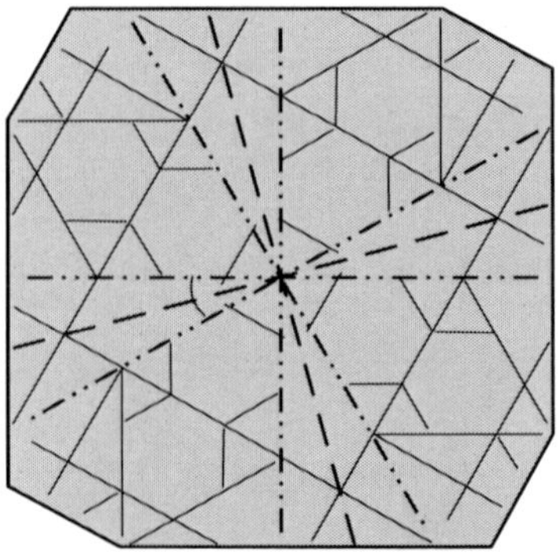

Fold and unfold to bisect the angles. Mountain-fold along the creases.

18

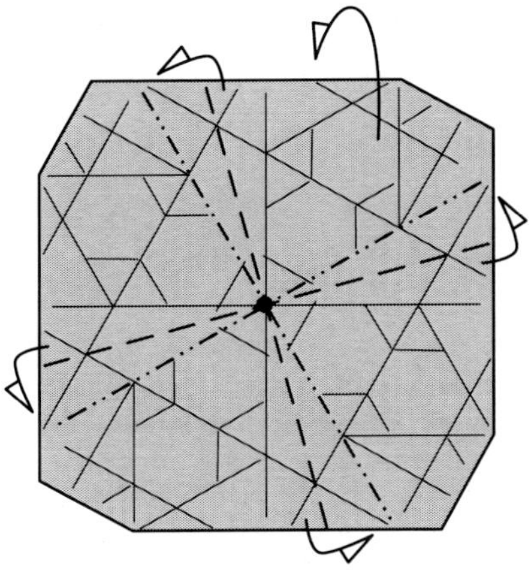

Puff out at the dot and fold along the creases.

19

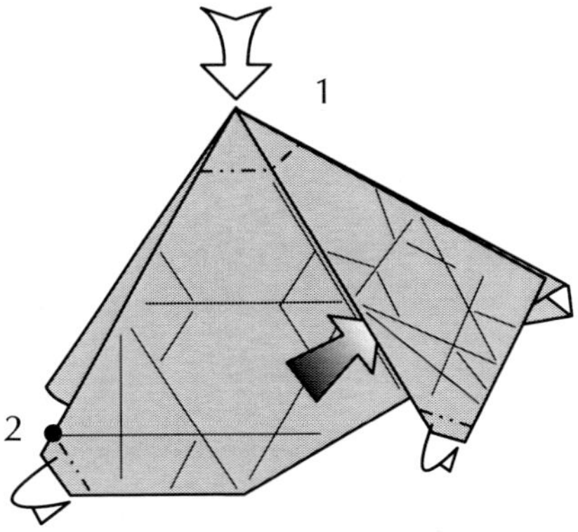

Note the dark arrow shows the direction of the layers.
1. Sink, keeping the ridges at the edges together.
2. Mountain-fold and repeat all around.

20

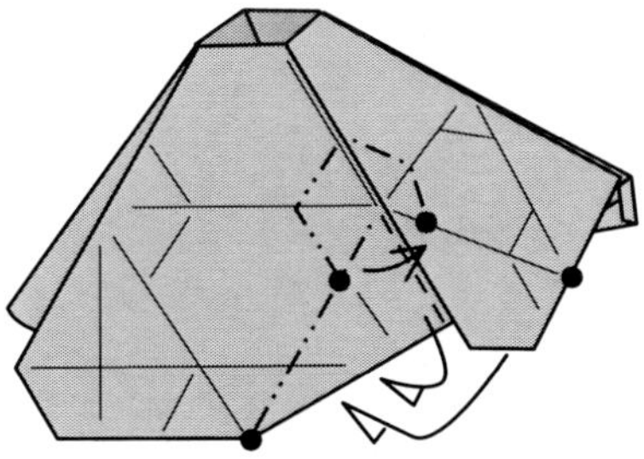

Form a sunken square by bringing the pair of dots together. Much of the paper will be tucked inside.

21

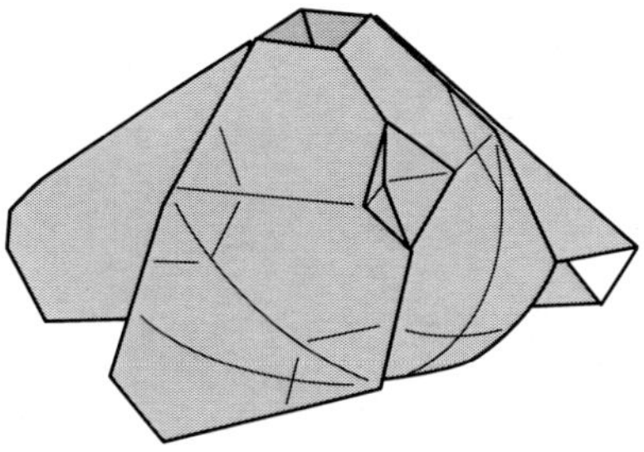

Repeat step 20 on the three other sides.

22

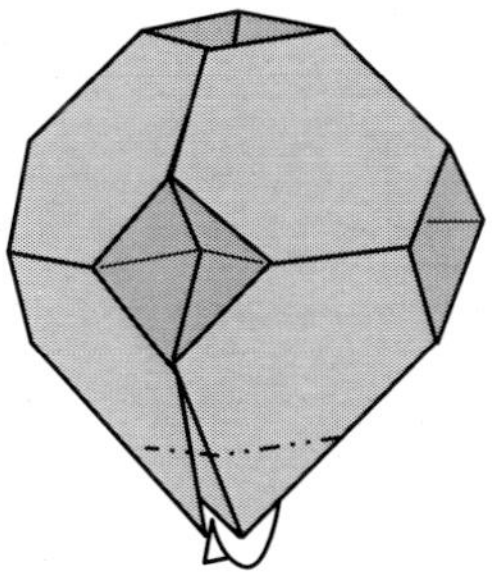

Reverse-fold one of the loose flaps on the bottom.

23

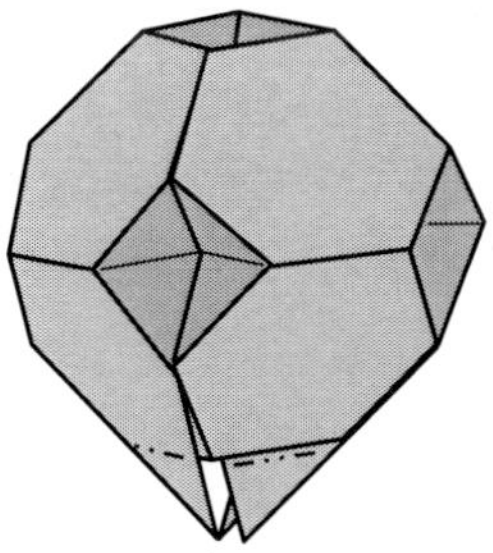

Continue reverse folding each of the layers inside each other to form the bottom sunken square. It is possible to inflate at the bottom to round out this polyhedron.

24

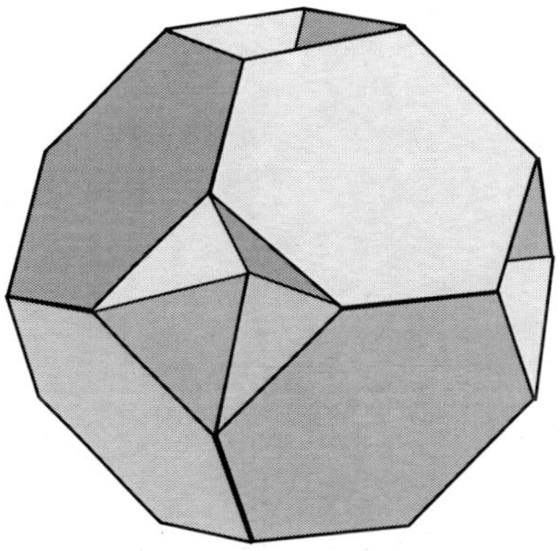

Dimpled Truncated Octahedron

Dimpled Hexagonal Dipyramid

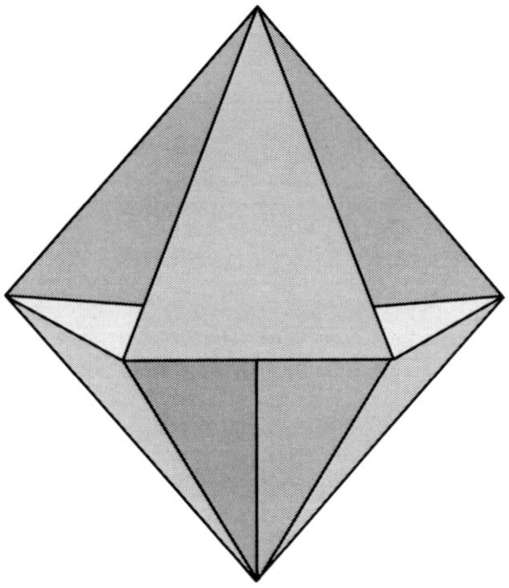

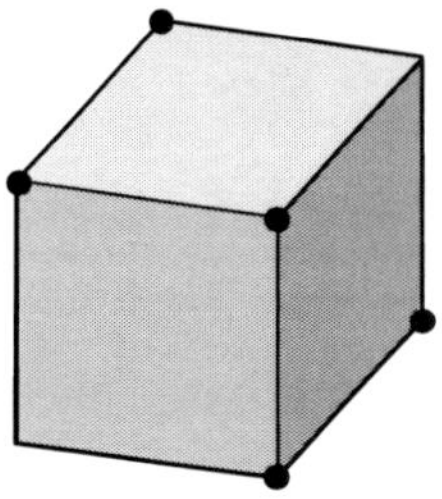

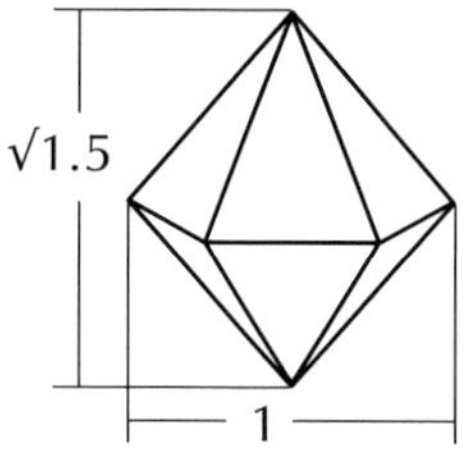

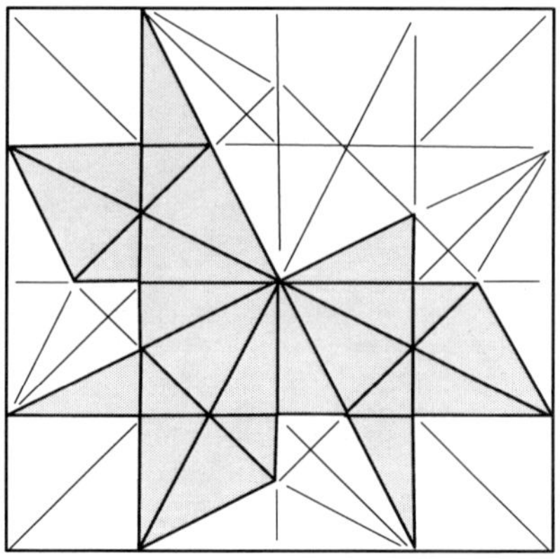

This model comes from a cube where six of the eight corners are sunken, as shown in the second picture. The angles in the nonsunken triangles are 36.87°, 71.565°, and 71.565°, and the ratio of the height to diameter is $\sqrt{1.5}$. = 1.225. The layout shows 3/4 square symmetry.

1

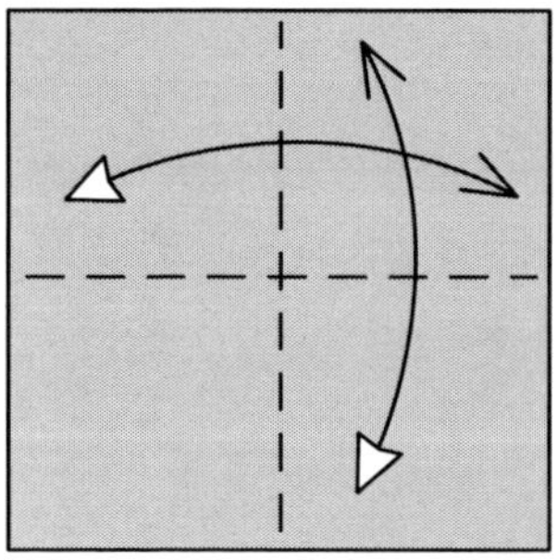

Fold and unfold.

2

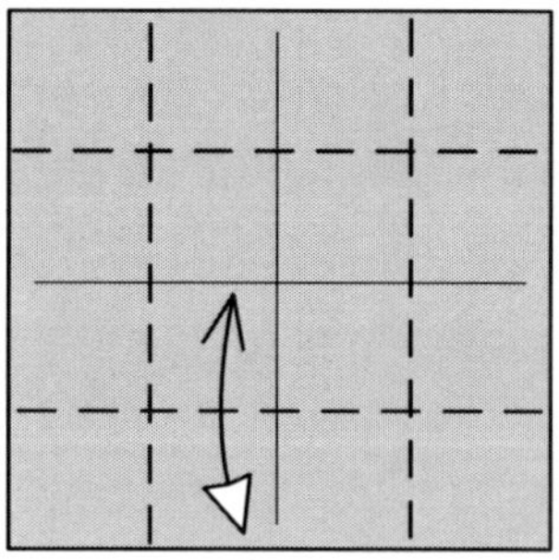

Fold and unfold four times.

3

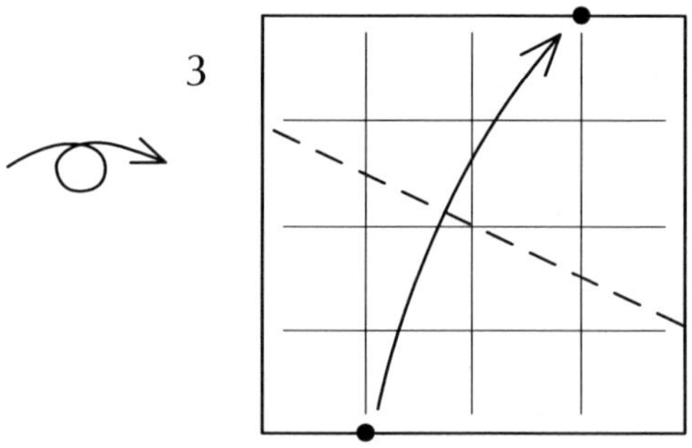

4

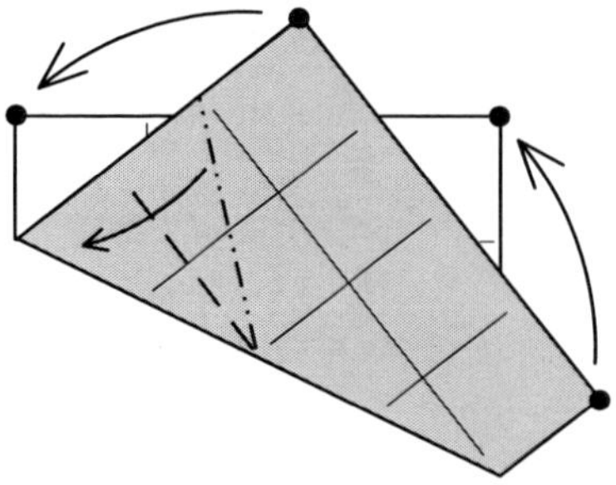

Valley-fold along the crease.
Turn over and repeat.

5

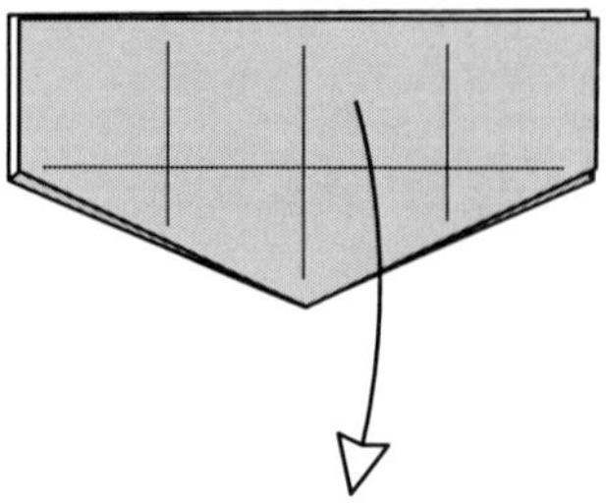

Unfold and rotate 90°.

6

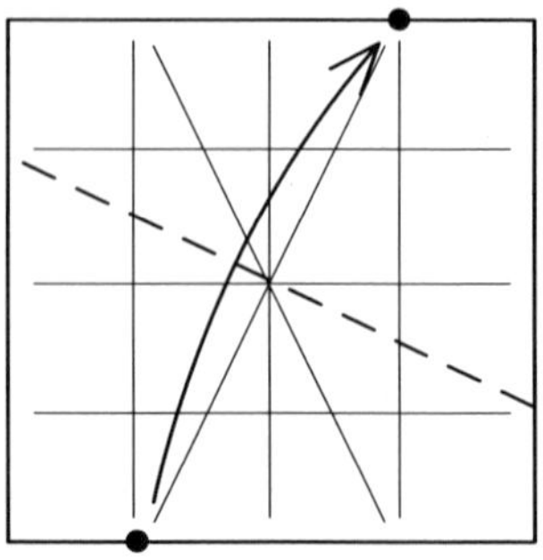

Repeat steps 3–5.

7

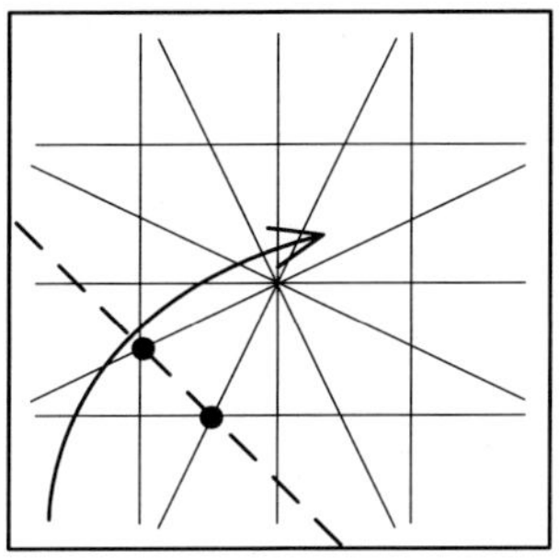

8

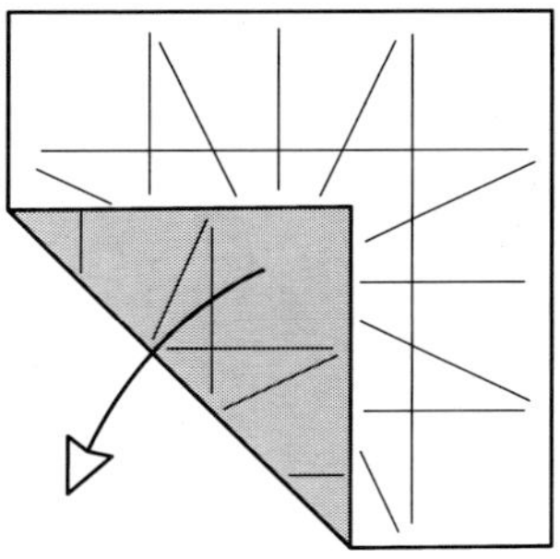

Unfold and rotate 90°.

9

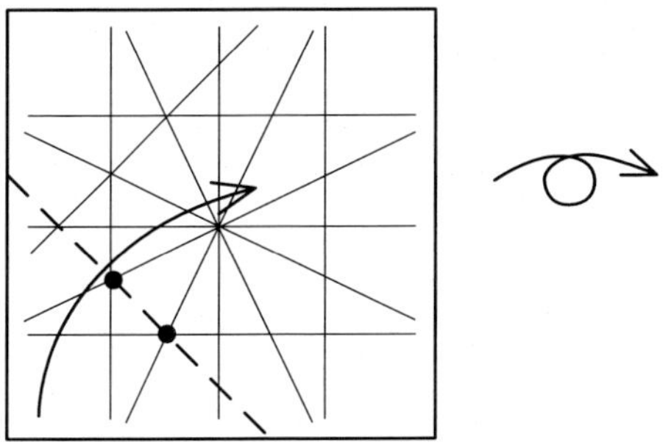

Repeat steps 7–8
three times.

10

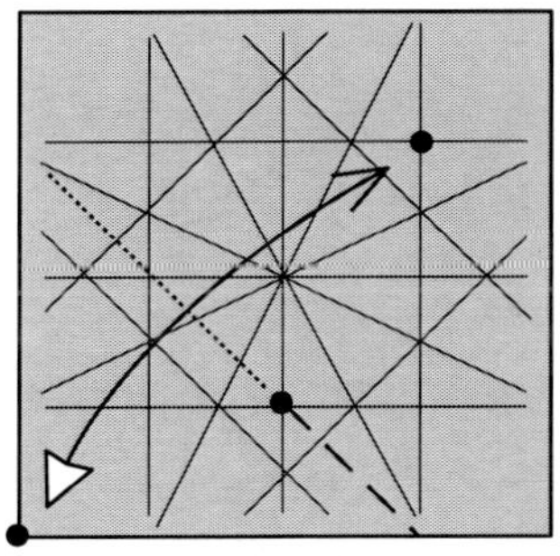

Fold and unfold. Rotate 90°.

11

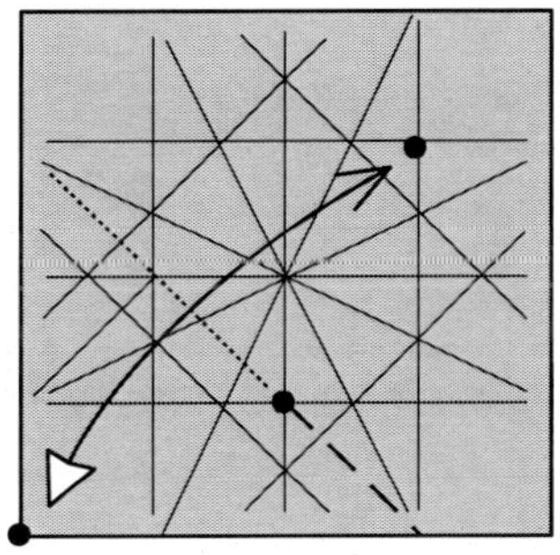

Repeat step 10
three times.

12

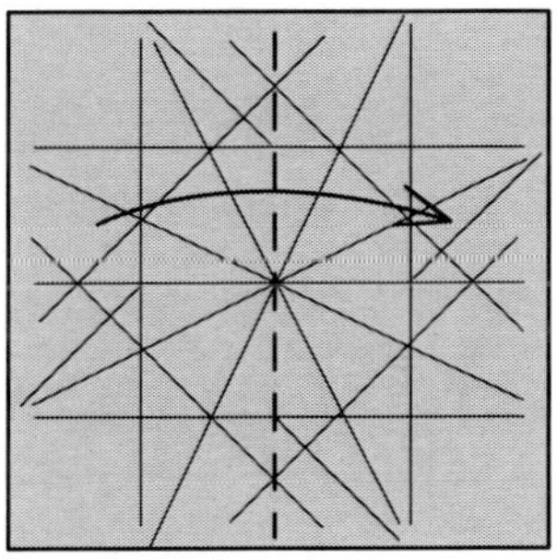

13

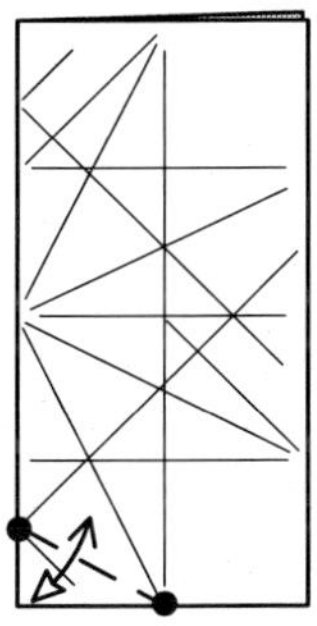

Fold and unfold.

14

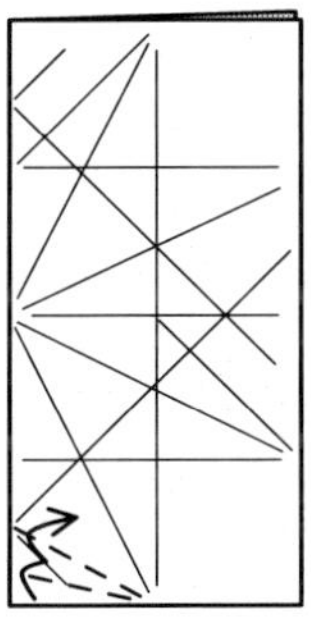

15

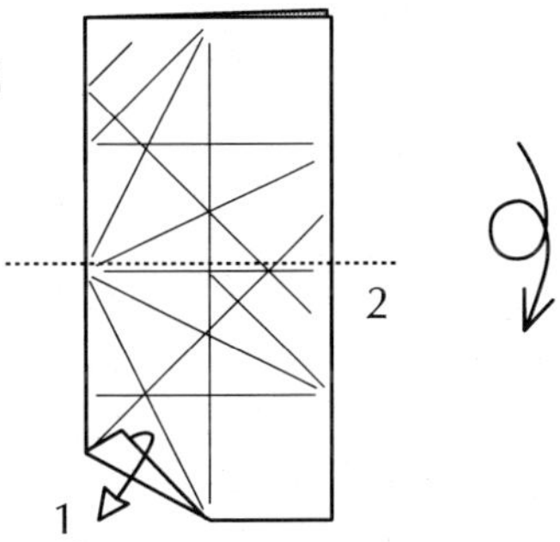

1. Unfold.
2. Turn over around the dotted line.

16

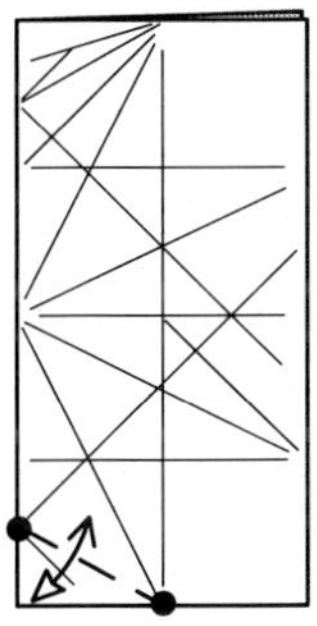

Repeat steps 13–15.

17

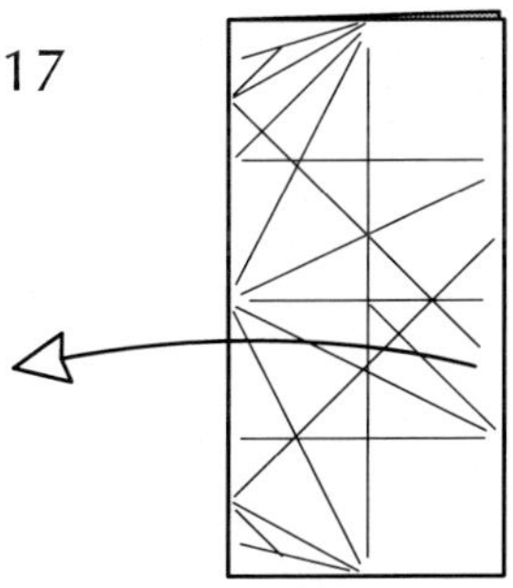

Unfold and rotate 90°.

18

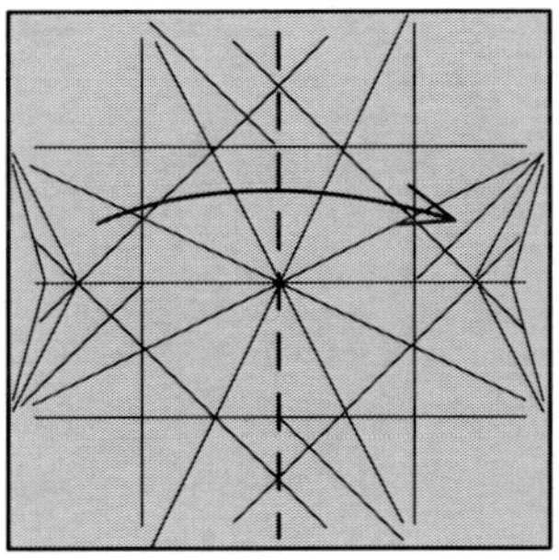

Repeat steps 12–17.

19

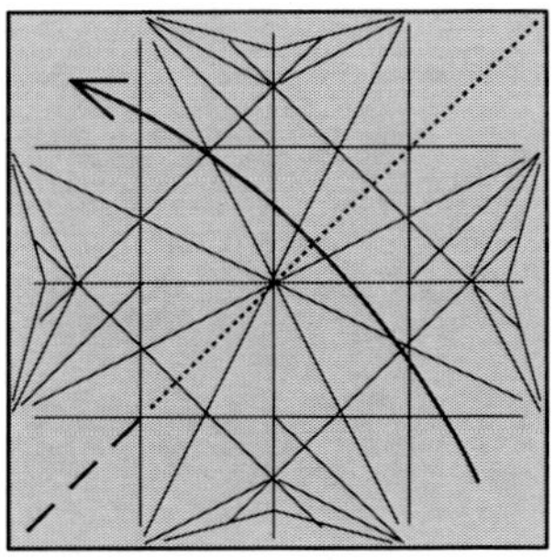

20

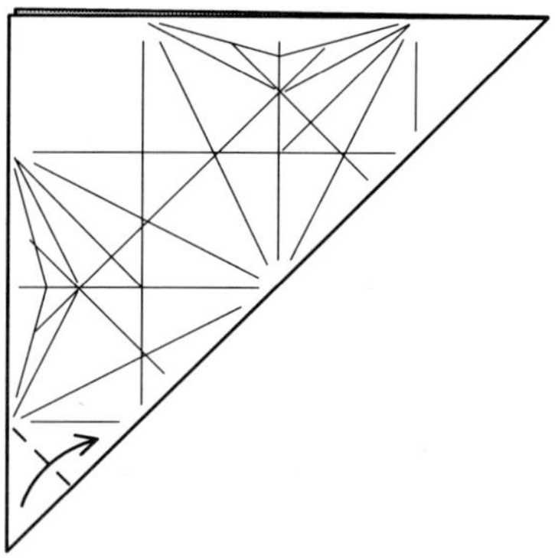

21

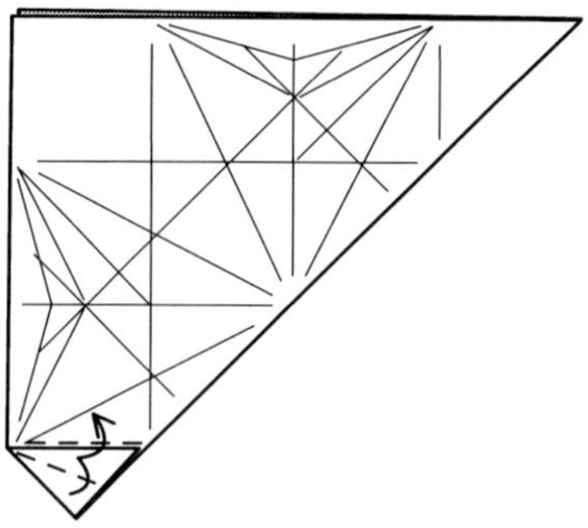

22

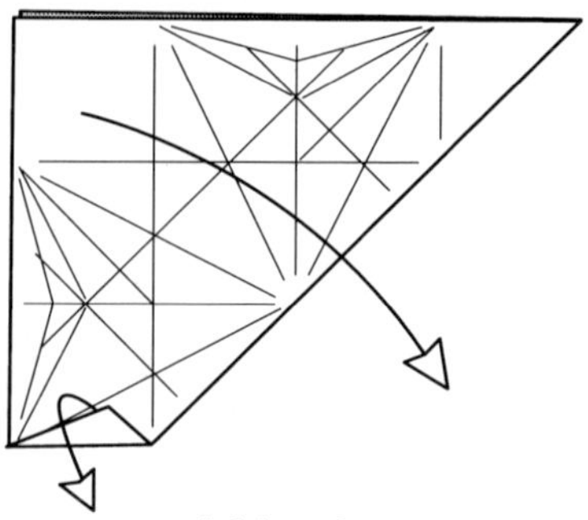

Unfold and rotate 90°.

23

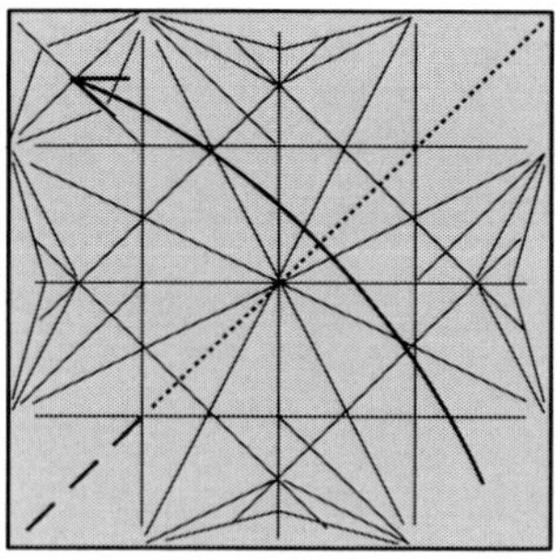

Repeat steps 19–22 three times.

24

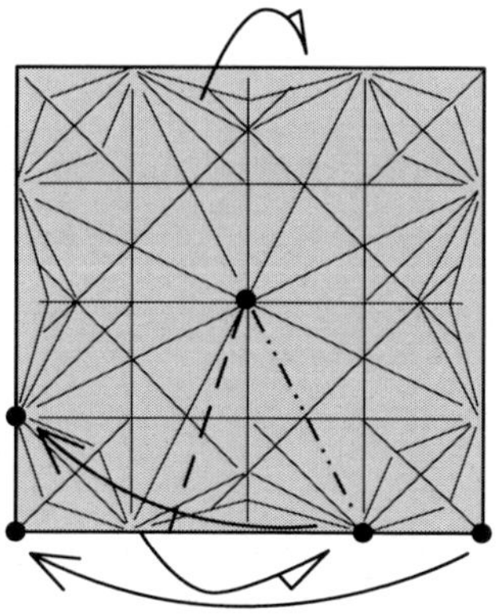

Puff out at the upper dot. The other pairs of dots will meet.

25

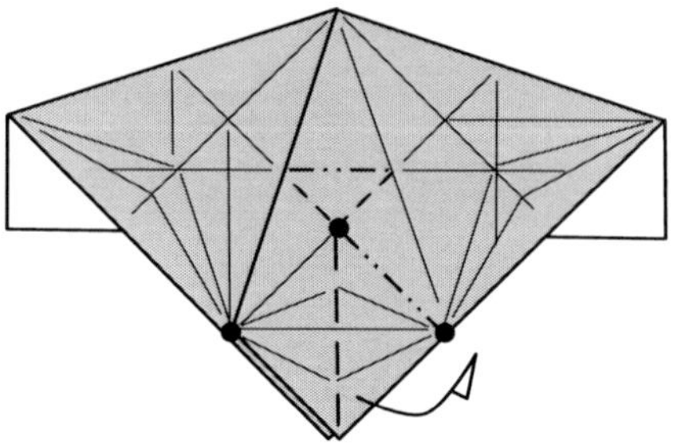

Push in at the upper dot. The other two dots will meet.

26

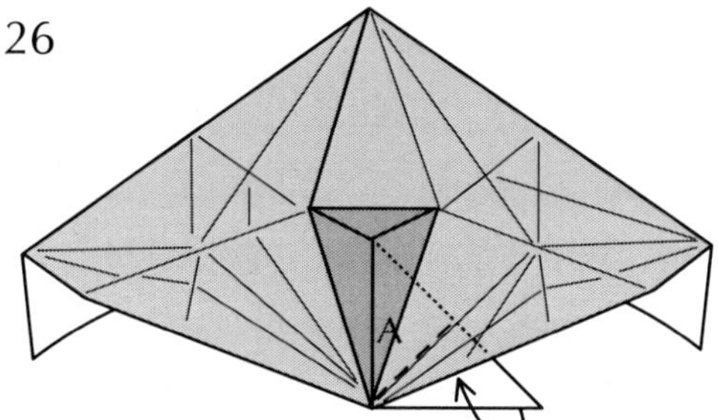

The folding is done on the inside. Refold the hidden flap as in steps 20–21 and flatten against the back of triangle A. Rotate.

27

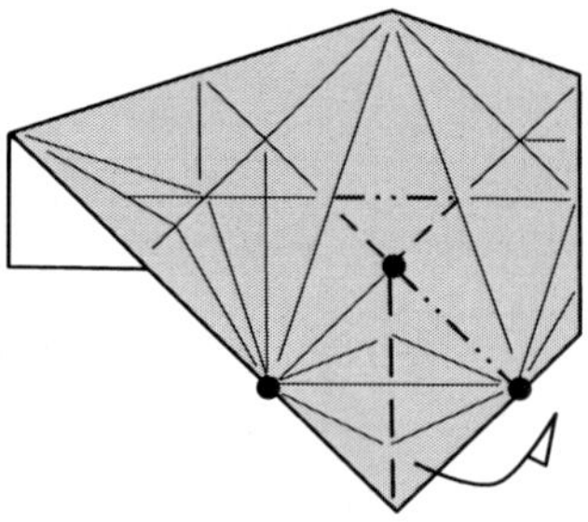

Repeat steps 25–26 two times. Rotate.

28

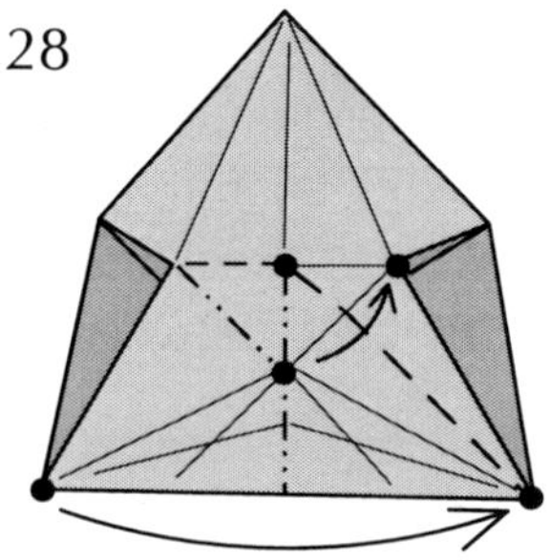

Push in at the upper dot. The other pairs of dots will meet.

29

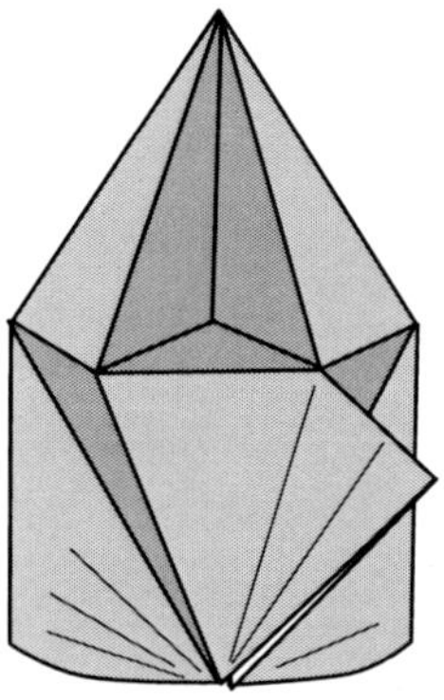

Repeat steps 28 two times.

30

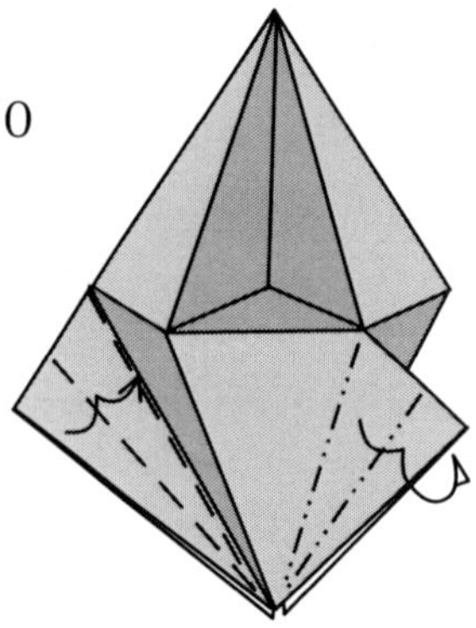

Fold along the creases and tuck inside. Do this on three sides, two are shown.

31

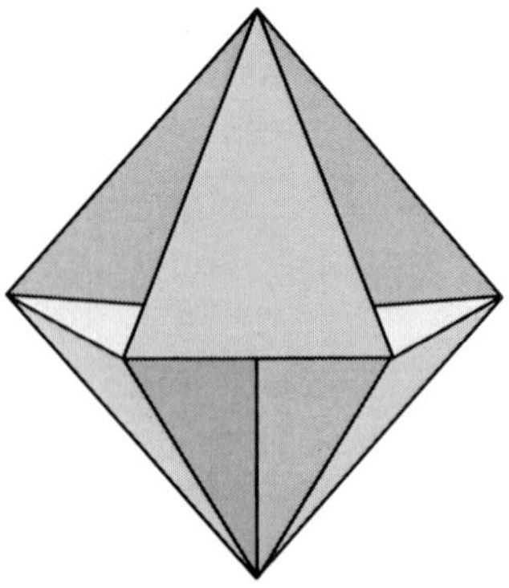

Dimpled Hexagonal Dipyramid

Dimpled Octagonal Dipyramid

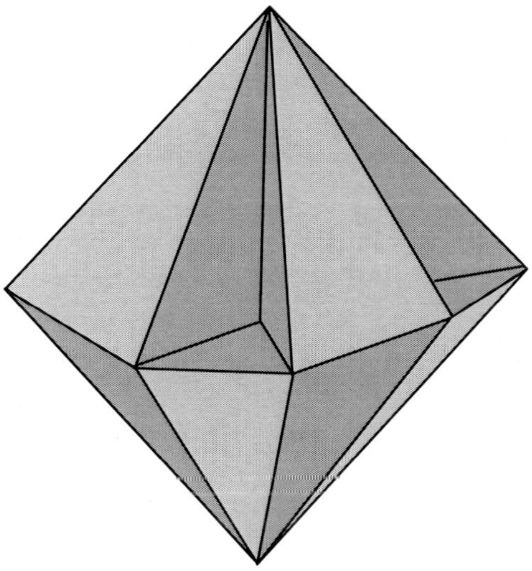

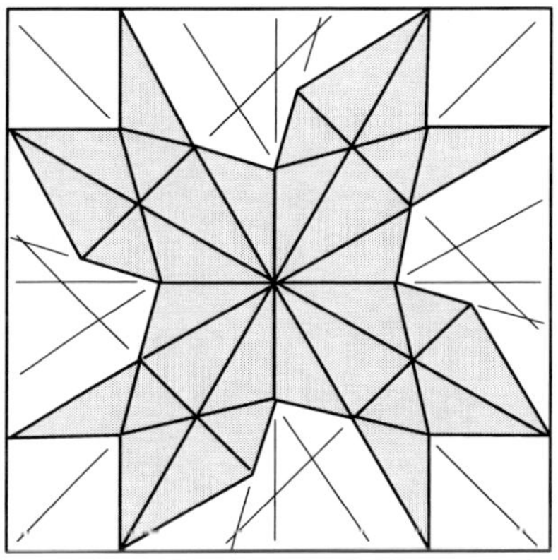

This dimpled dipyramid has eight sunken sides. The angles of each of the eight nonsunken triangles are 30°, 75°, and 75°.

1

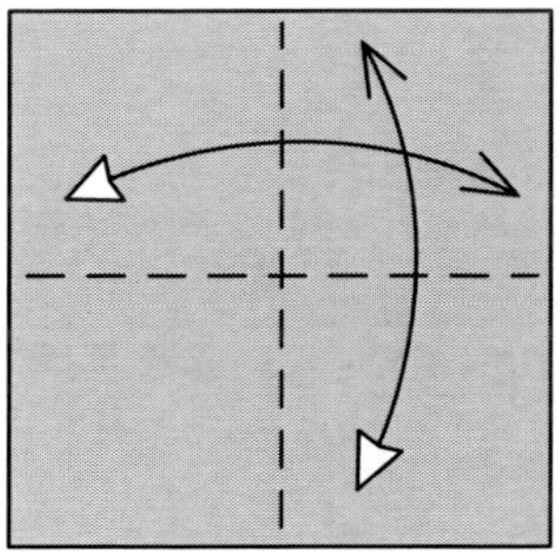

Fold and unfold.

2

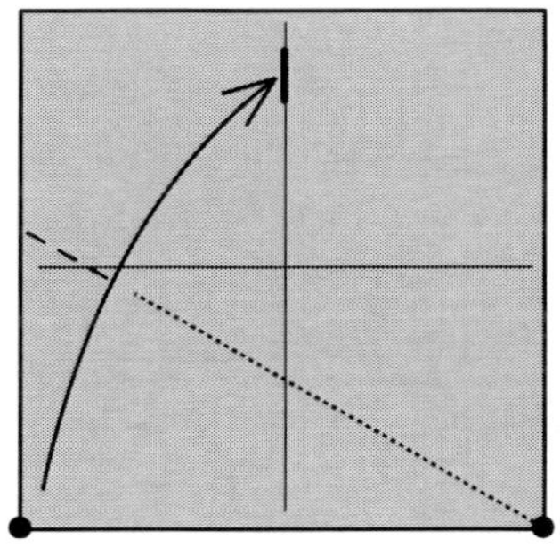

Bring the corner to the line. Fold on the left.

3

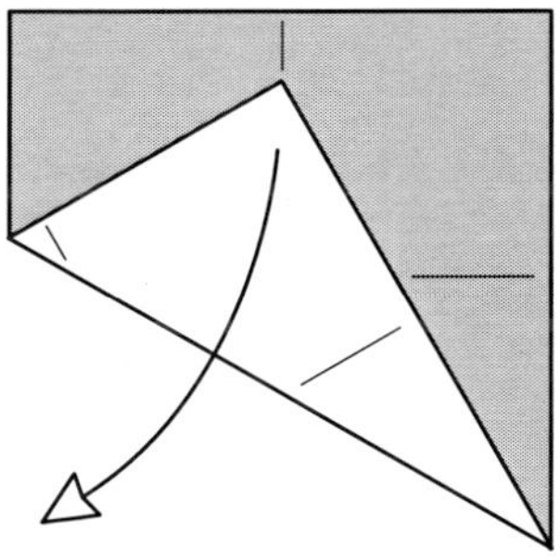

Unfold.

4

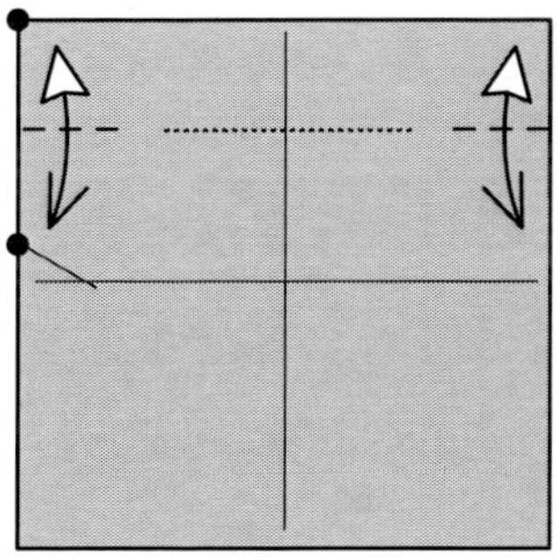

Fold and unfold on the left and right. Rotate 180°.

5

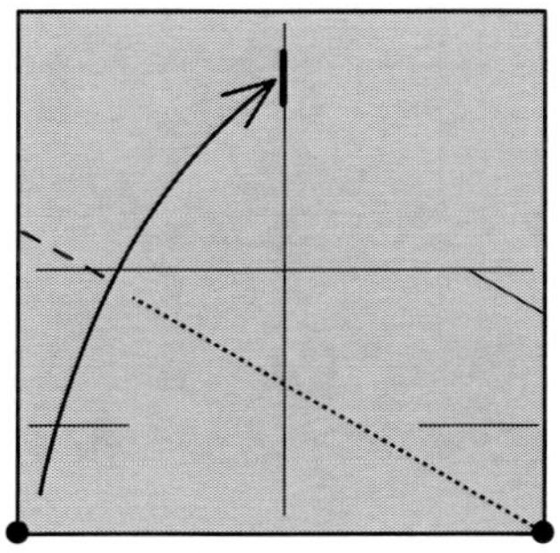

Repeat steps 2–4.

6

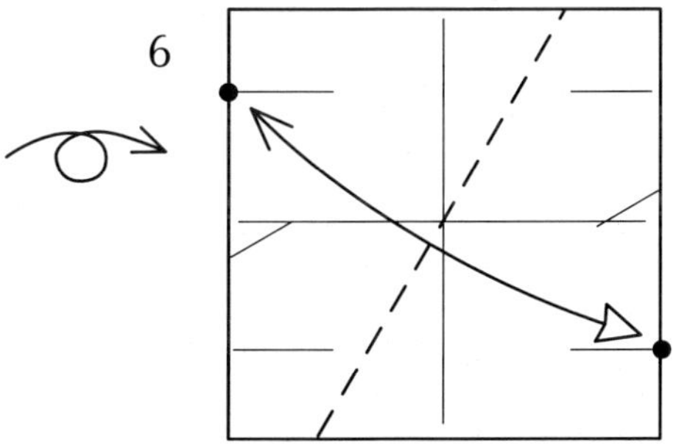

Fold and unfold.

7

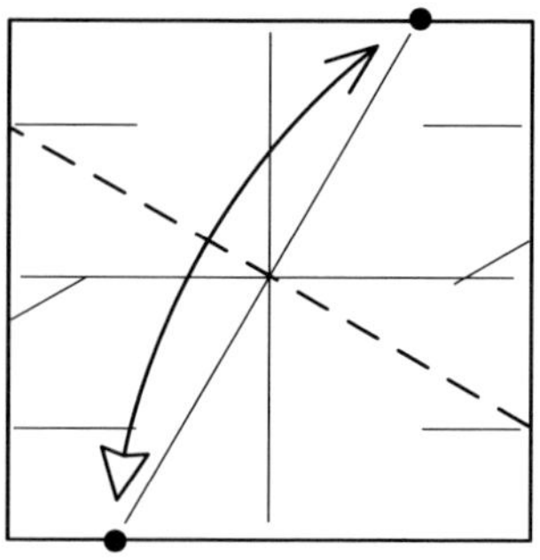

Fold and unfold.

8

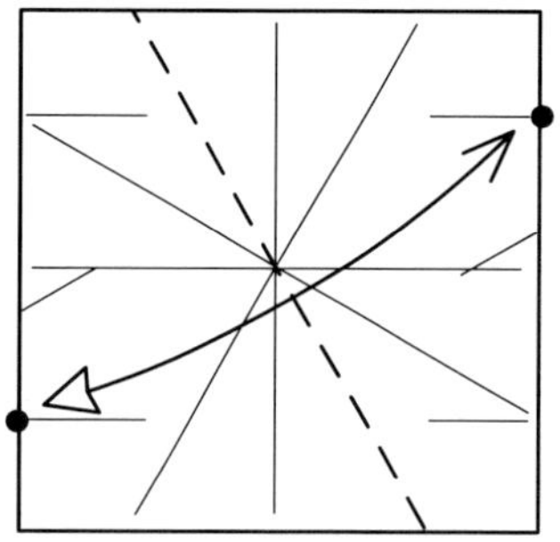

Repeat steps 6–7 in the opposite direction.

9

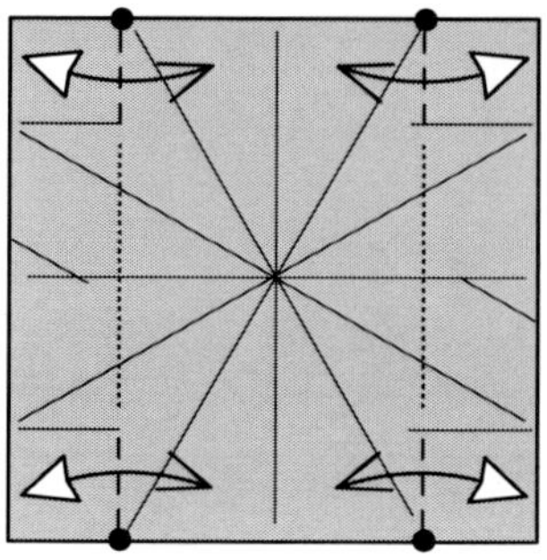

Fold and unfold.

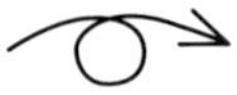

10

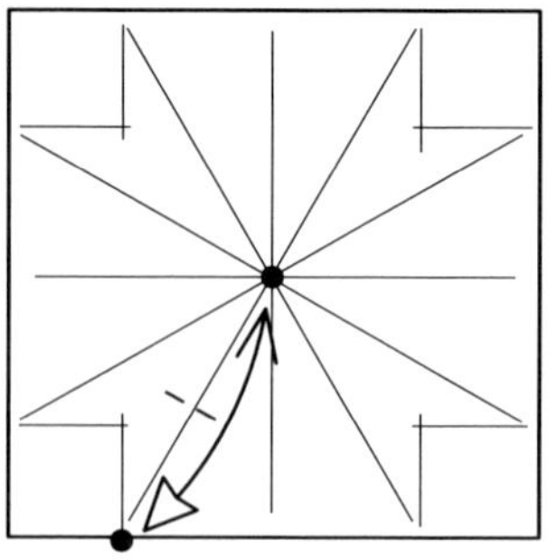

Fold and unfold.

11

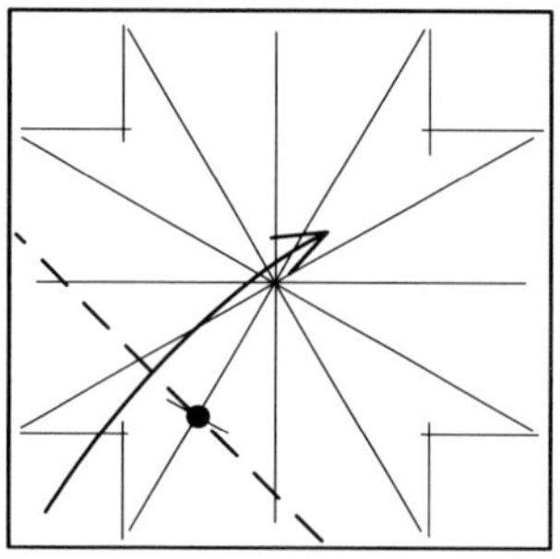

Note the right angle in the next step.

12

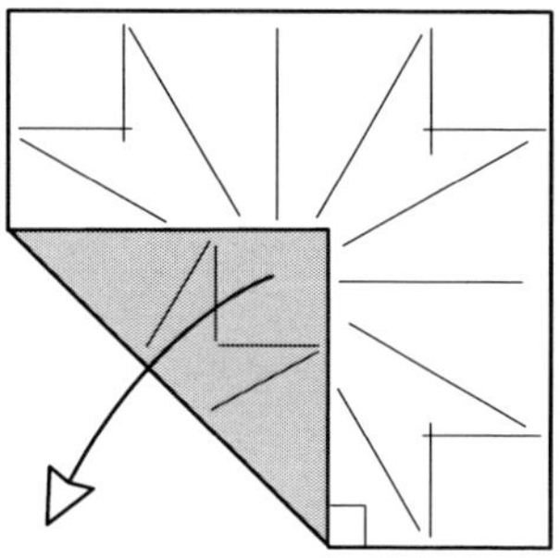

Unfold and rotate 90°.

13

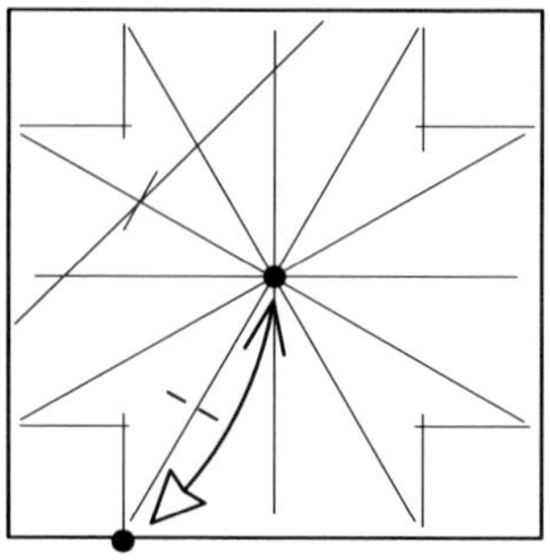

Repeat steps 10–12 three times.

14

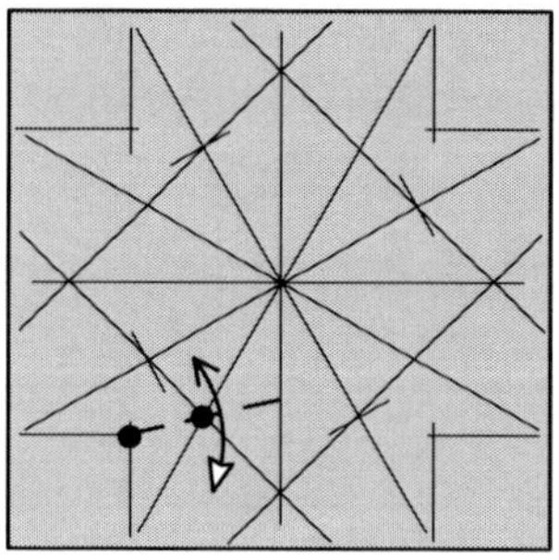

Fold and unfold.

15

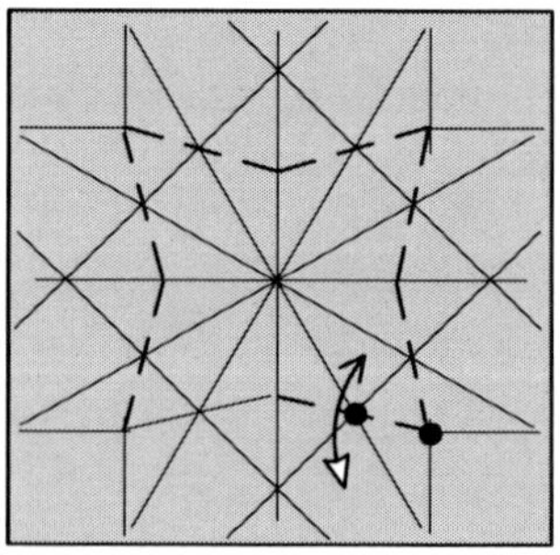

Fold and unfold seven times.

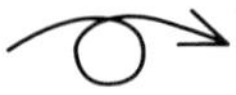

16

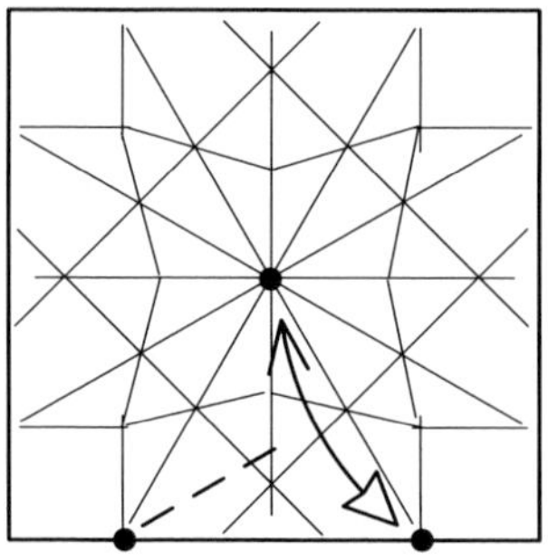

Fold and unfold.

17

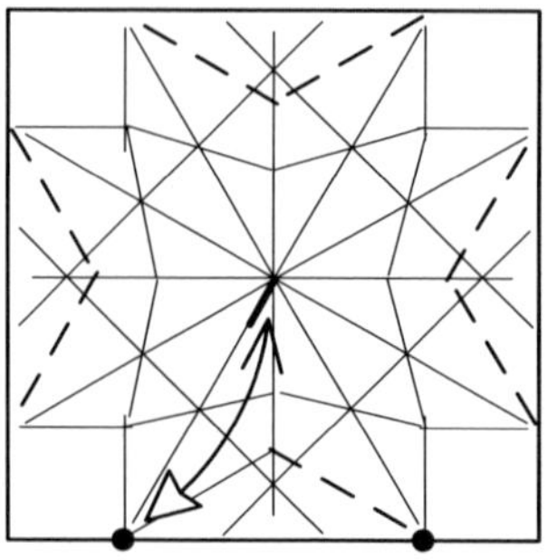

Fold and unfold.

18

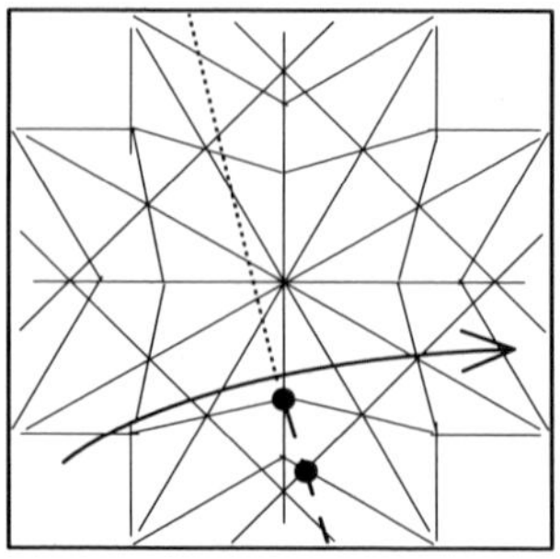

19

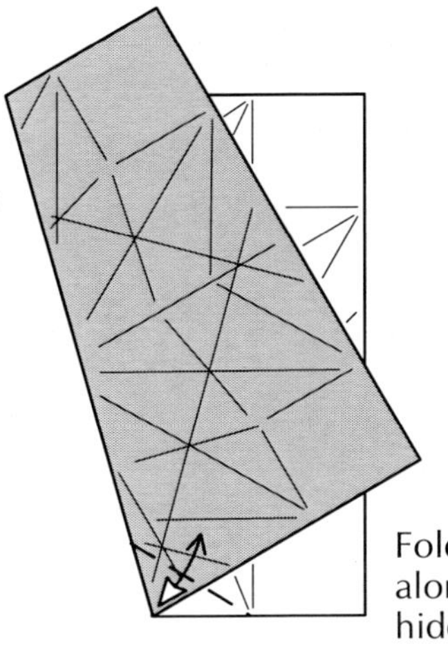

Fold and unfold along a partially hidden crease.

20

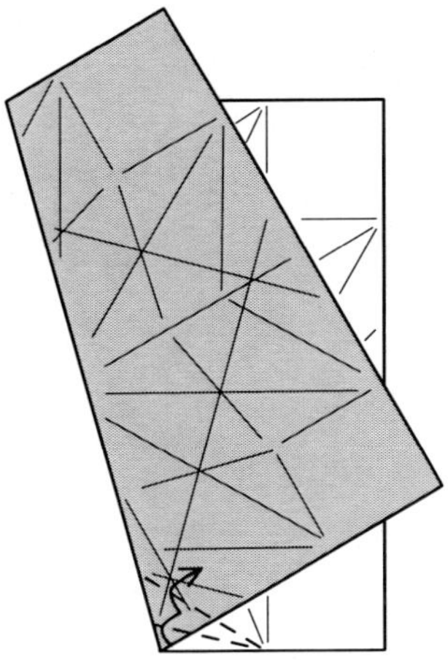

21

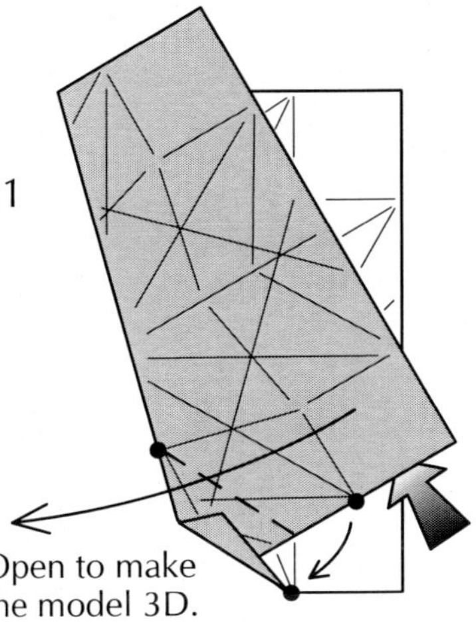

Open to make the model 3D.

22

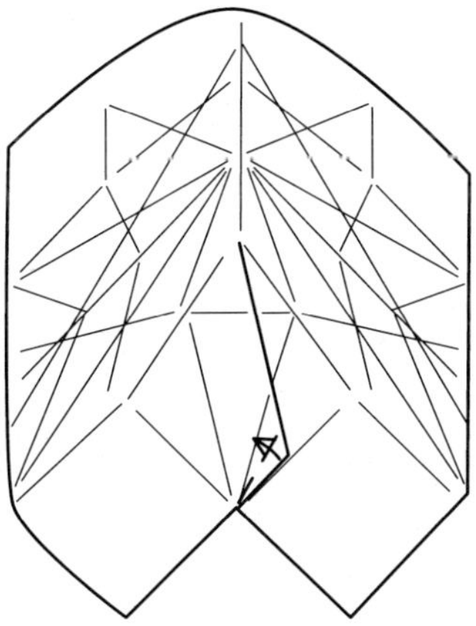

23

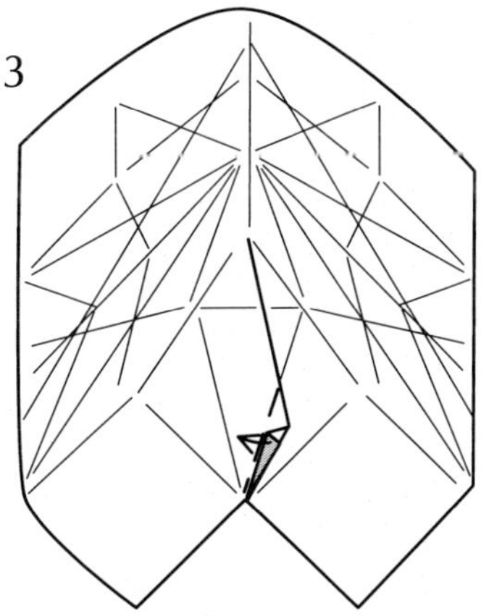

Fold along the crease.

24

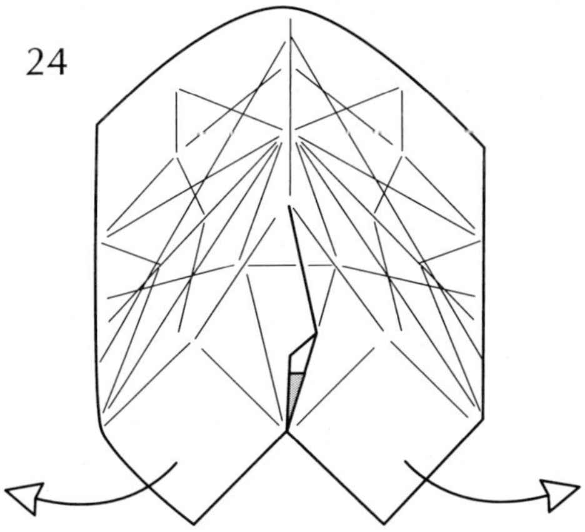

Unfold and rotate 90°.

25

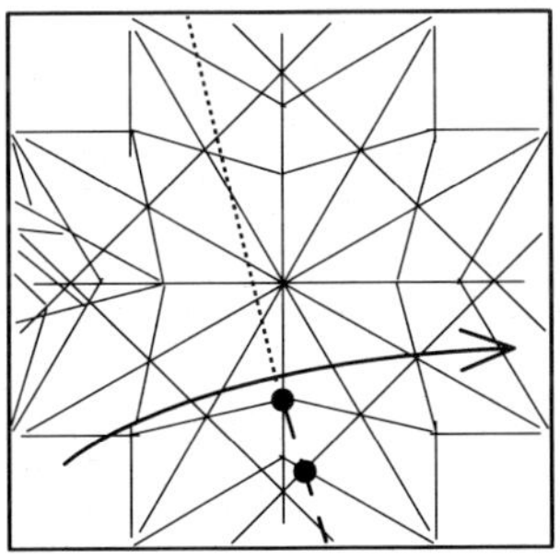

Repeat steps 18–24 three times.

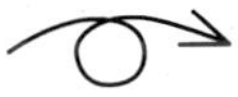

26

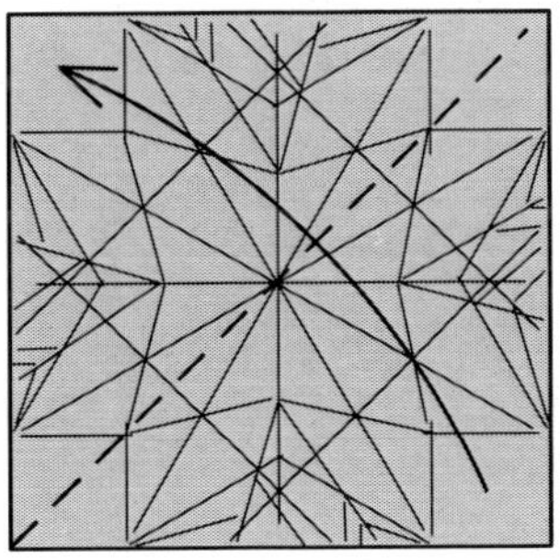

27

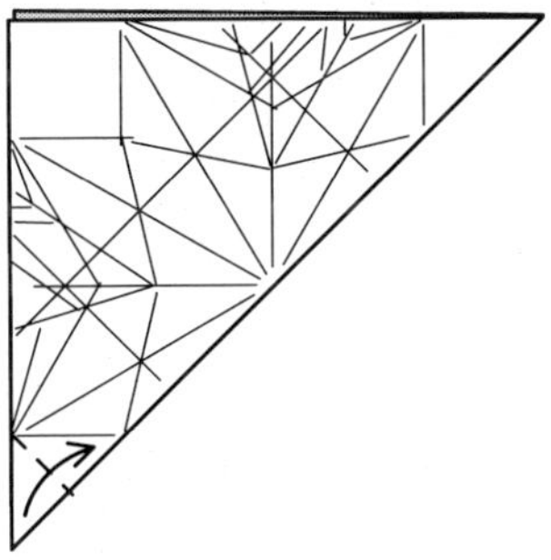

28

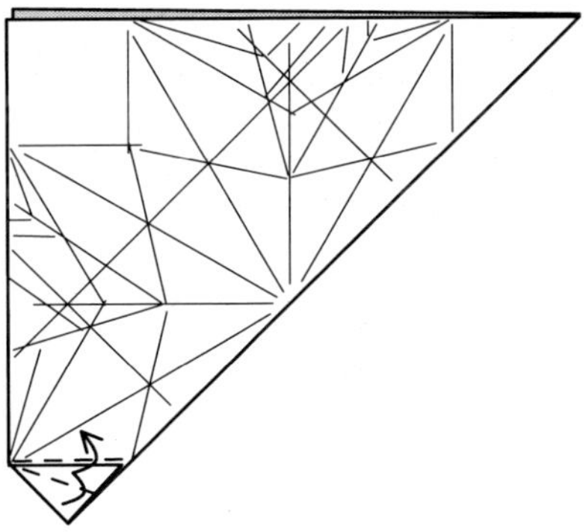

29

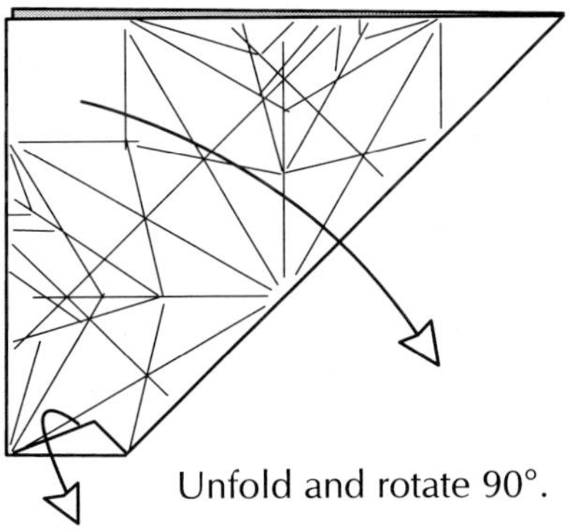

Unfold and rotate 90°.

30

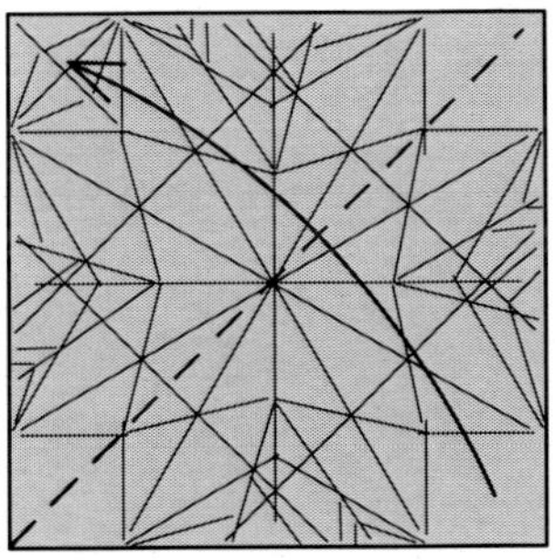

Repeat steps 26–29 three times.

31

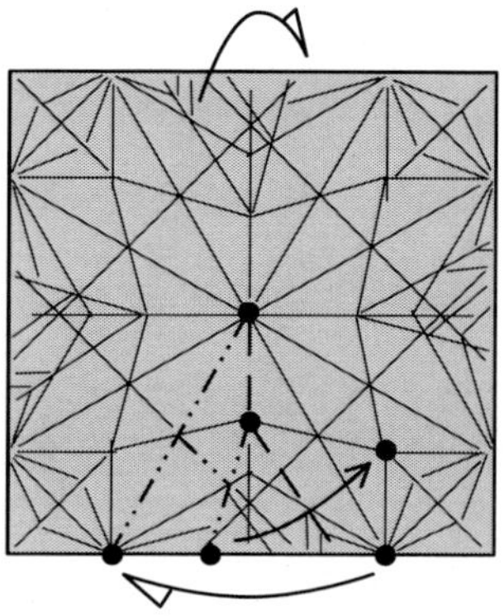

Puff out at the upper dot and push in at the center dot to form a sunken triangle. The other pairs of dots will meet.

32

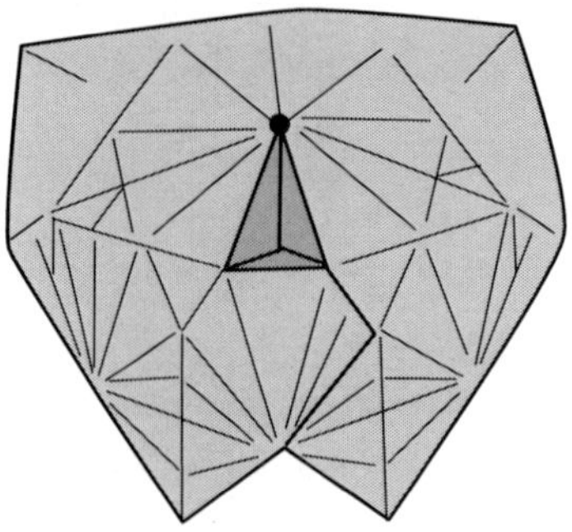

Repeat step 31 three times, with the dot here macthing the upper dot in step 31.

33

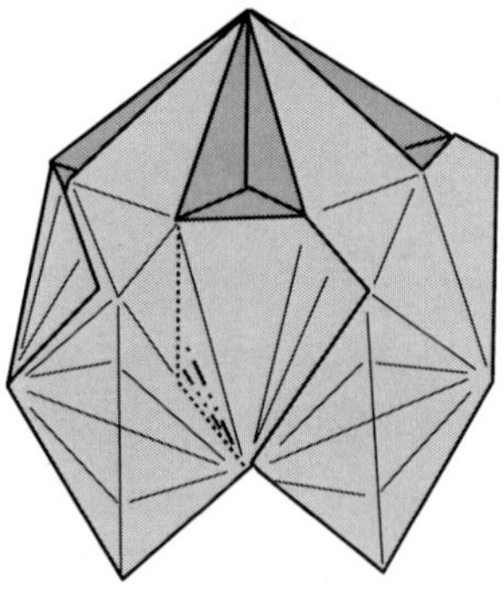

The folding is done on the inside. Refold the hidden flap as in steps 22–23.

34

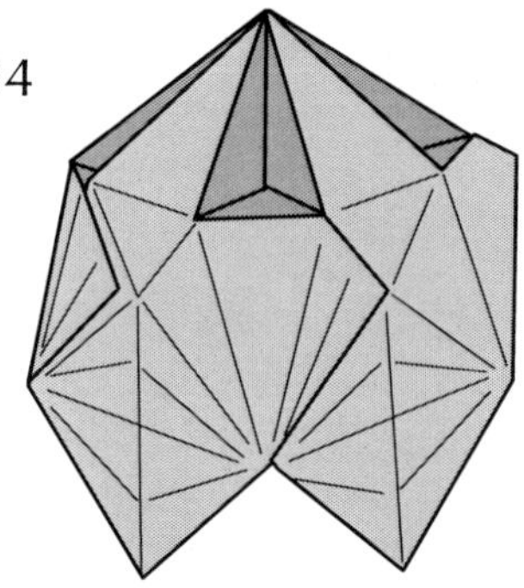

Repeat step 33 three times.

35

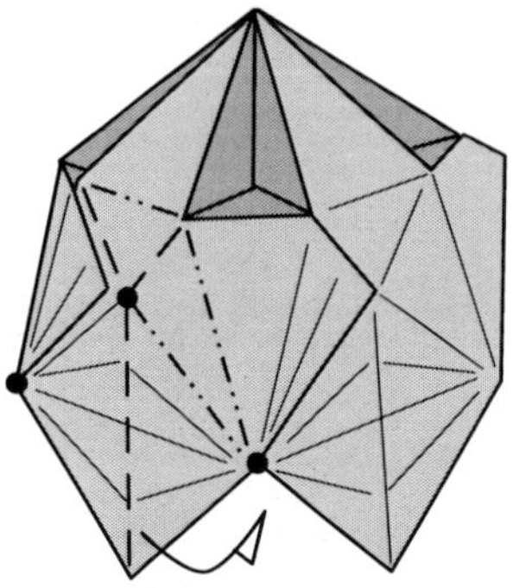

Push in at the upper dot. The other two dots will meet.

36

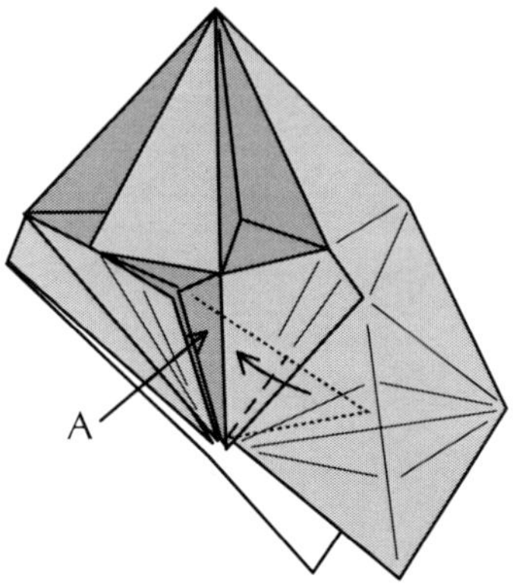

The folding is done on the inside. Refold the hidden flap as in steps 27–28 and flatten against the back of triangle A.

37

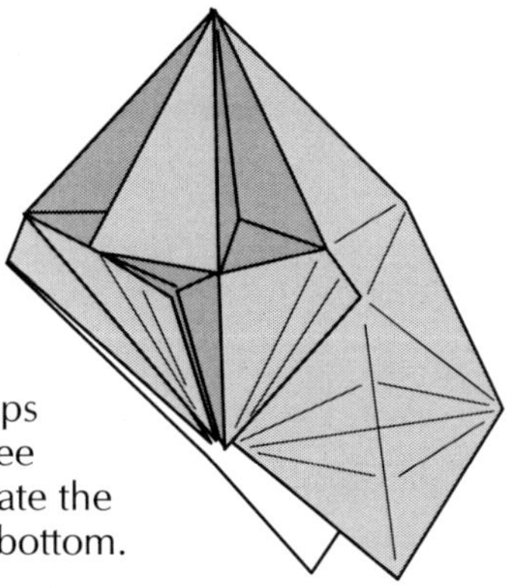

Repeat steps 35–36 three times. Rotate the top to the bottom.

38

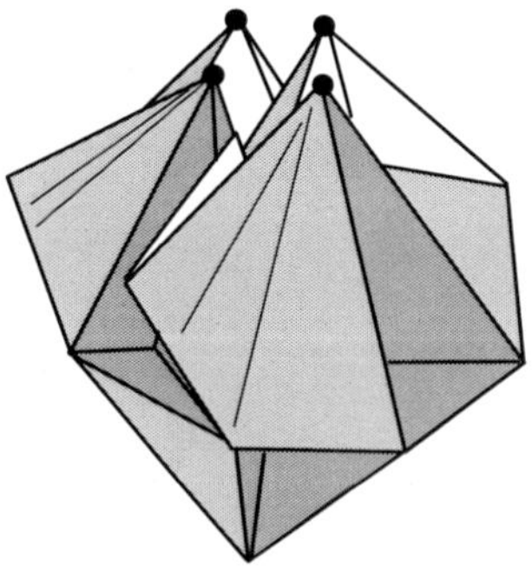

Bring the four dots together.

39

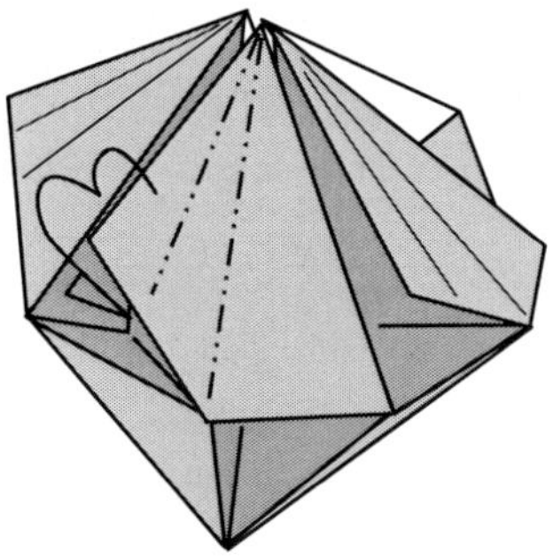

Fold along the creases and tuck inside. Repeat three times.

40

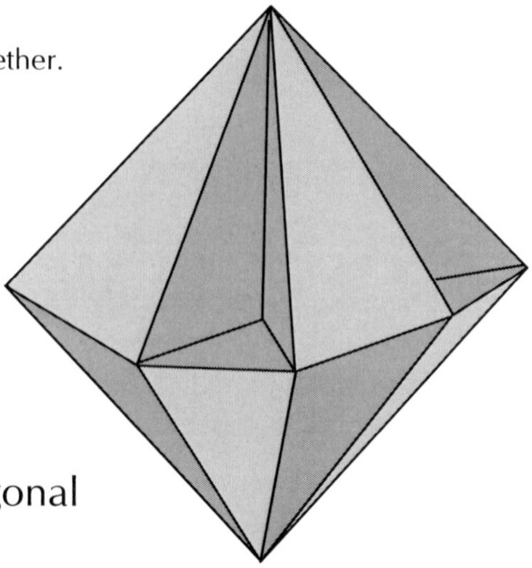

Dimpled Octagonal Dipyramid

Sunken Polyhedra

These sunken polyhedra become more challenging and add magic to origami.

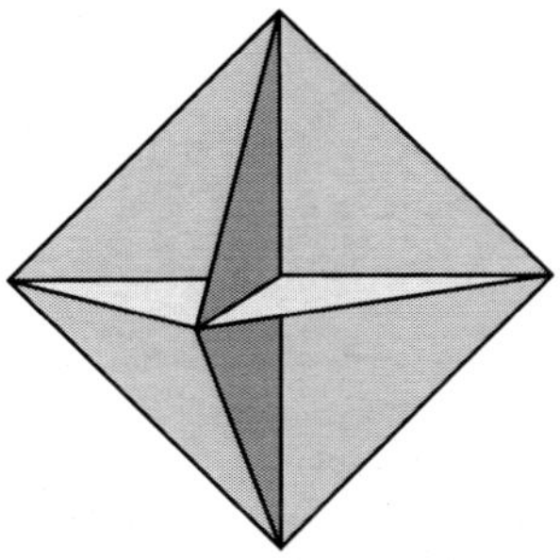

Sunken Octahedron

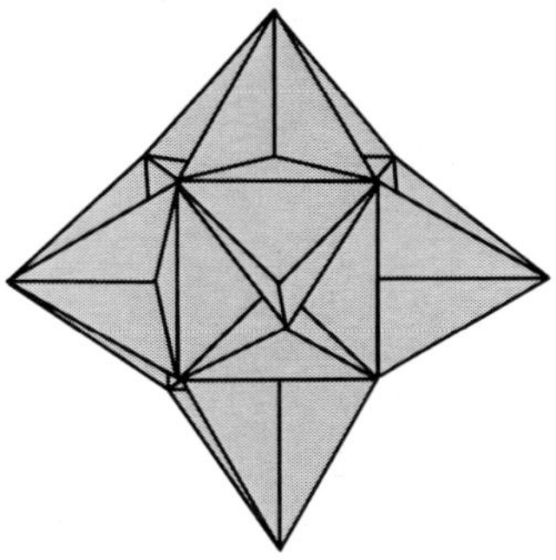

Sunken Stellated Cube

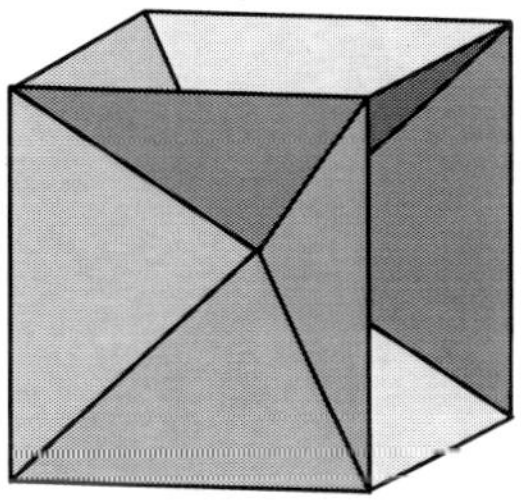

Sunken Cube

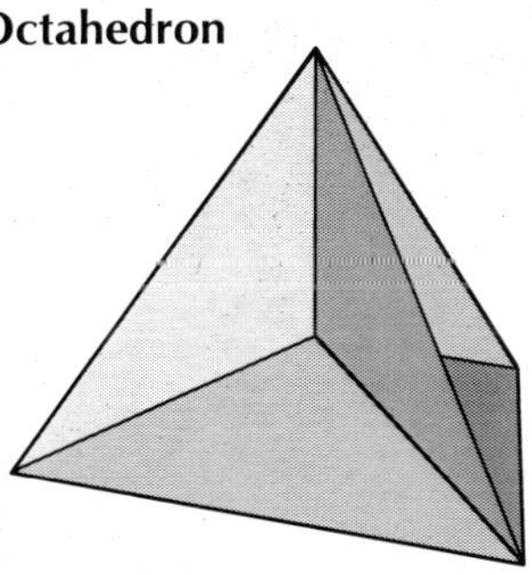

Sunken Tetrahedron

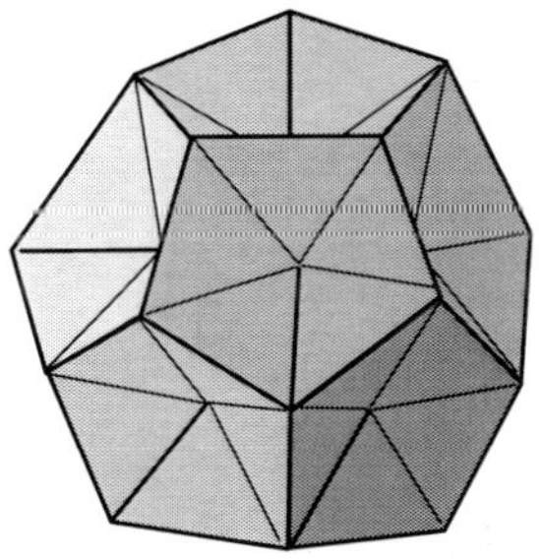

Sunken Dodecahedron

Sunken Octahedron

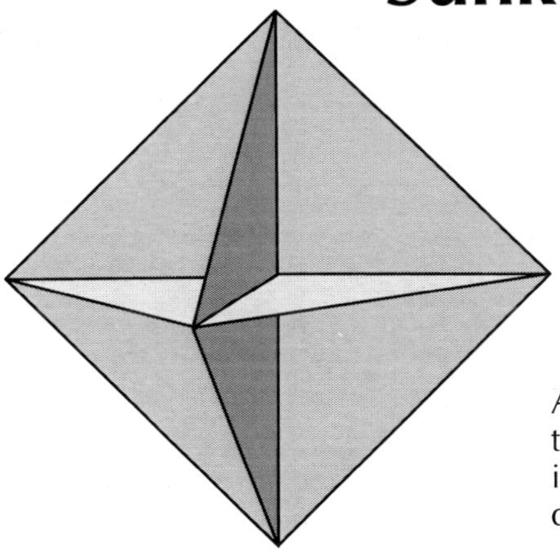

All the faces meet at the center, and this model can be viewed as three intersecting squares in three dimensions. Square symmetry is used.

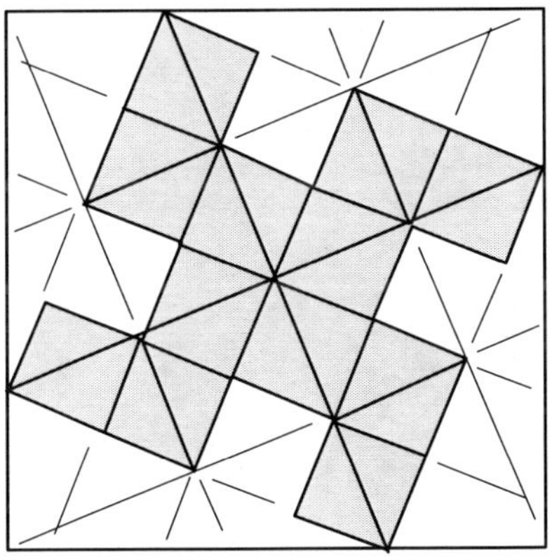

1

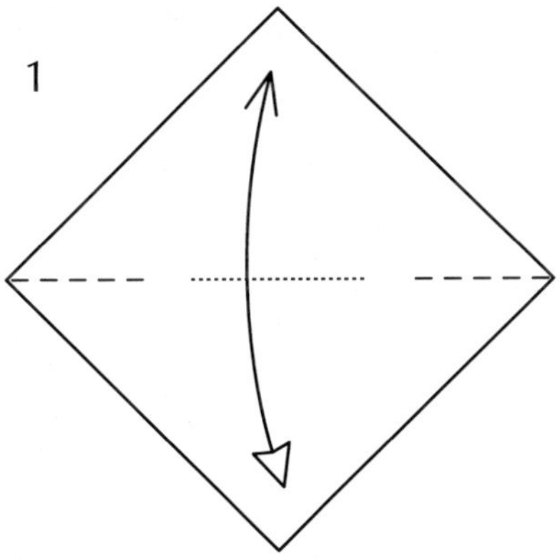

Fold and unfold.

2

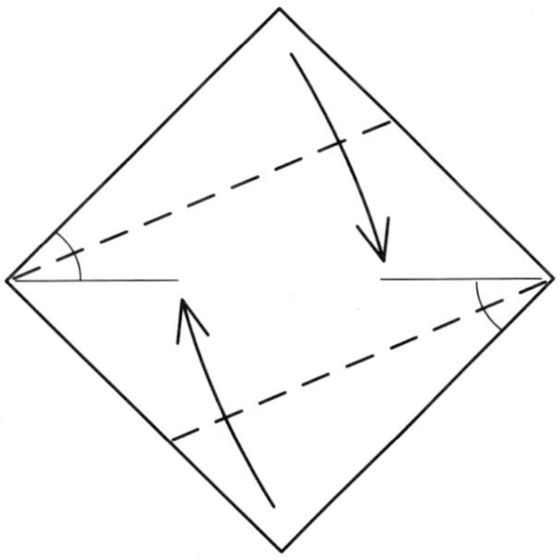

Fold to the center.

3

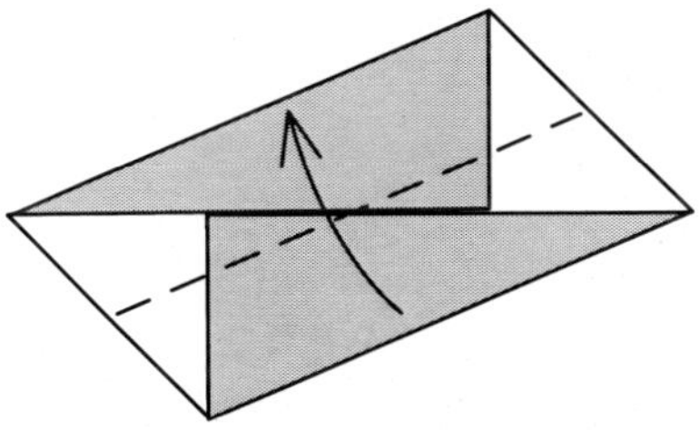

Fold in half.

4

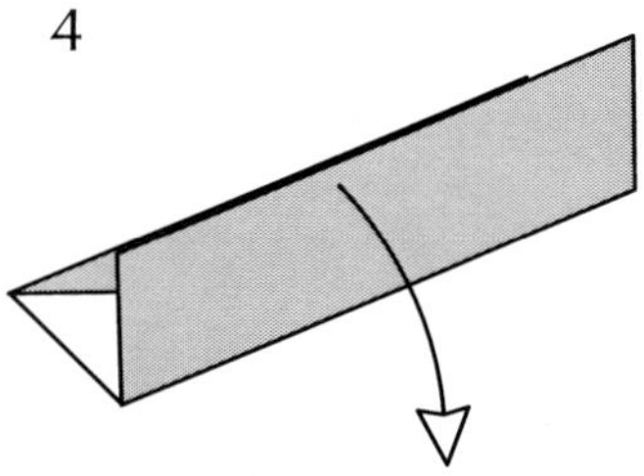

Unfold and rotate 90°.

5

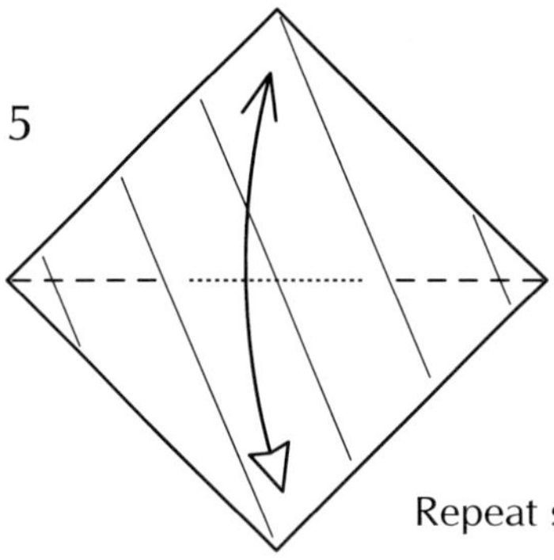

Repeat steps 1–4.

6

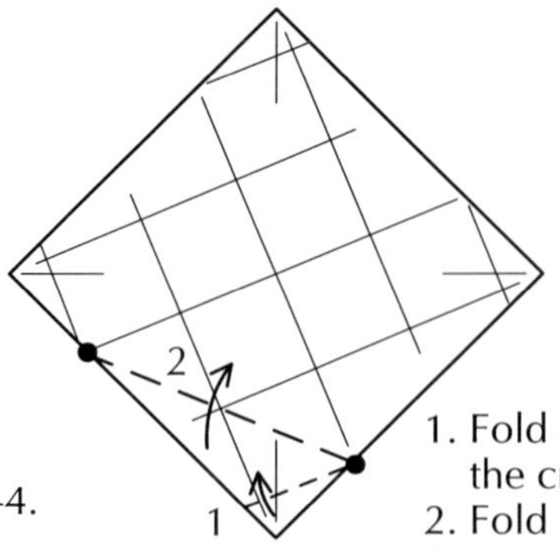

1. Fold along the crease.
2. Fold by the dots.
Rotate 90°.

7

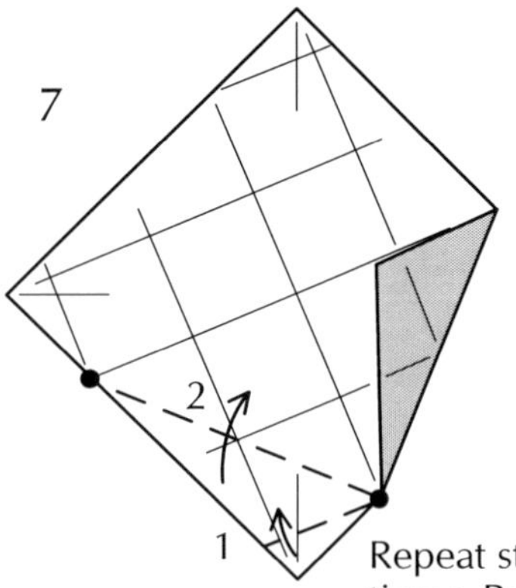

Repeat step 6 three times. Rotate.

8

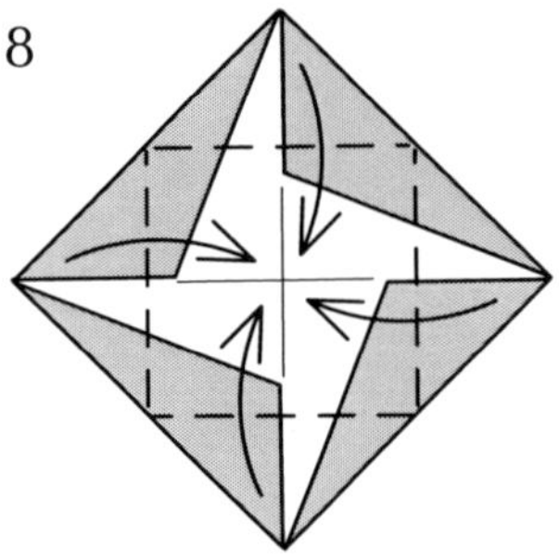

Fold along the creases.

9

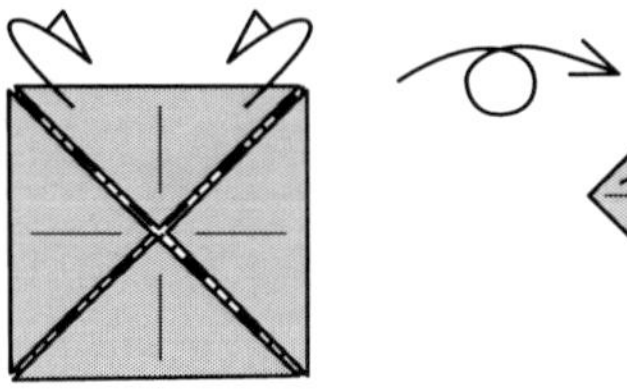

Fold and unfold. Rotate.

10

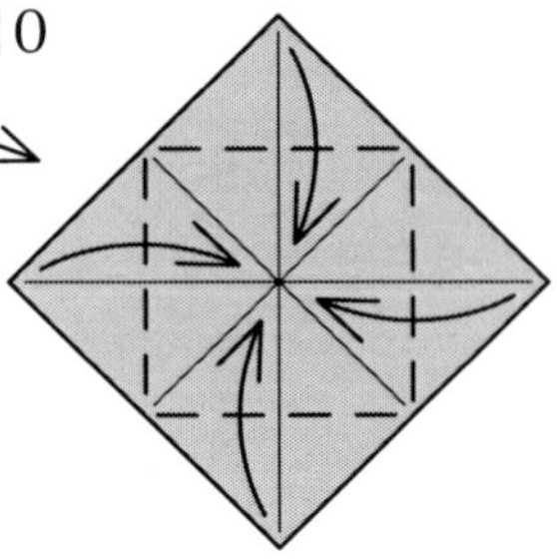

Fold to the center.

11

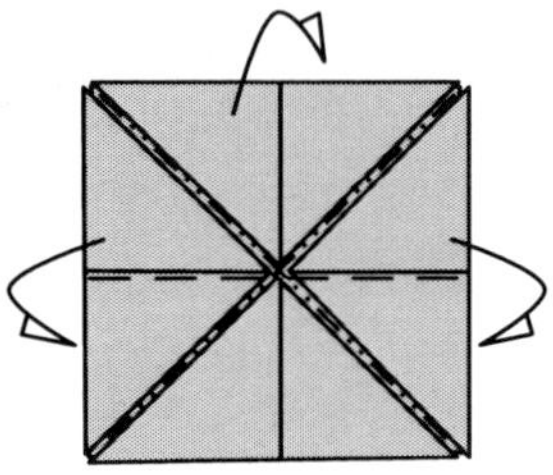

Fold along the creases. It is easier to do steps 11 and 12 as one step.

12

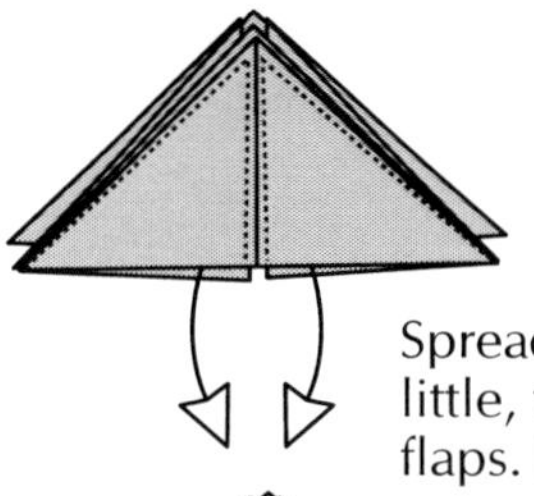

Spread the paper a little, to pull out the flaps. Repeat behind.

13

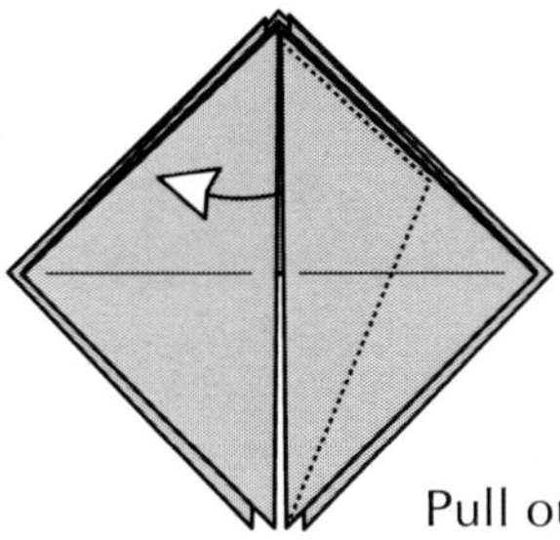

Pull out.

14

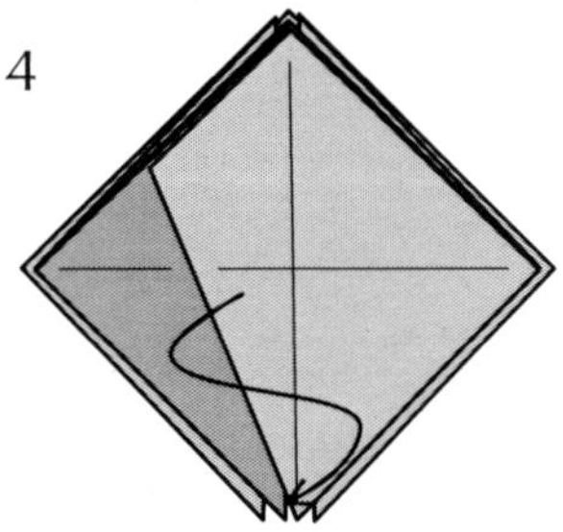

Tuck under the dark layer.

15

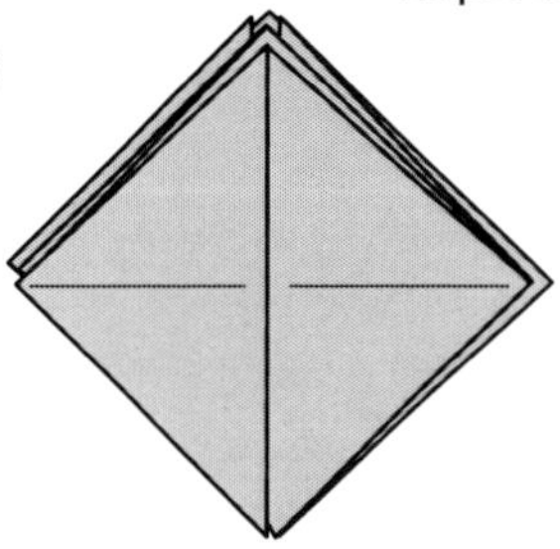

Repeat steps 13–14 three times, on the back and sides.

16

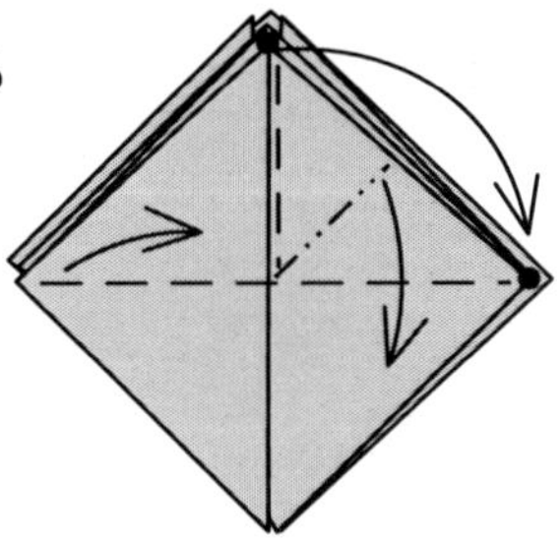

Bring the dots together and lift up on the left. The model will become 3D.

17

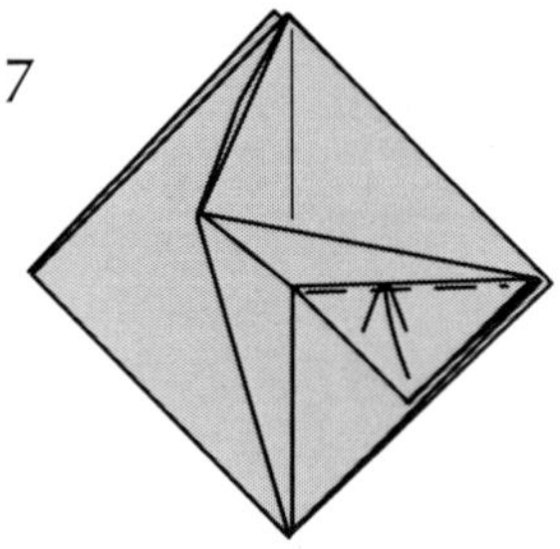

Tuck inside.

18

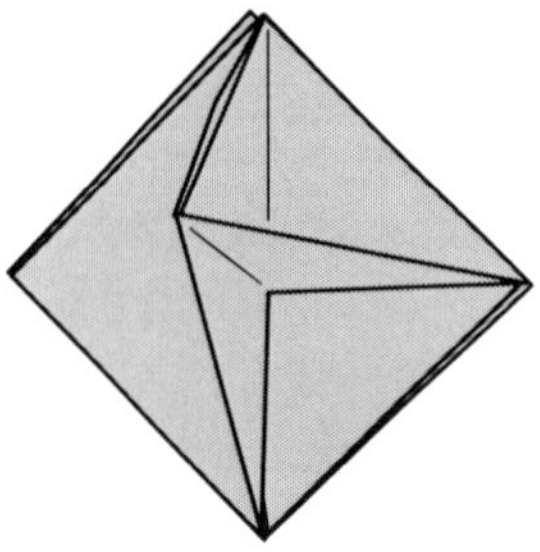

Repeat steps 16–17 on the three remaining sides.

19

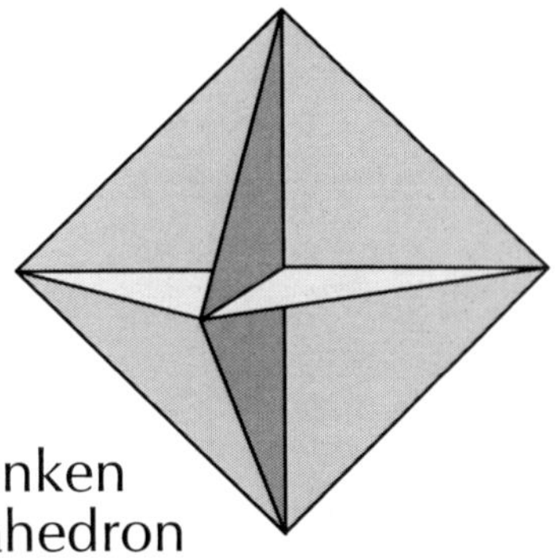

Sunken Octahedron

Sunken Tetrahedron

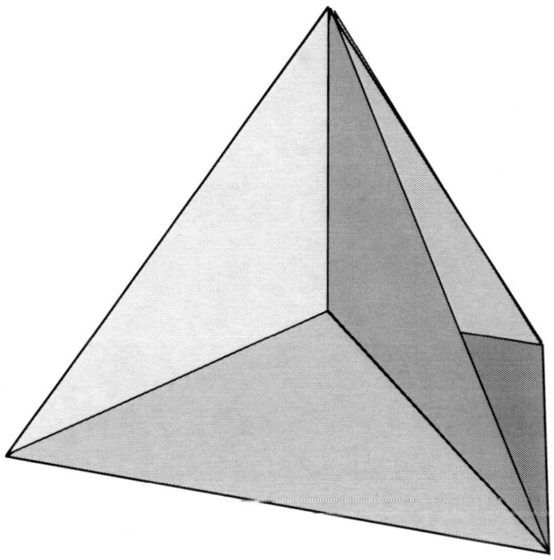

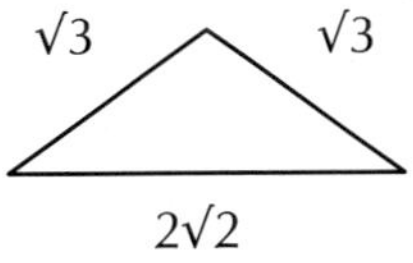

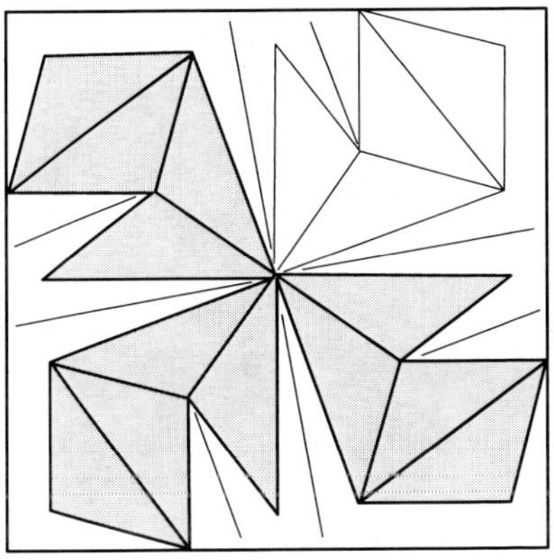

The twelve faces of the sunken tetrahedron are isosceles triangles. The sides of each triangular face are proportional to $2\sqrt{2}$, $\sqrt{3}$, and $\sqrt{3}$. The crease pattern shows 3/4 square symmetry.

1

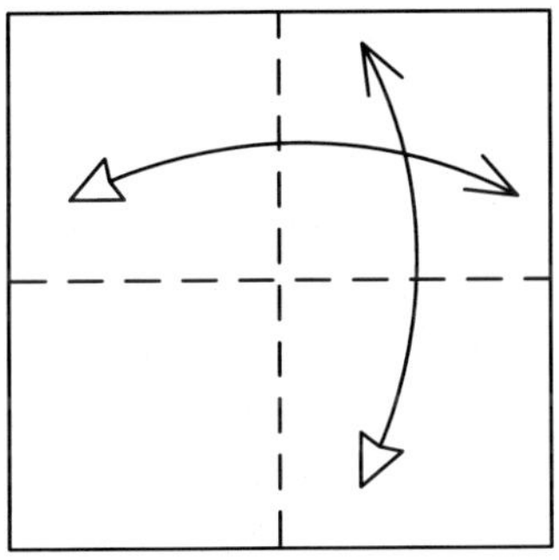

Fold and unfold.

2

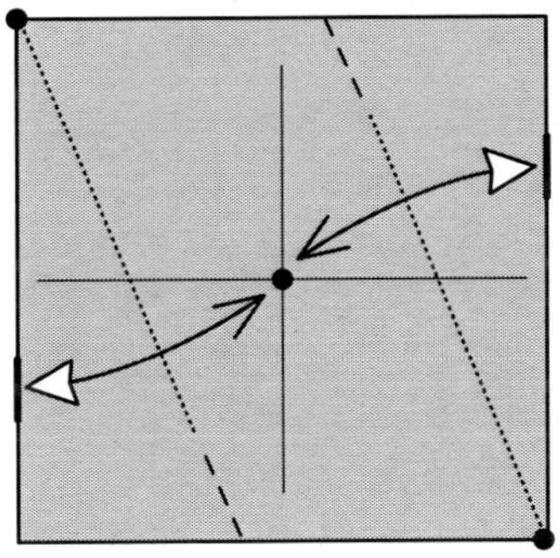

Bring the edges to the center creasing on the edges. Fold and unfold.

3

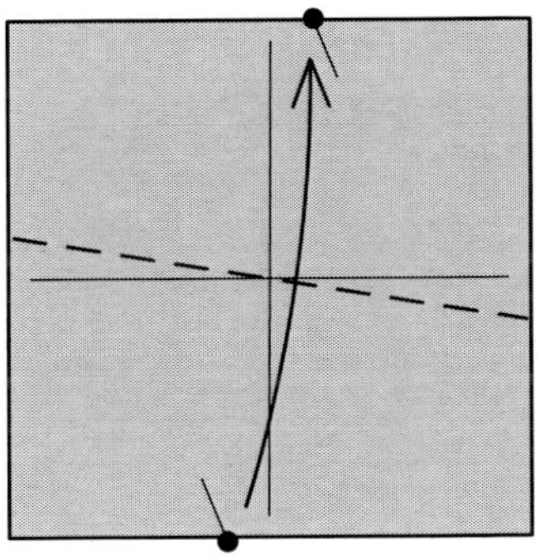

4

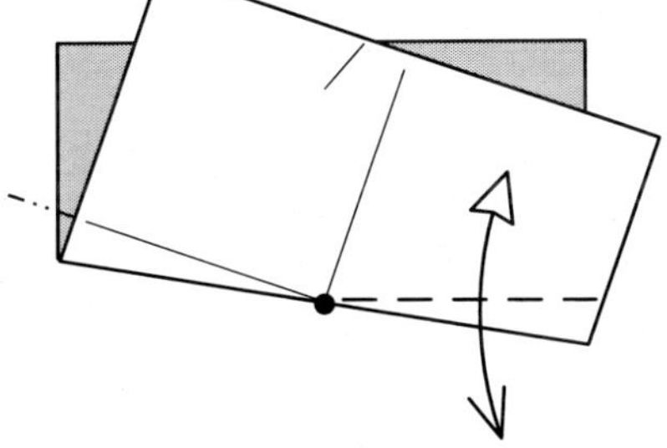

Fold and unfold the top layer along a hidden crease. Turn over and repeat.

5

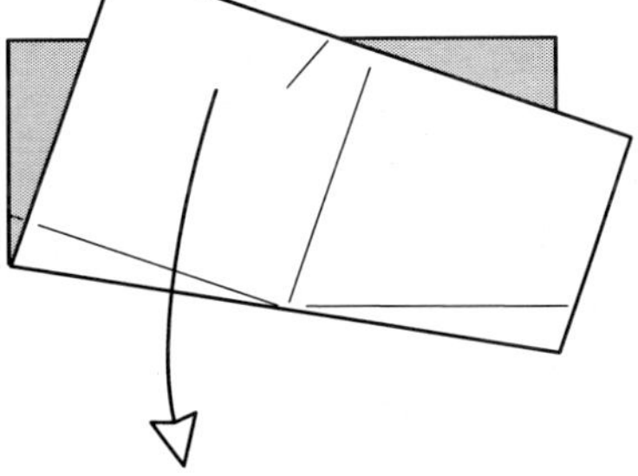

Unfold and rotate 90°.

6

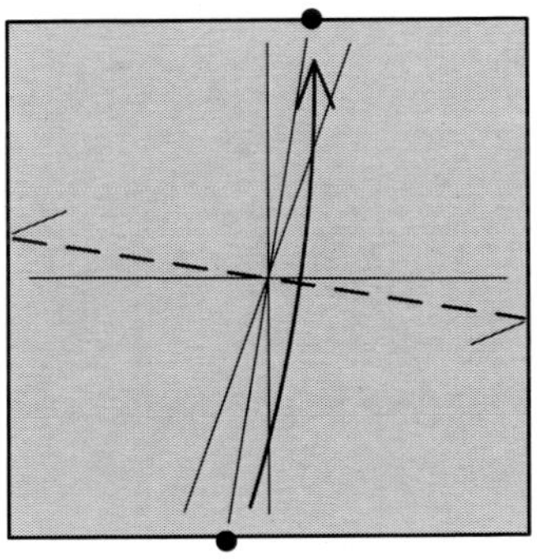

Repeat steps 3–6.

7

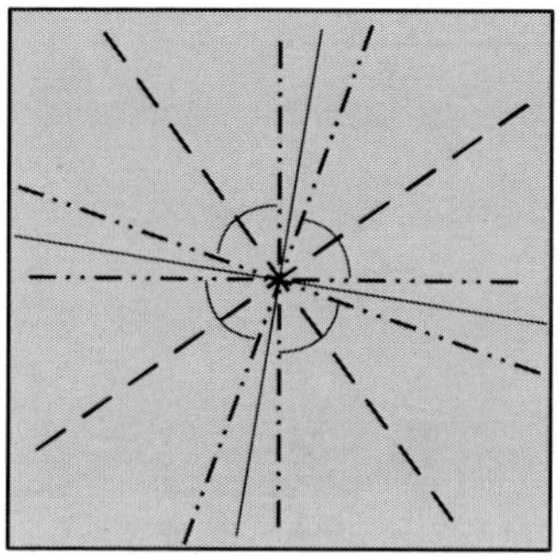

Fold and unfold to bisect the angles. Mountain-fold along the creases. Rotate 45°.

8

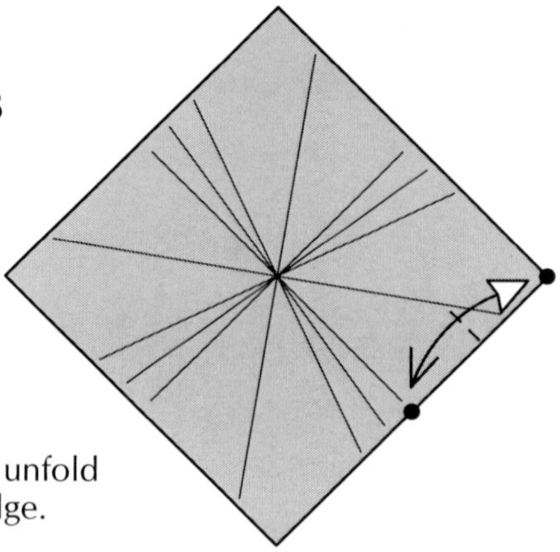

Fold and unfold on the edge.

9

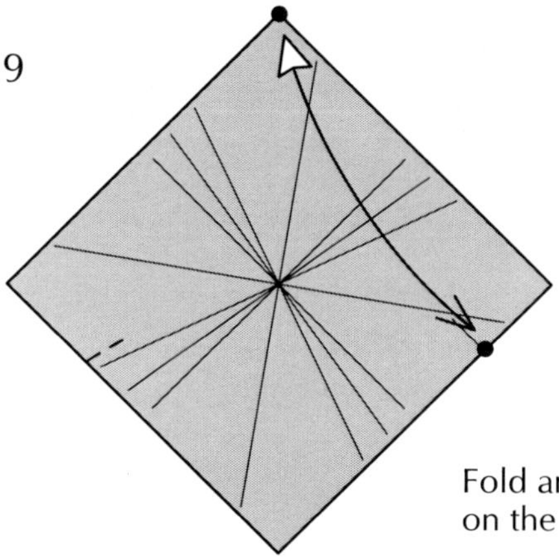

Fold and unfold on the edge.

10

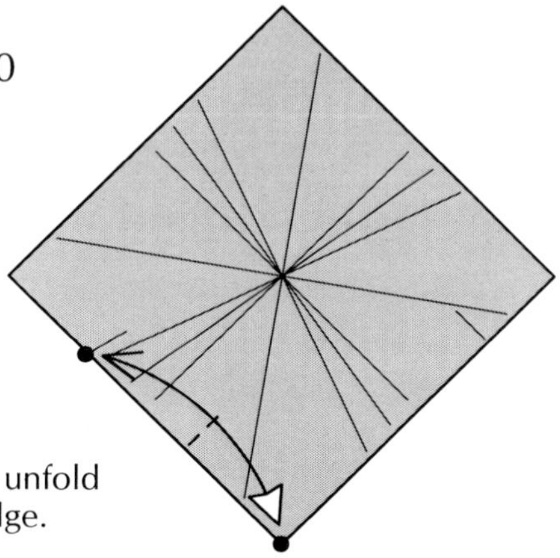

Fold and unfold on the edge.

11

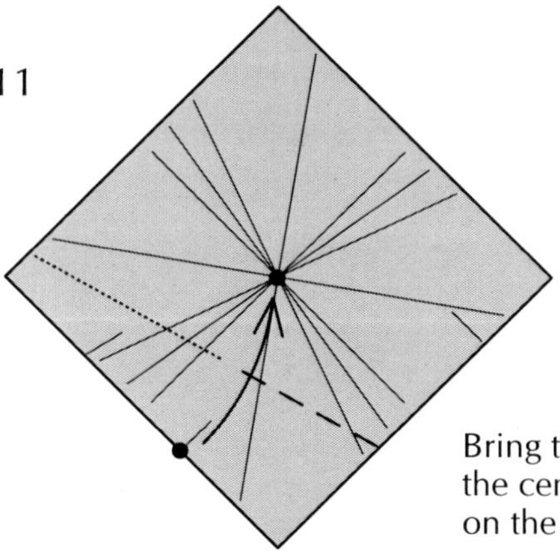

Bring the lower dot to the center and crease on the right side.

12

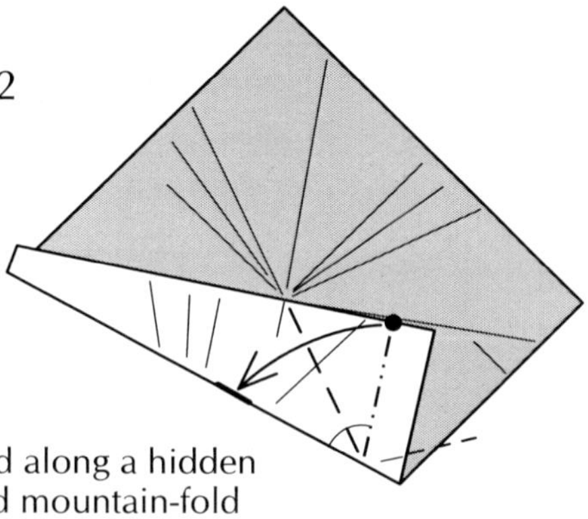

Valley-fold along a hidden crease and mountain-fold at the edge.

13

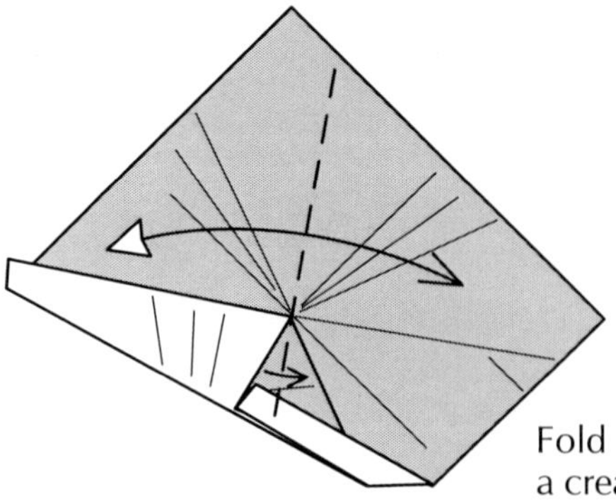

Fold and unfold along a crease. Do not unfold at the bottom.

14

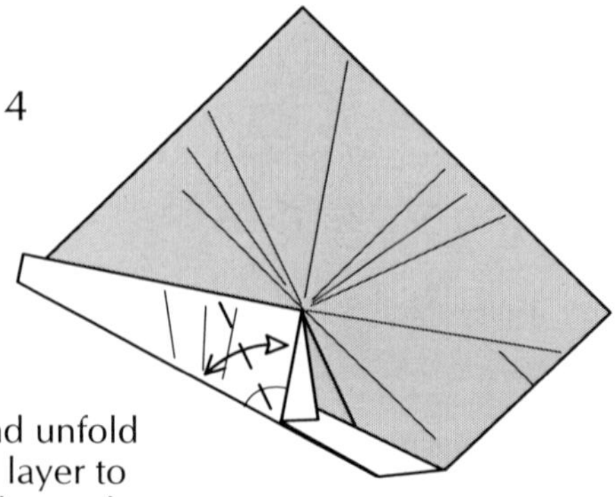

Fold and unfold the top layer to bisect the angle.

15

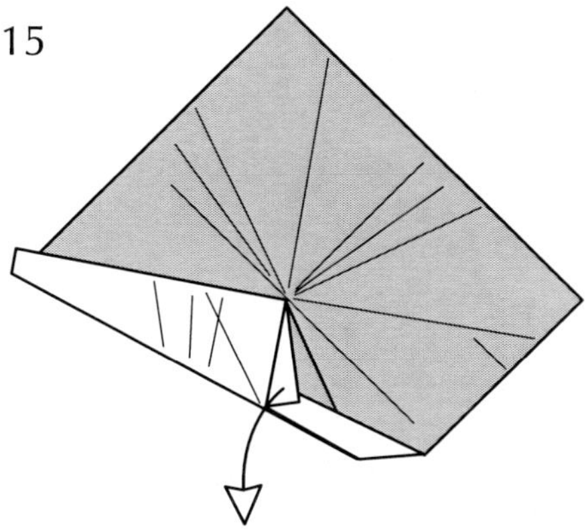

Unfold and
rotate 90°.

16

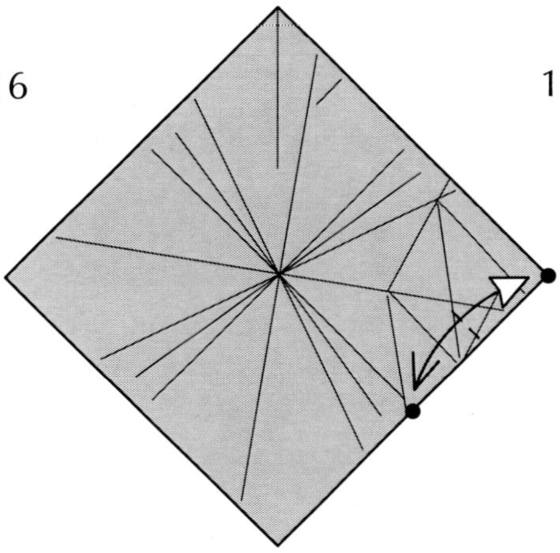

Repeat steps 8–15 three
times. Rotate 45°.

17

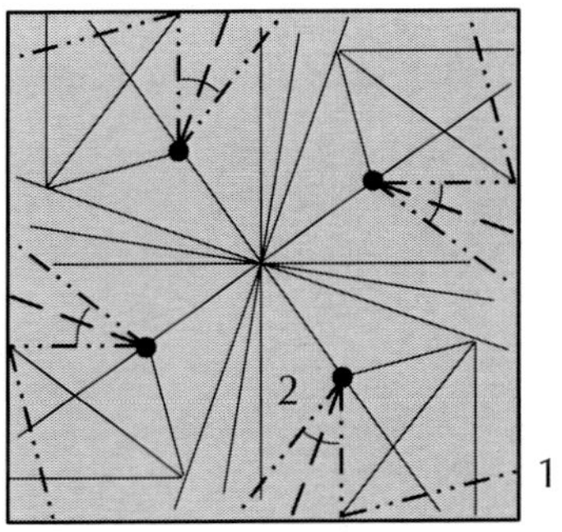

1. Fold and unfold along the crease.
2. Mountain-fold along the creases to bisect the angle.

Repeat all around.

18

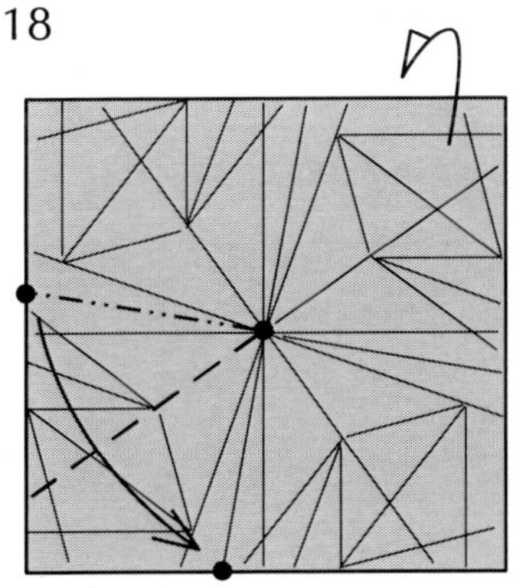

Fold along the creases and puff out at the dot in the center. The other two dots will meet.

19

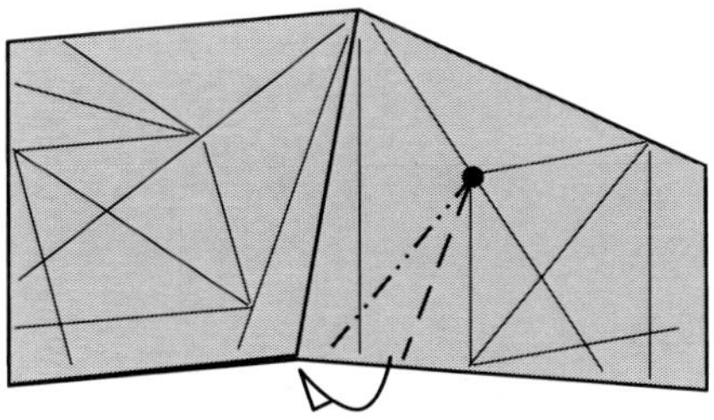

Fold along the creases
and push in at the dot.

20

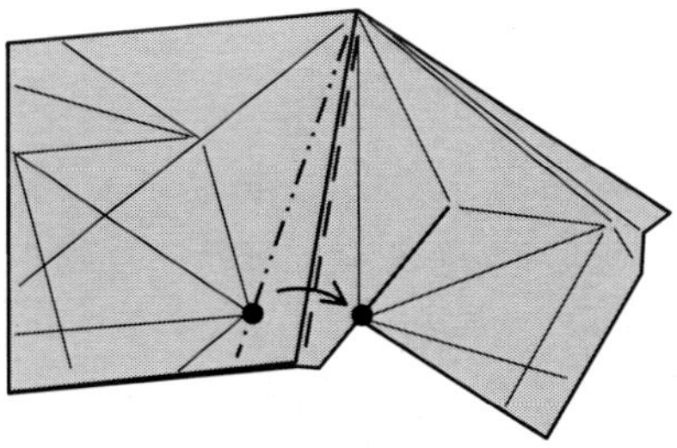

Fold along the creases.
The dots will meet.

21

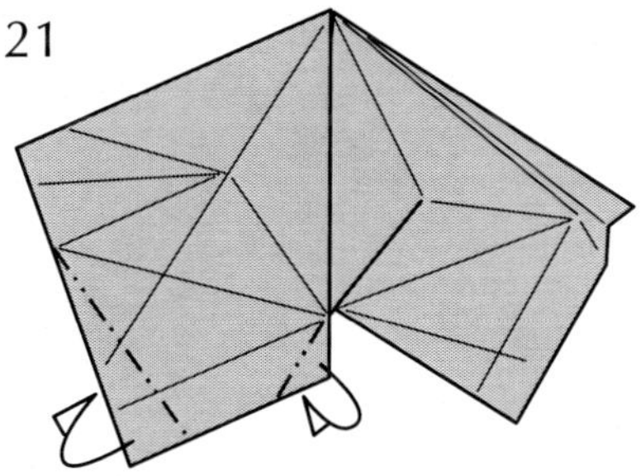

Fold along the
crease on the left.

22

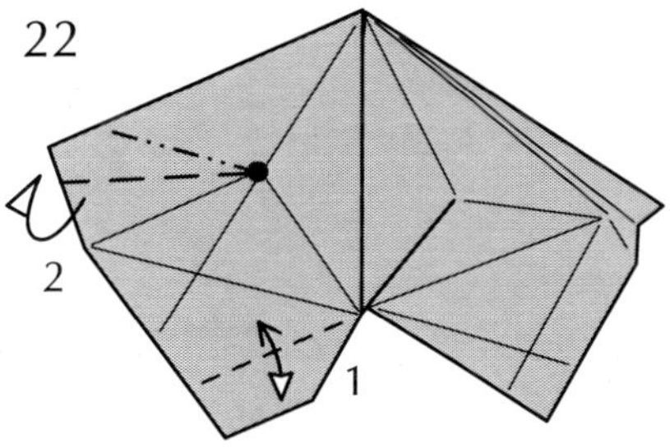

1. Fold and unfold.
2. Repeat steps 19–22 two times.

Rotate to view the bottom.

23

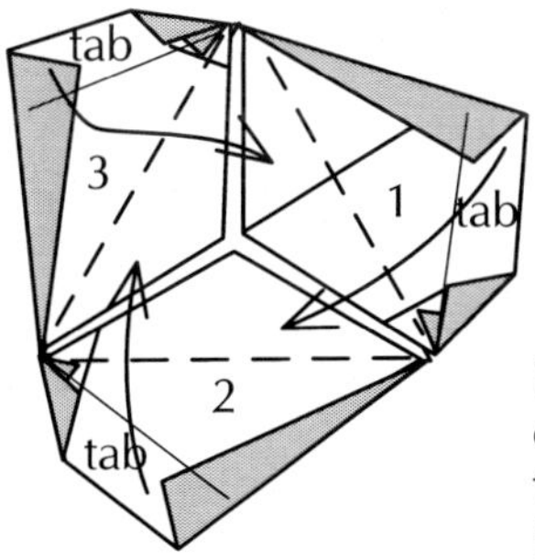

Fold paper 1, then fold 2 which covers tab 1. Then fold 3 to cover tab 2 and tuck tab 3 under 1. Push in at the center.

24

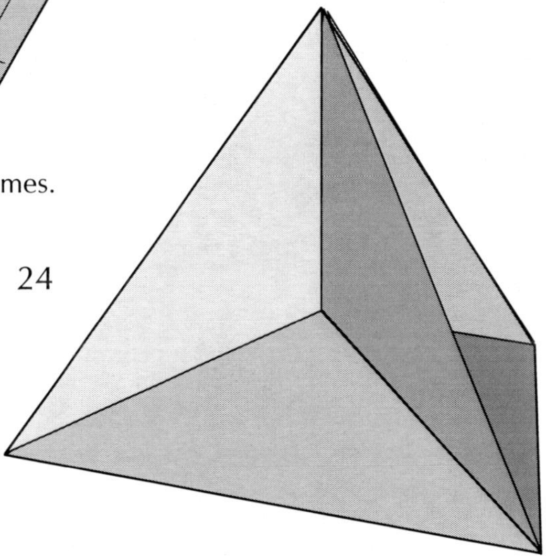

Sunken Tetrahedron

Sunken Cube

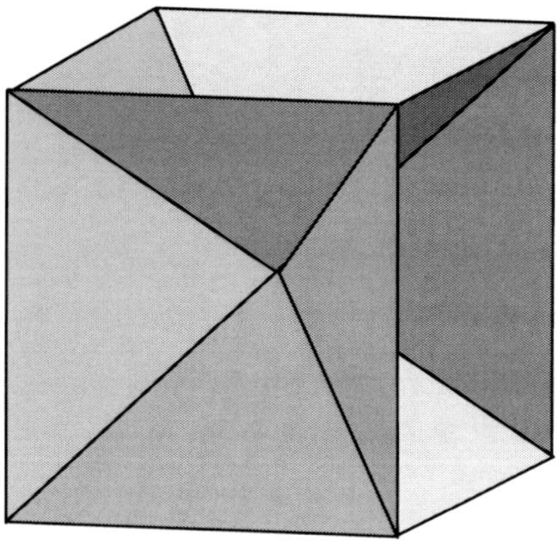

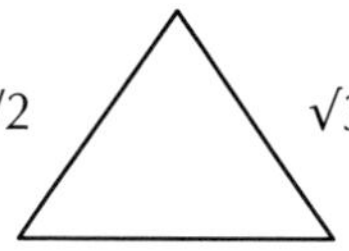

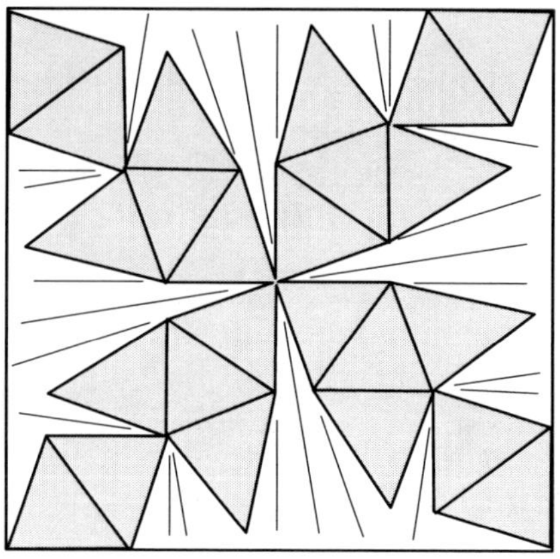

The sunken cube is composed of 24 isosceles triangles all meeting at the center. The sides of each triangle are proportional to 1, √3/2, √3/2. This beautiful shape can also be called a hexahedral nolid, that is, it comes from the cube but has no volume. The crease pattern shows square symmetry.

1

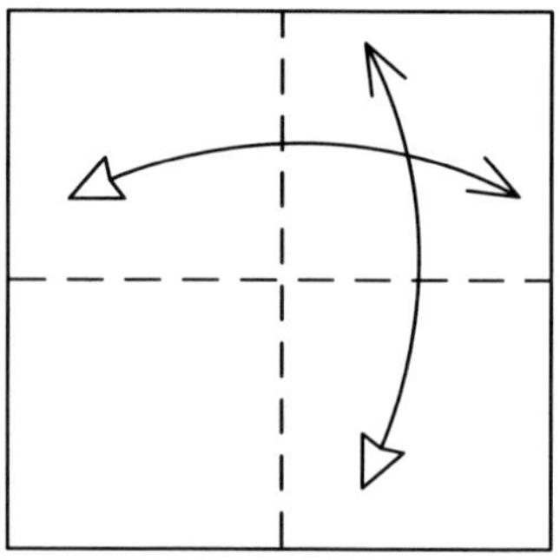

Fold and unfold.

2

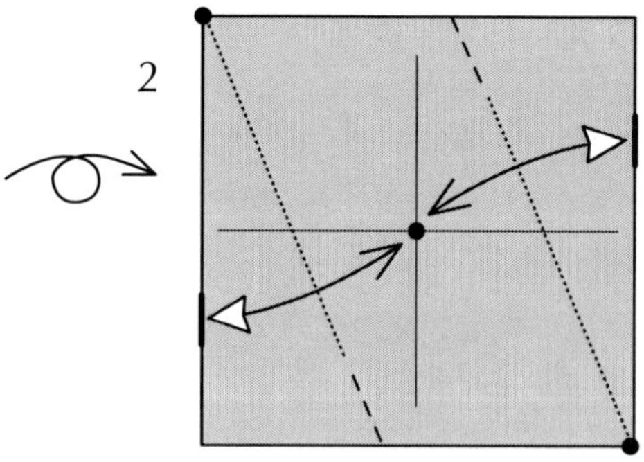

Bring the edges to the center creasing on the edges. Fold and unfold.

3

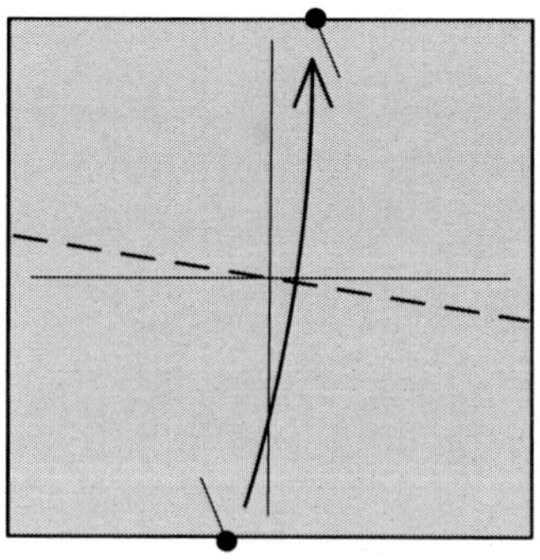

4

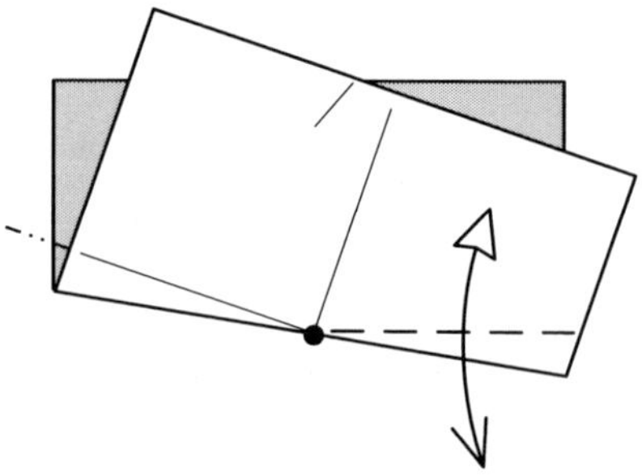

Fold and unfold the top layer along a hidden crease. Turn over and repeat.

5

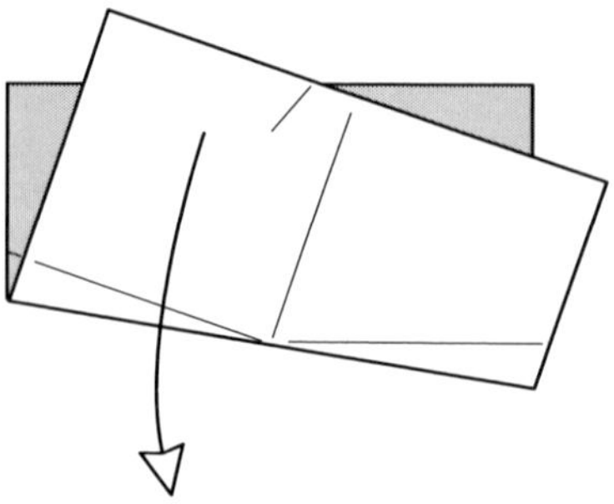

Unfold and rotate 90°.

6

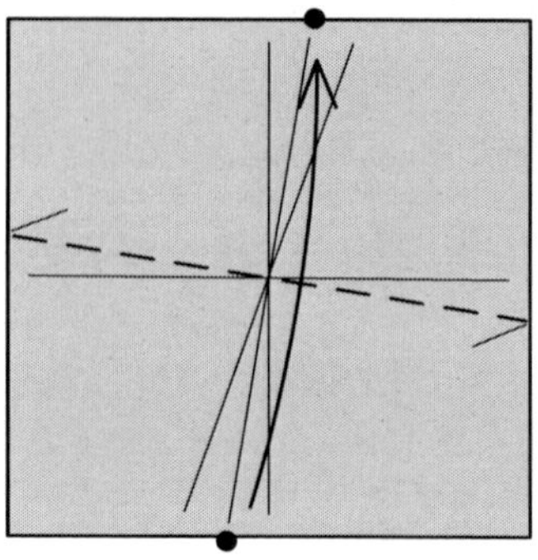

Repeat steps 3–6.

7

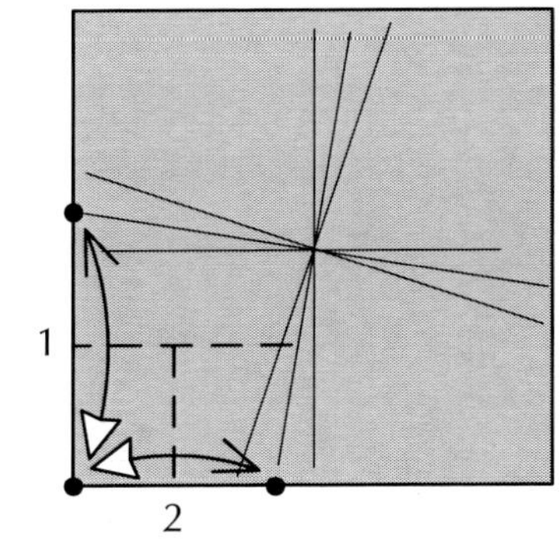

Fold and unfold.

8

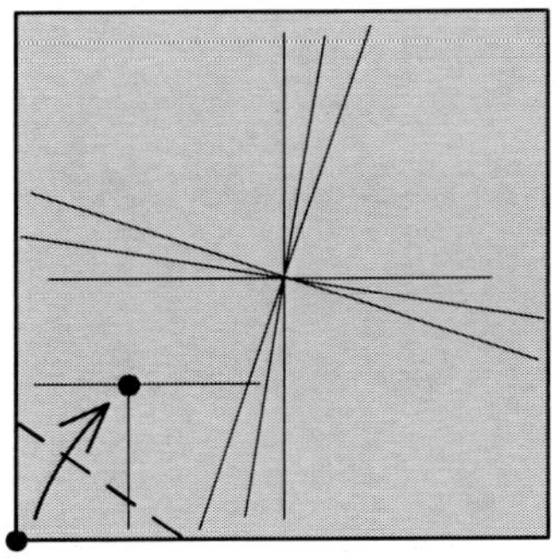

9

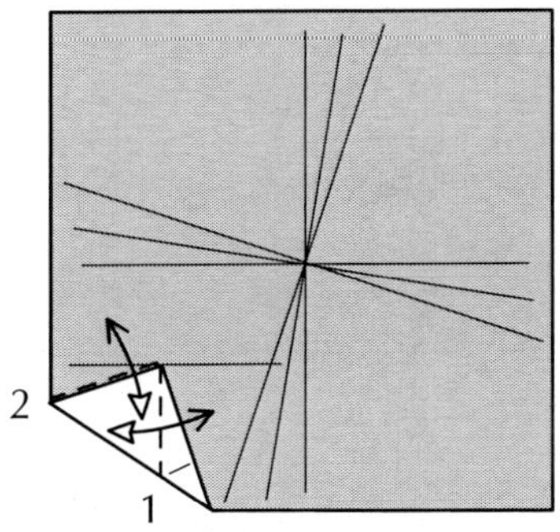

1. Fold and unfold along a hidden crease.
2. Fold and unfold along the top edge.

10

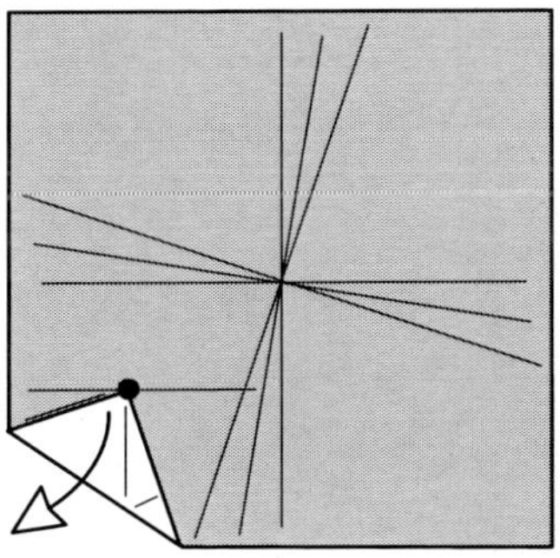

Unfold and rotate so the dot stays at the bottom left corner.

11

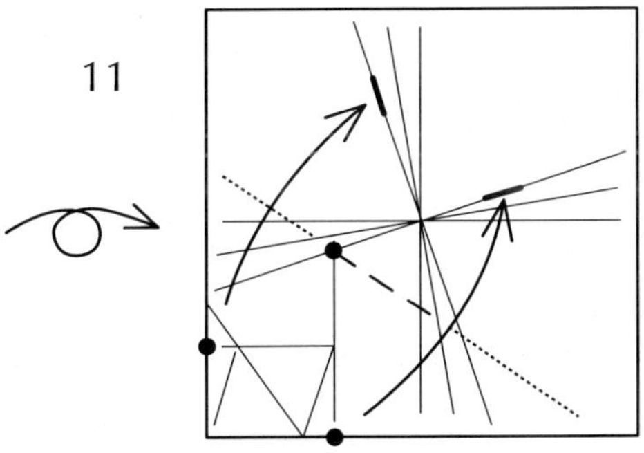

Crease in the center.

12

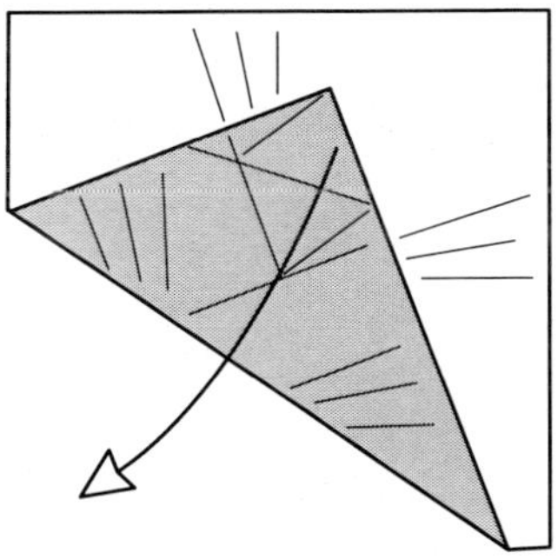

Unfold.

13

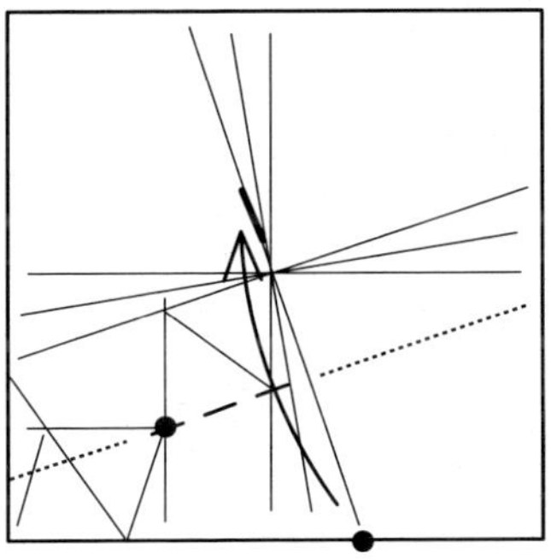

14

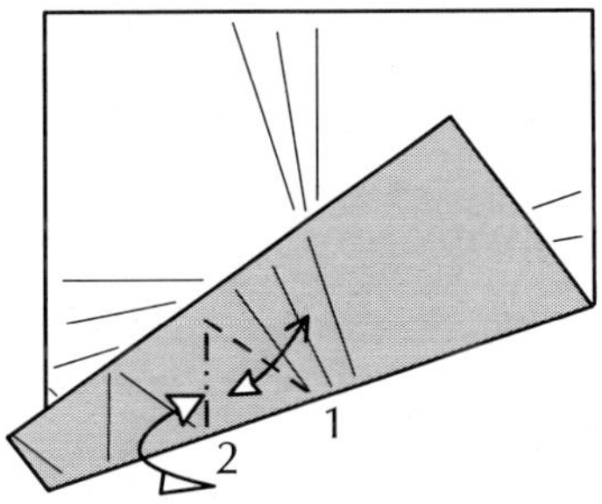

Fold and unfold along hidden creases.

15

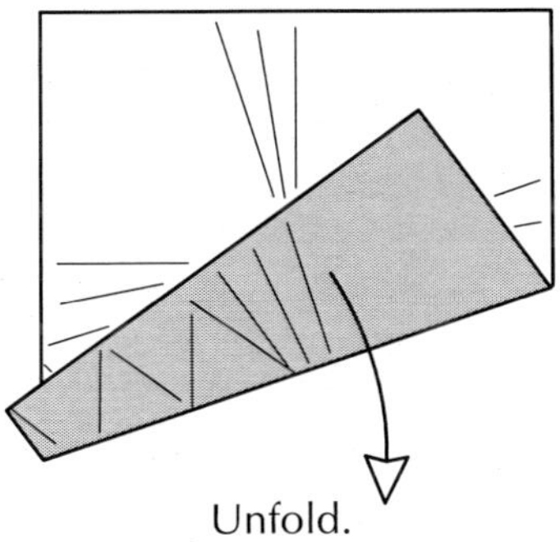

Unfold.

16

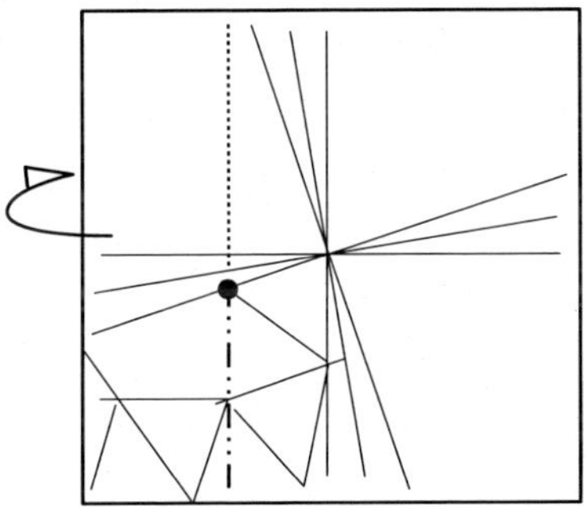

Fold along the crease on the bottom half.

17

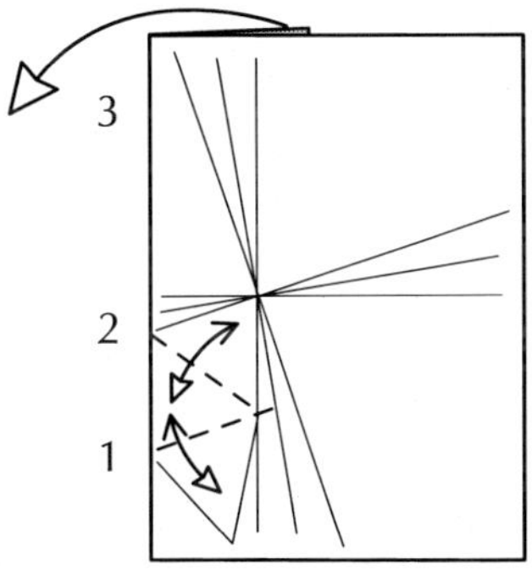

1,2. Fold both layers along hidden creases.
3. Unfold and rotate 90°.

18

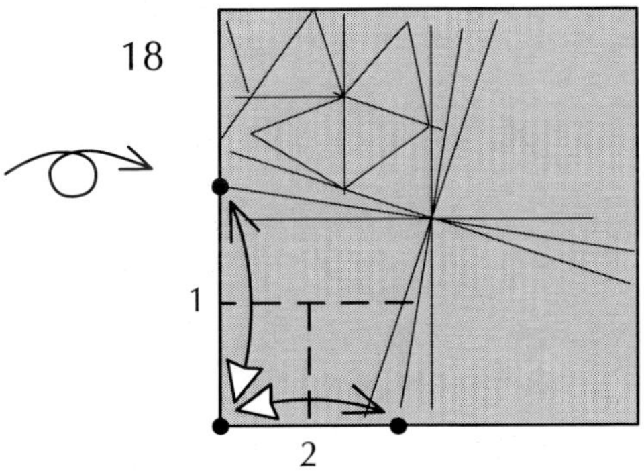

Repeat steps 7–17 three times.

19

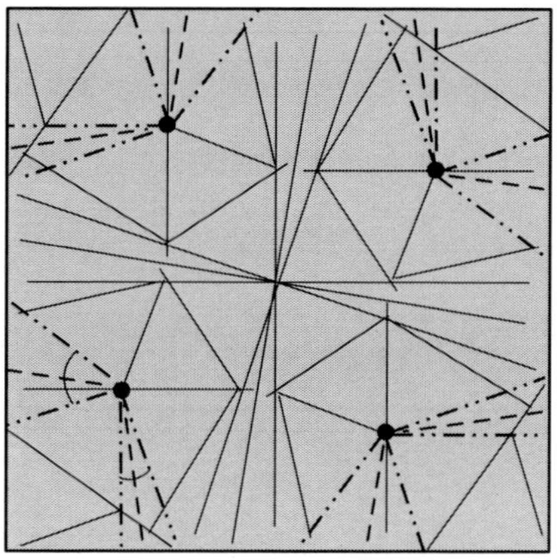

Fold and unfold to bisect the angles. Mountain-fold along the creases.

20

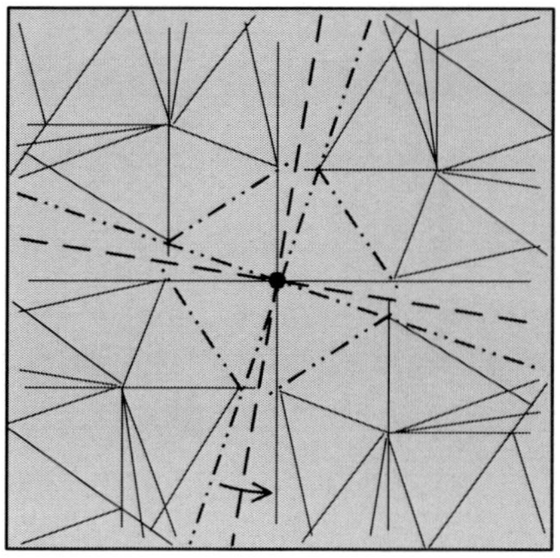

Push in at the dot to make the first sunken square. Rotate the center to the bottom.

21

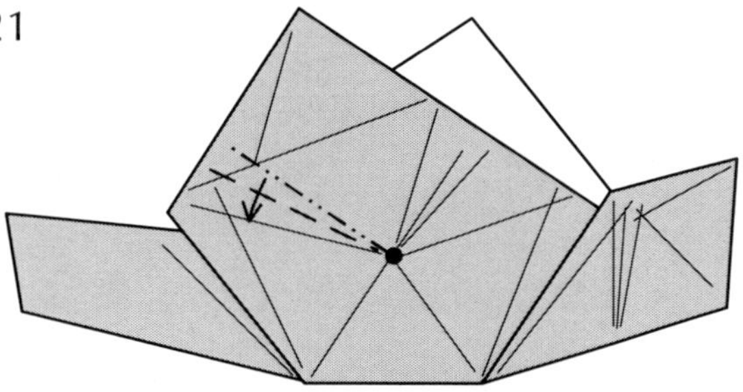

Push in at the dot.

22

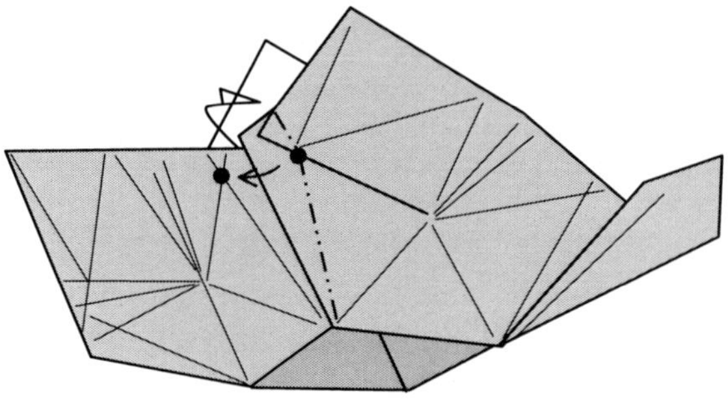

Slide the paper. There is a small squash fold at the top.

23

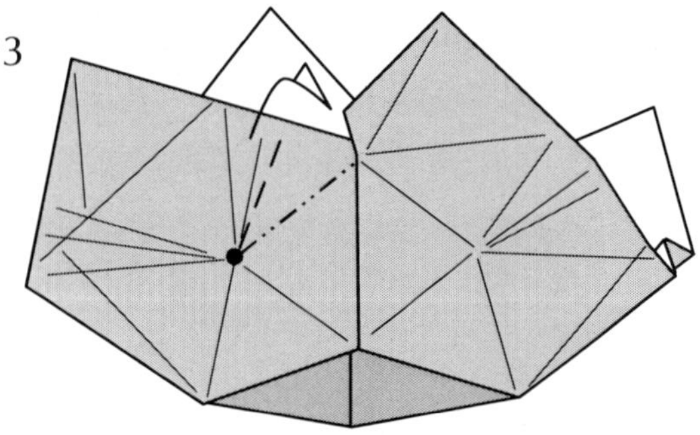

Push in at the dot.

24

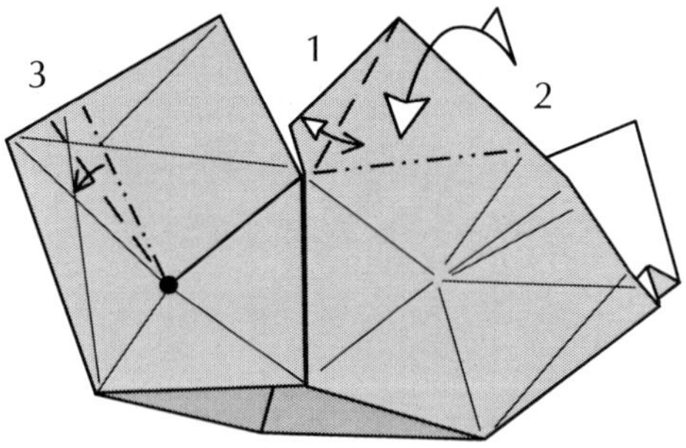

1. Fold and unfold.
2. Fold and unfold.
3. Repeat steps 21–24 three times.

25

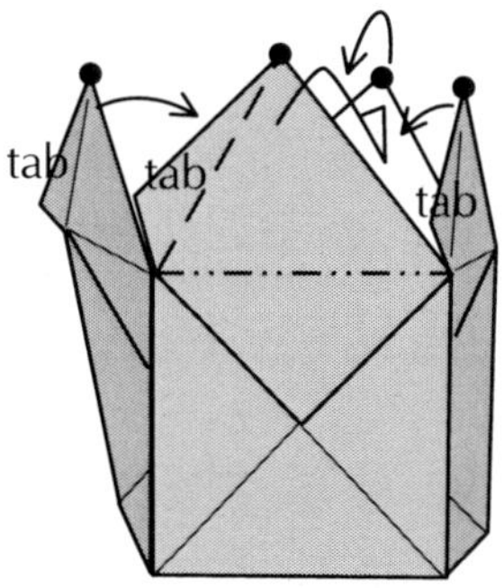

Fold the four corners inside to form the last sunken square. Interlock to cover the tabs.

26

Sunken Cube

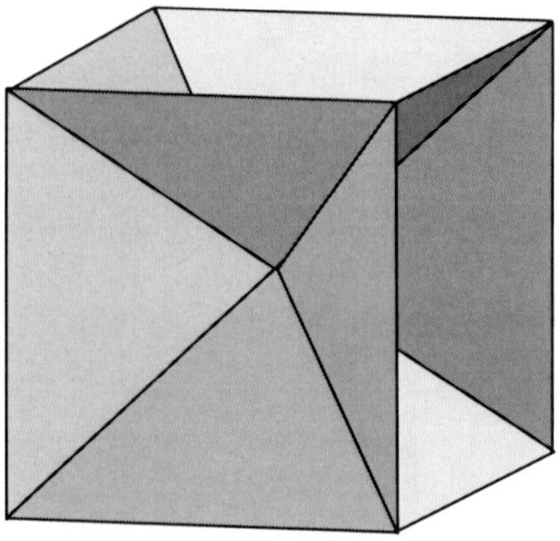

Sunken Dodecahedron

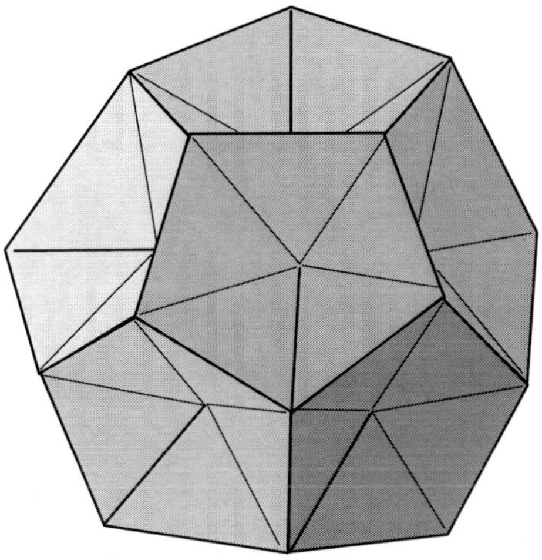

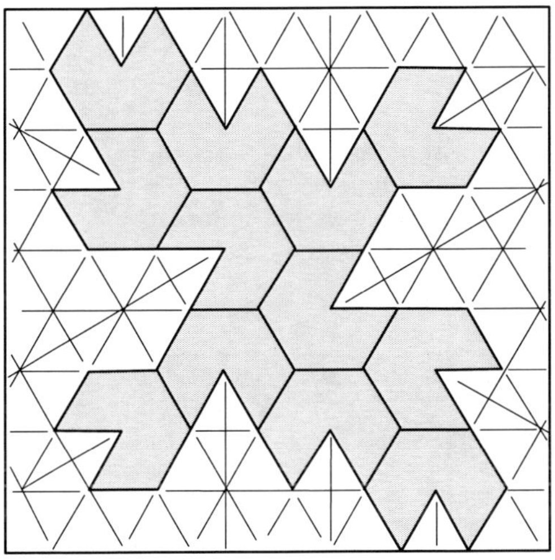

This sunken dodecahedron, one of several stellated icosahedrons, is composed of 60 equilateral triangles. This version has the same surface as an icosahedron. The crease pattern shows odd symmetry.

1

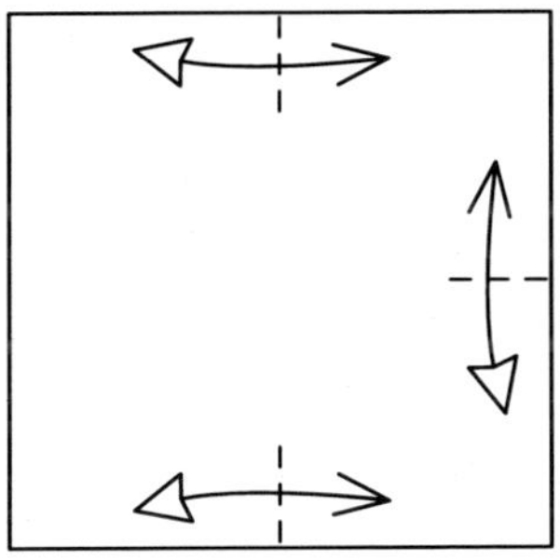

Fold and unfold on three edges.

2

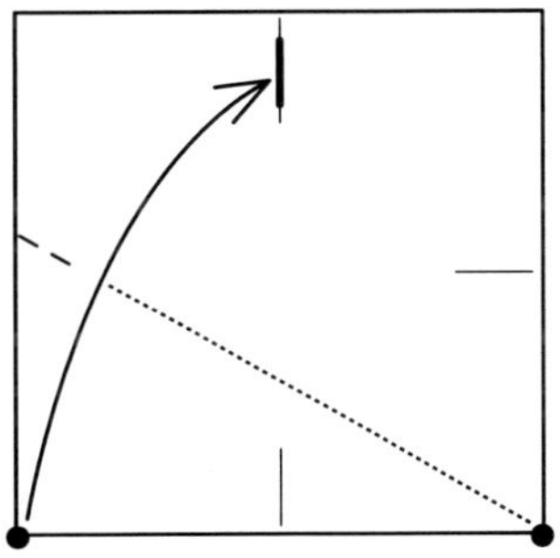

Bring the left dot to the line. Crease on the left.

3

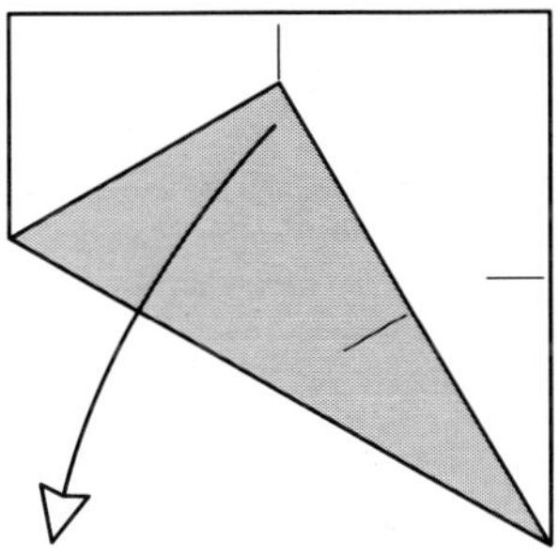

Unfold and rotate 180°.

4

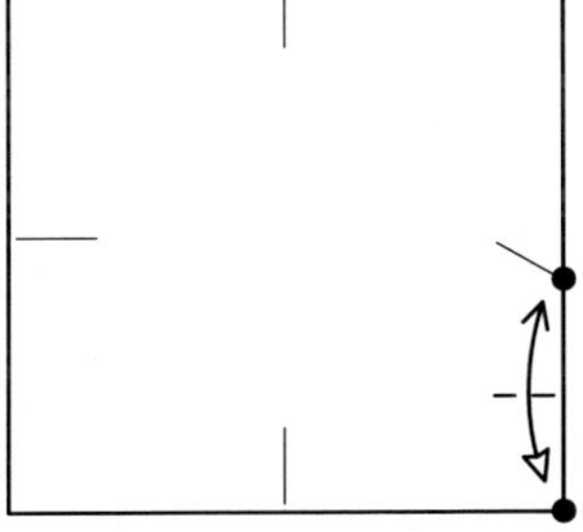

Fold and unfold on the right.

5

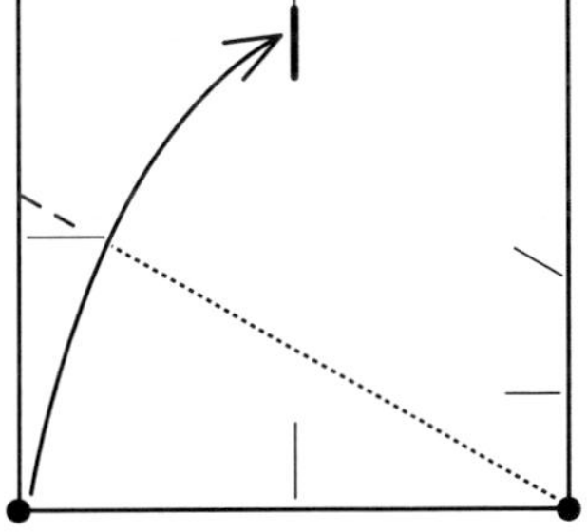

Repeat steps 2–4.

6

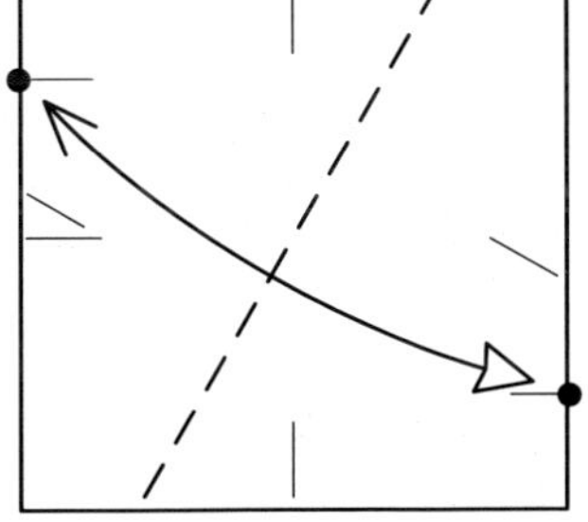

Fold and unfold.

7

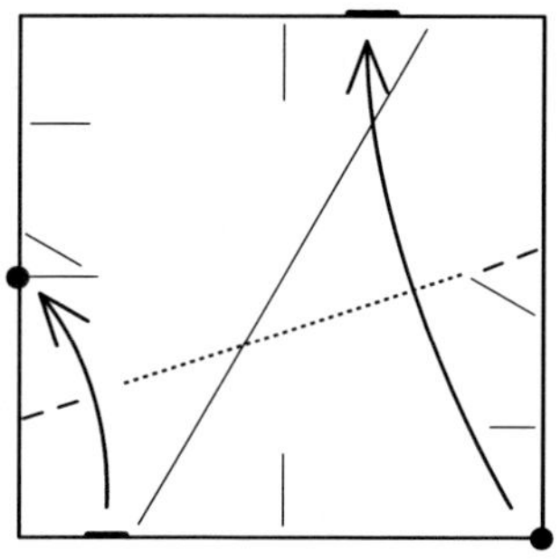

Bring the bottom edge to the left dot and the bottom right corner to the top. Crease on the edges.

8

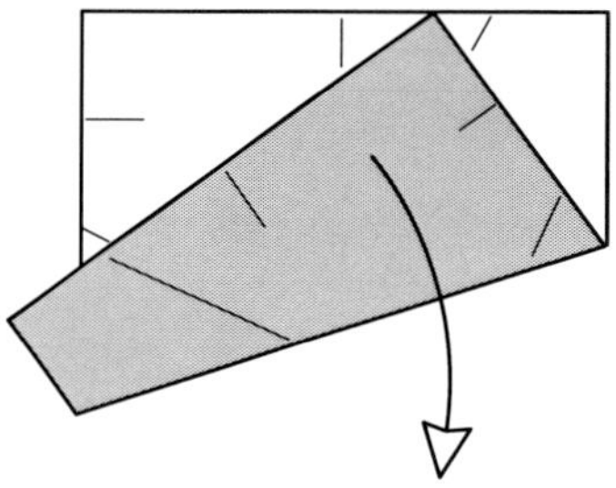

Unfold.

9

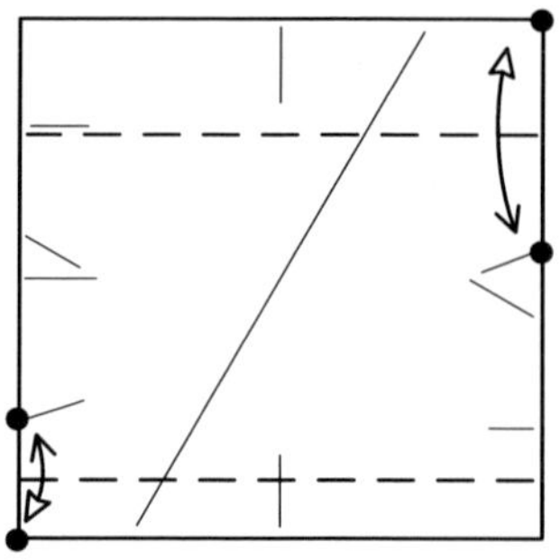

Fold and unfold.

10

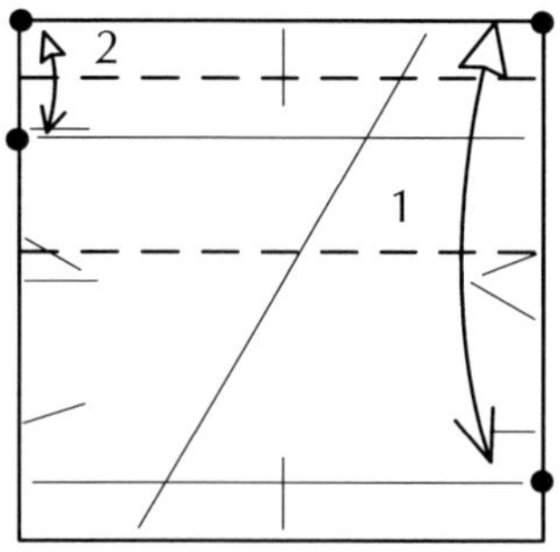

Fold and unfold.

11

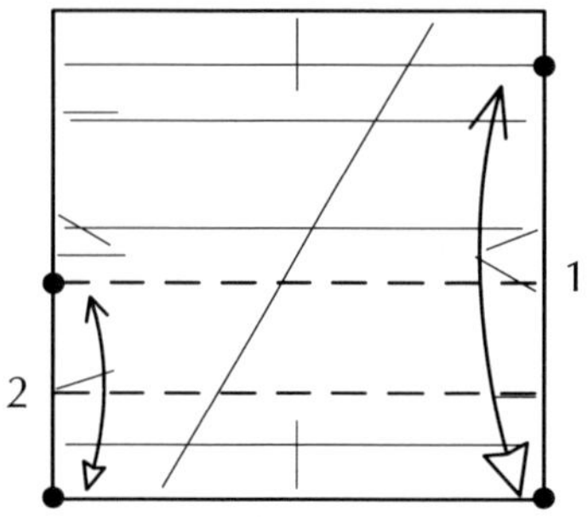

Fold and unfold.

12

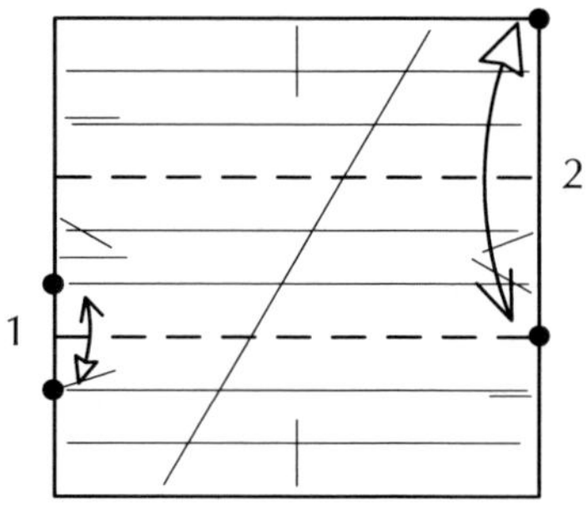

Fold and unfold. The paper is divided into 9ths.

13

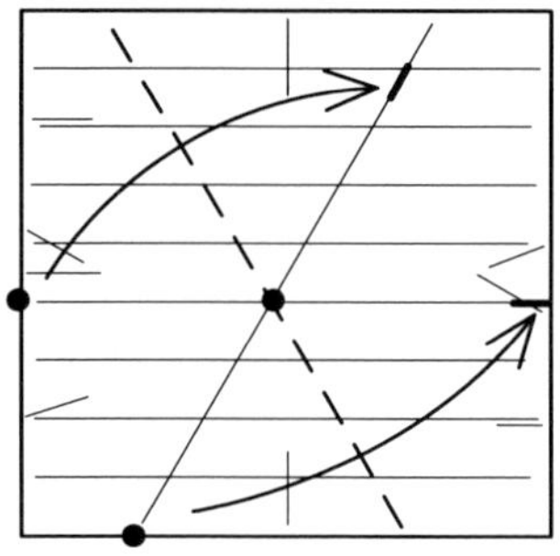

14

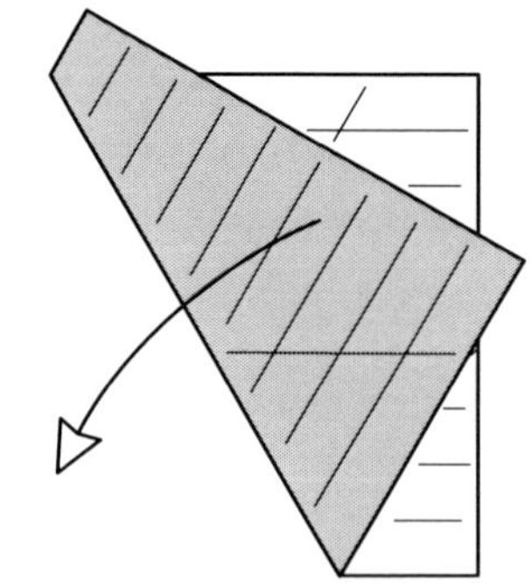

Unfold.

15

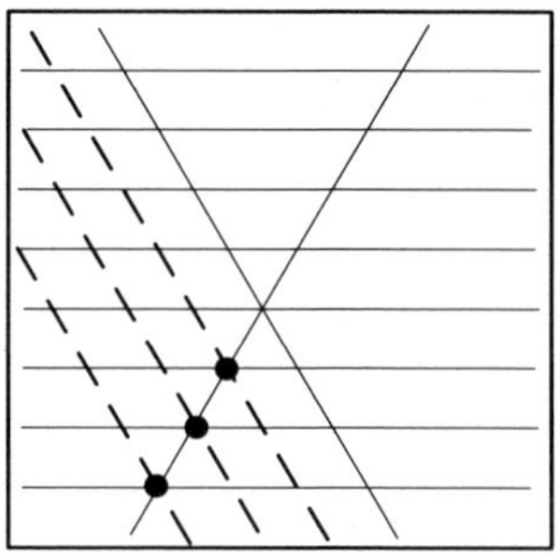

Fold and unfold.
Rotate 180°.

16

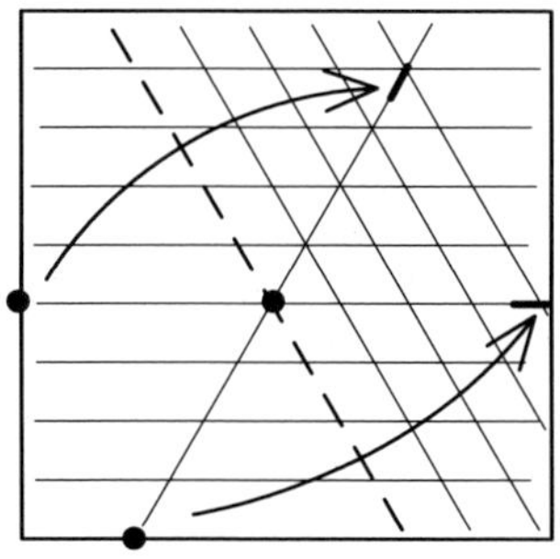

Repeat steps 13–15.

17

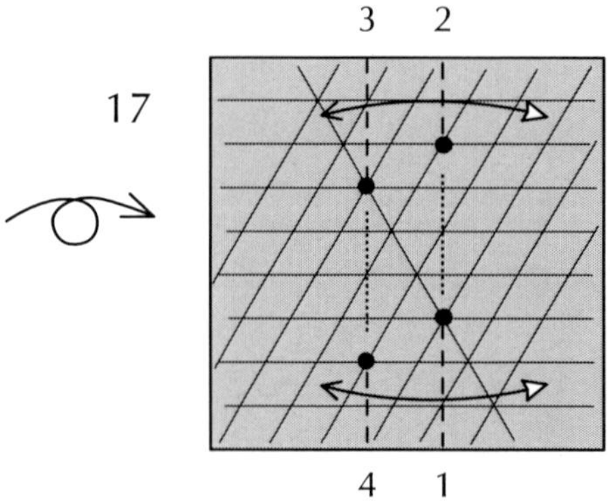

Fold and unfold.

18

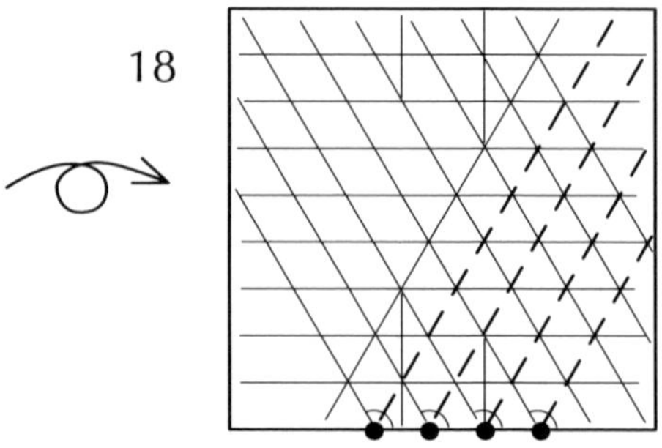

Fold and unfold.
Rotate 180°.

19

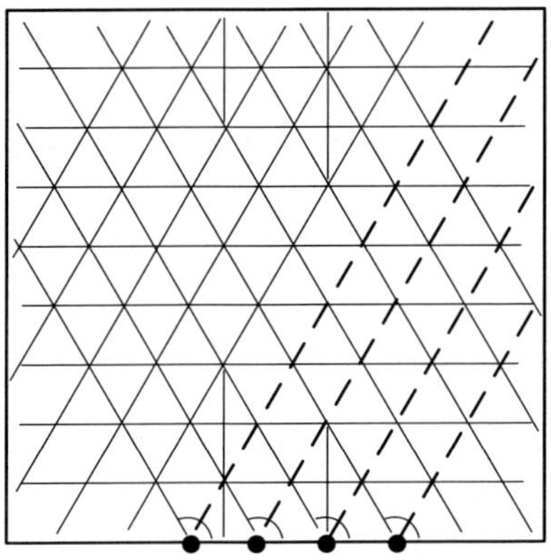

Repeat step 18.

20

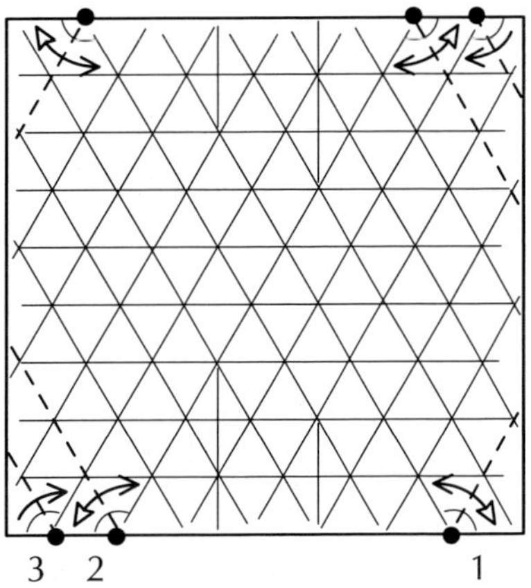

1. Fold and unfold.
2. Fold and unfold.
3. Valley-fold.
Rotate 180° and repeat, shown at the top.

21

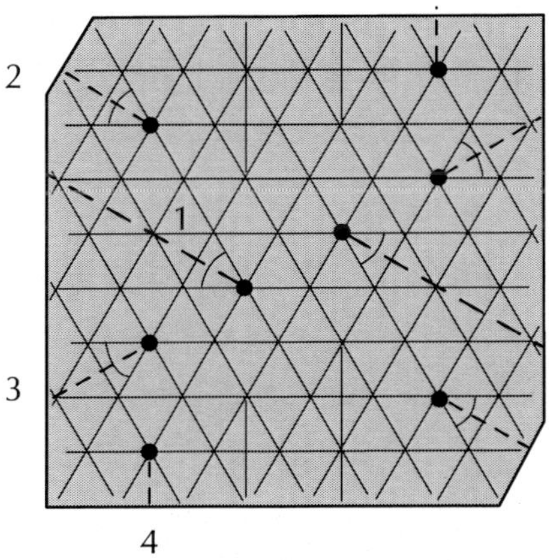

Fold and unfold on four lines on the left. Rotate 180° and repeat. Rotate 90°.

22

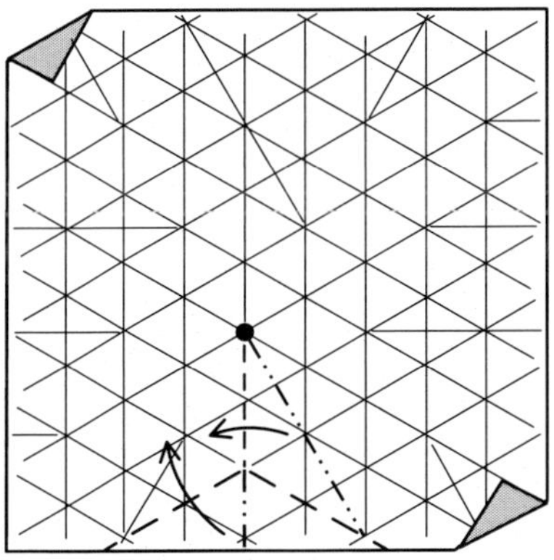

Push in at the dot.

23

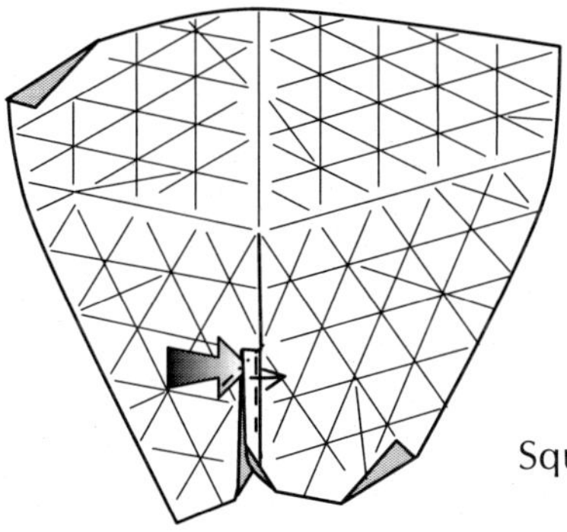

Squash-fold.

24

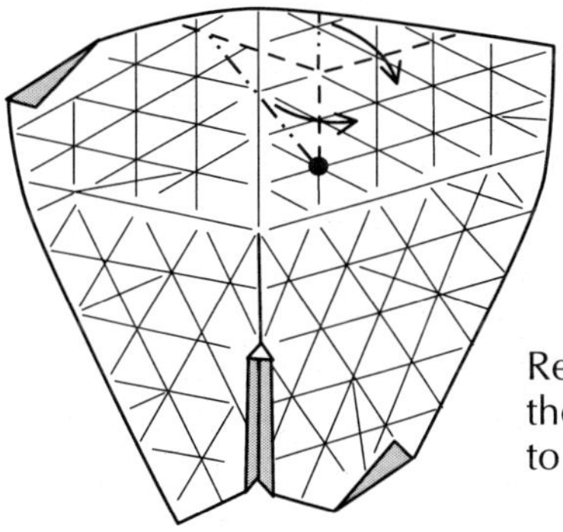

Repeat steps 22–23 on the other side. Turn over to view the outside.

25

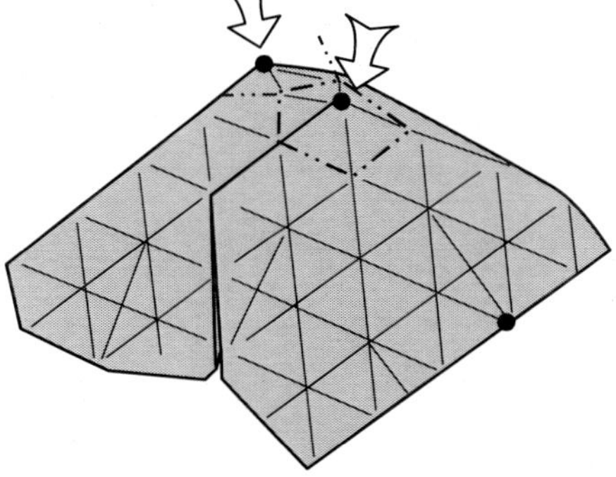

Push in at the upper dots to form two sunken pentagons. Rotate the lower dot to the center.

26

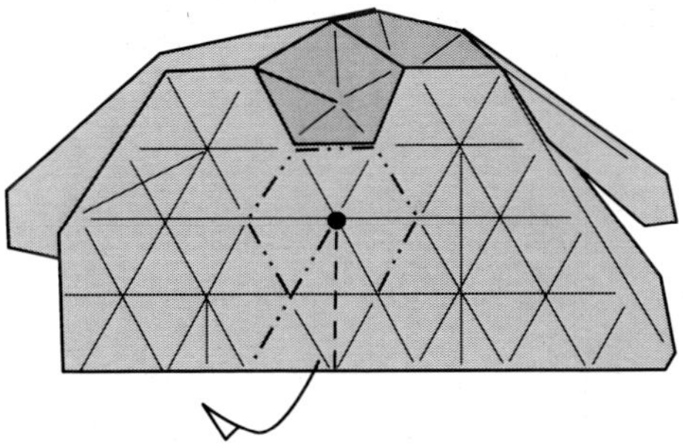

Push in at the dot to form a sunken pentagon. Turn over and repeat.

27

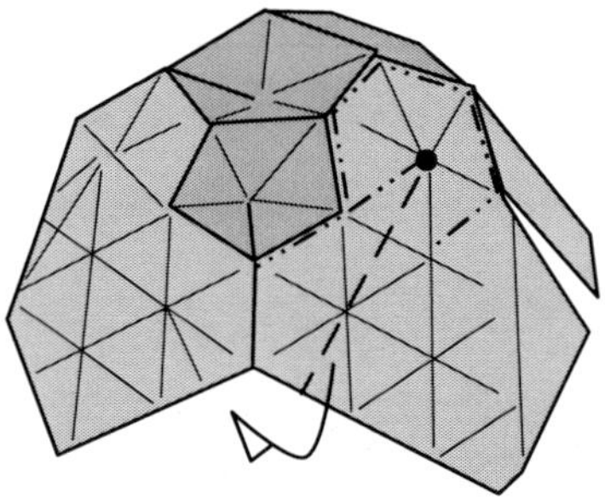

Push in at the dot and tuck under the left side to form a sunken pentagon. Turn over and repeat.

28

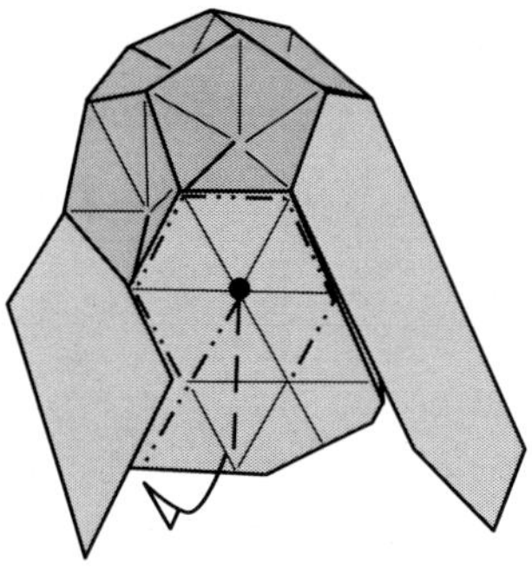

Push in at the dot to form a sunken pentagon. Turn over and repeat.

29

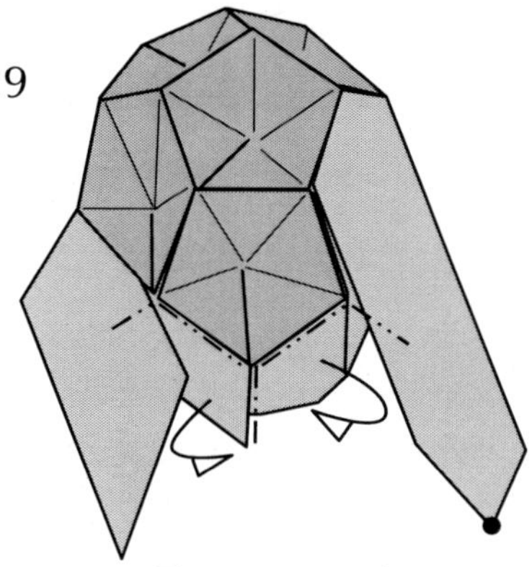

Turn over and repeat. Rotate to view the dot at the top, center.

30

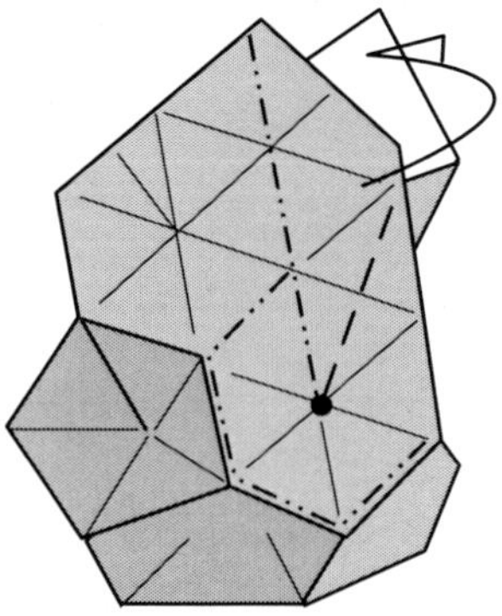

Push in at the dot to form a sunken pentagon. Turn over and repeat.

31

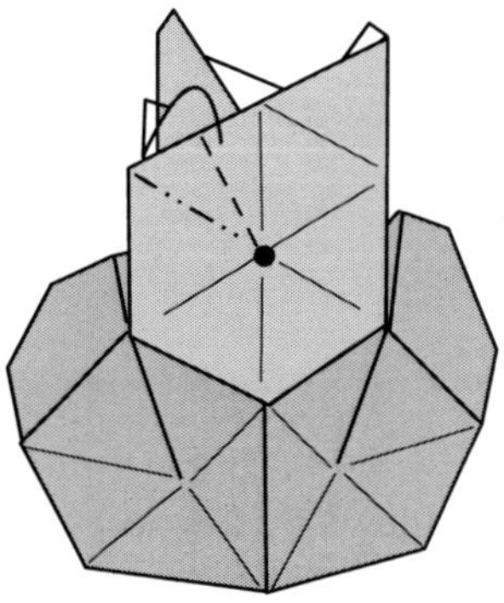

Push in at the dot and tuck into a hidden pocket. Turn over and repeat.

32

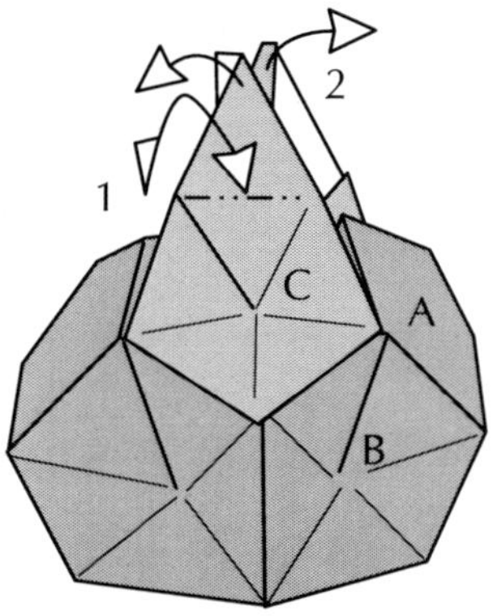

1. Fold and unfold. Turn over and repeat.
2. Spread at the top to view the inside.

Follow pentagons A, B, and C into the next step.

33

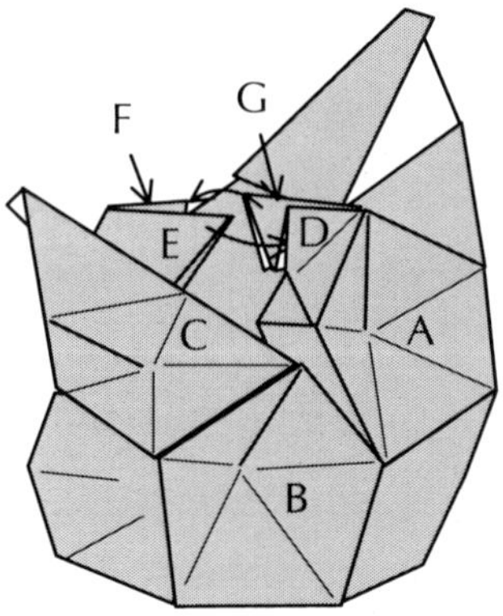

There are four flaps at the top, flaps D, E, F, and G. Flaps E and G larger than D and F. Tuck E into D, and G into F. This will create a pocket at the top.

34

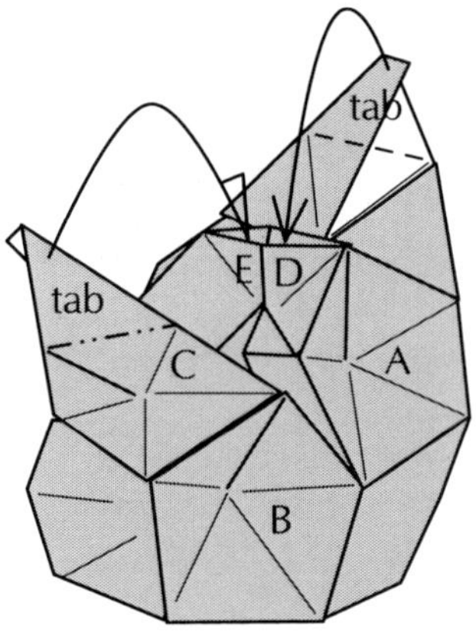

Tuck the tabs into the pocket.

35

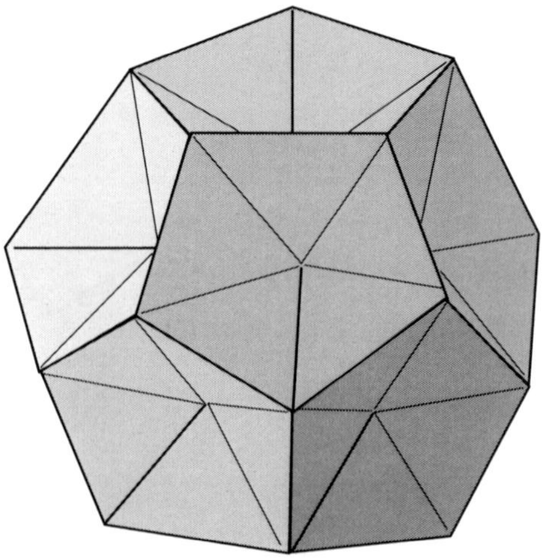

Sunken Dodecahedron

Sunken Stellated Cube

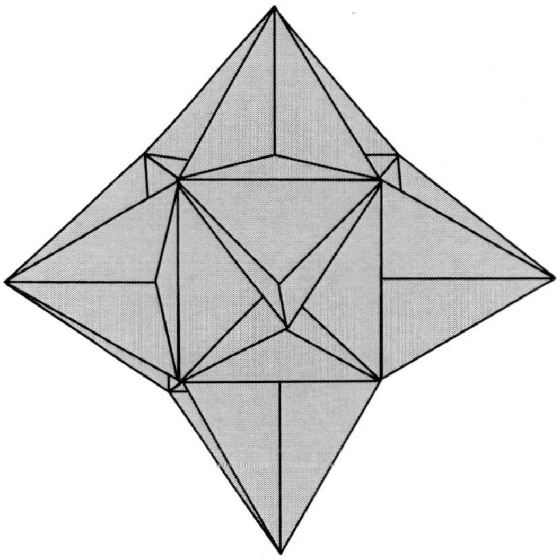

Each face of a central cube is stellated to produce this star. All the sides are triangles, with angles 45°, 45°, and 90°. Square symmetry is used. The four corners of the square meet at the bottom to form a vertex and lock the star.

1

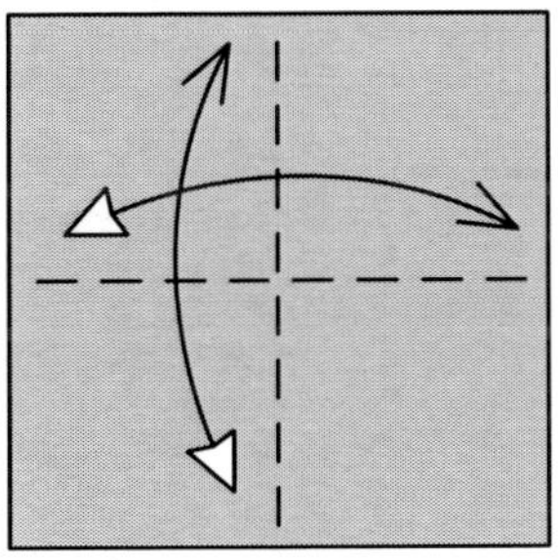

Fold and unfold.

2

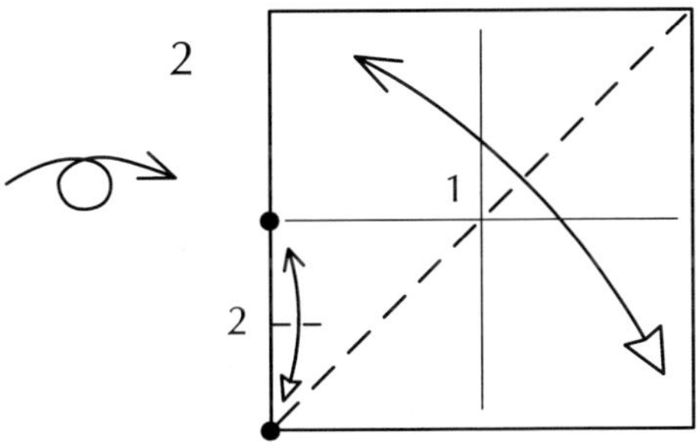

1. Fold and unfold.
2. Fold and unfold on the left.

3

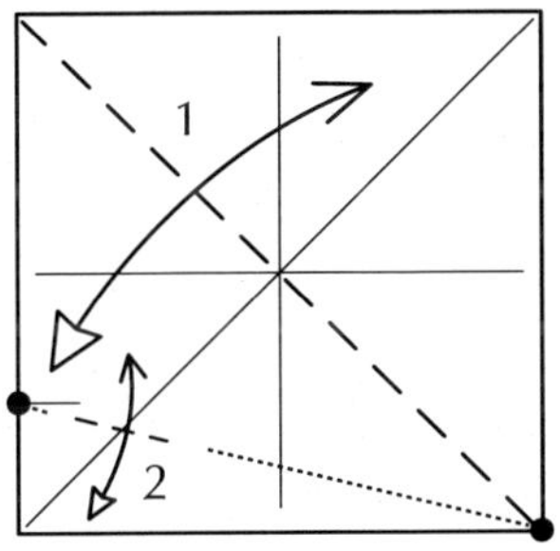

1. Fold and unfold.
2. Fold and unfold on the diagonal.

Rotate 180°.

4

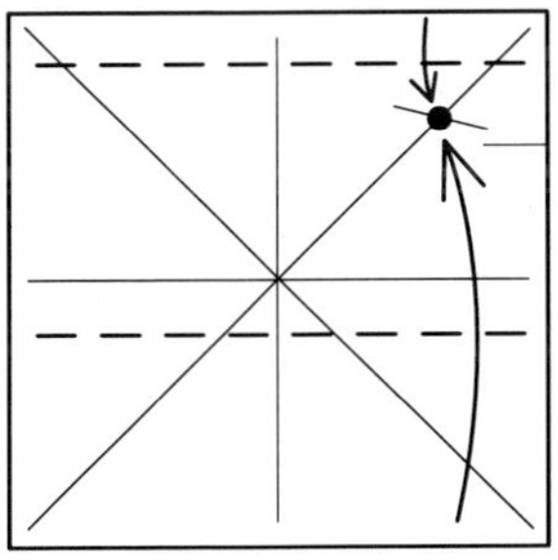

5

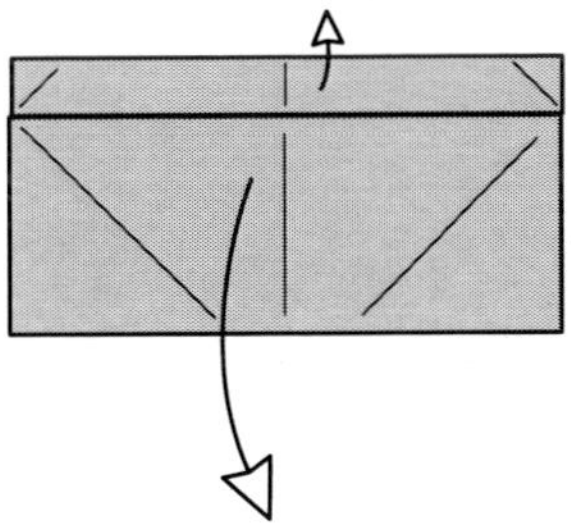

Unfold.

6

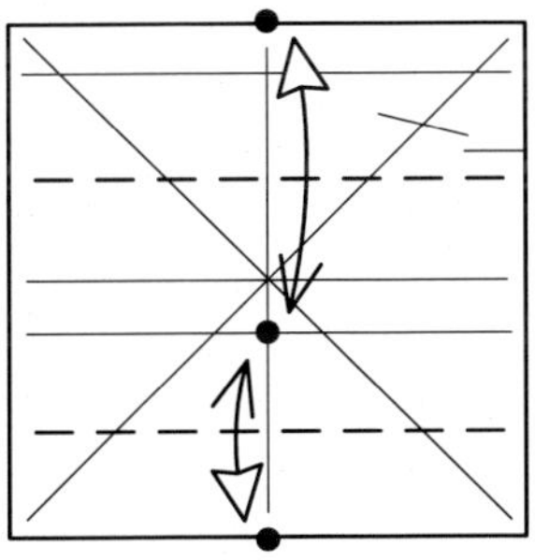

Fold and unfold.

7

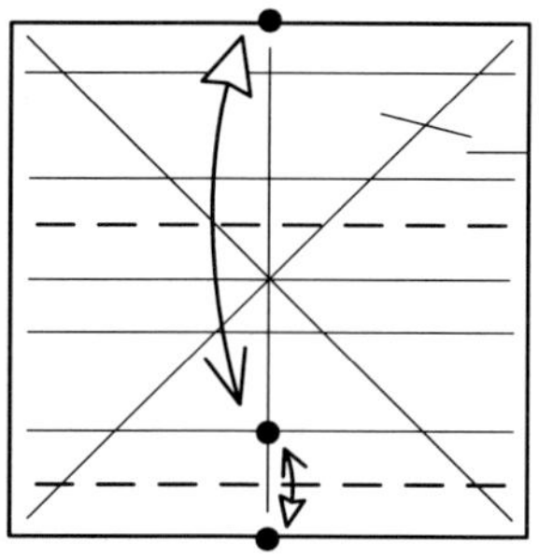

Fold and unfold.

8

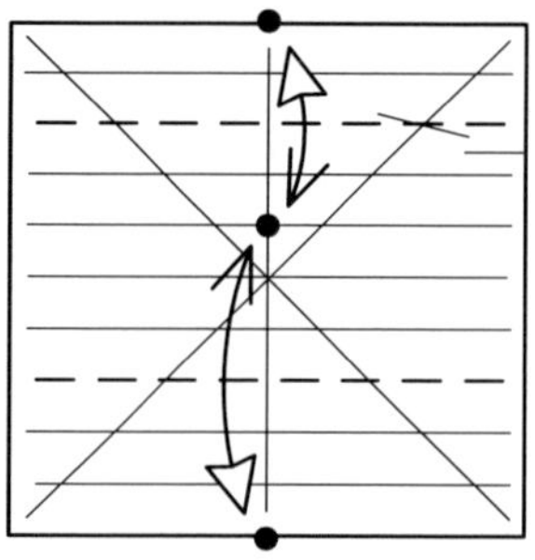

Fold and unfold.
Rotate 90°.

9

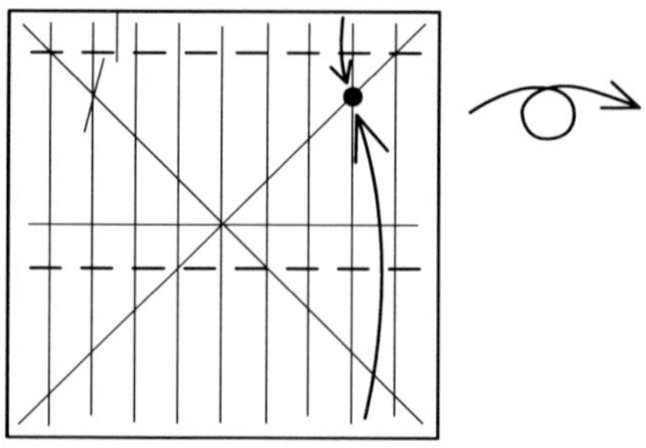

Repeat steps 4–8.
Rotate 45°.

10

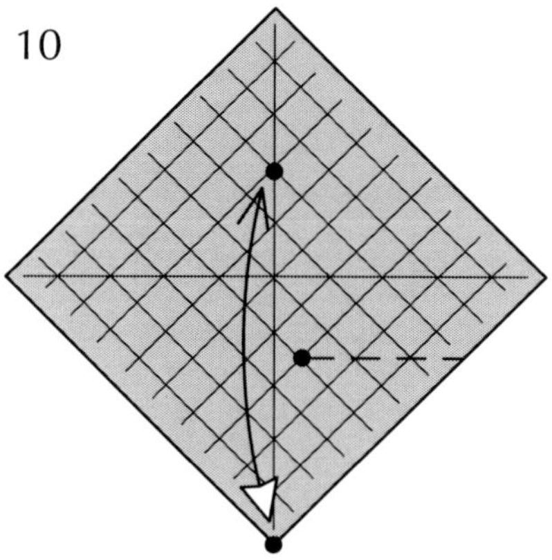

Fold and unfold.

11

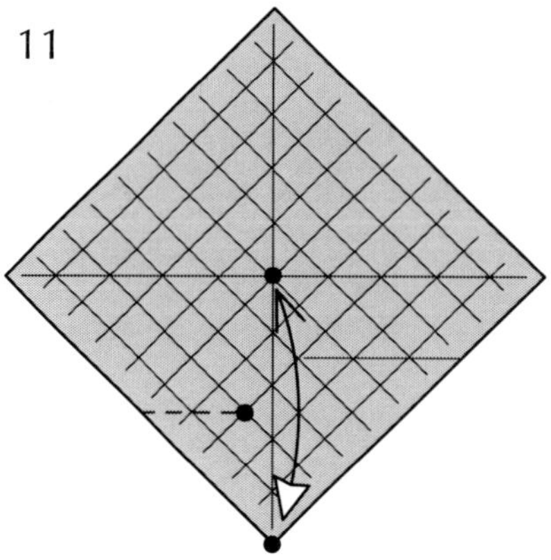

Fold and unfold.
Rotate 90°.

12

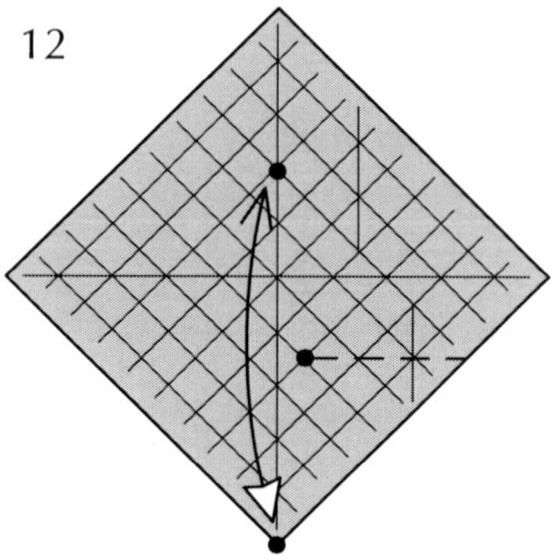

Repeat steps 10–11 three times. Rotate 45°.

13

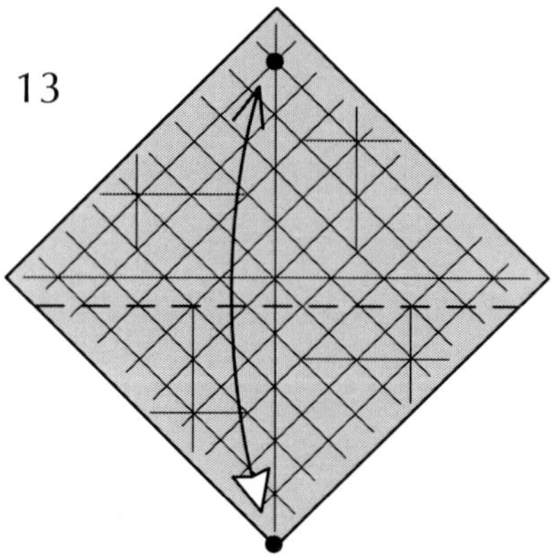

Fold and unfold.
Rotate 90°.

14

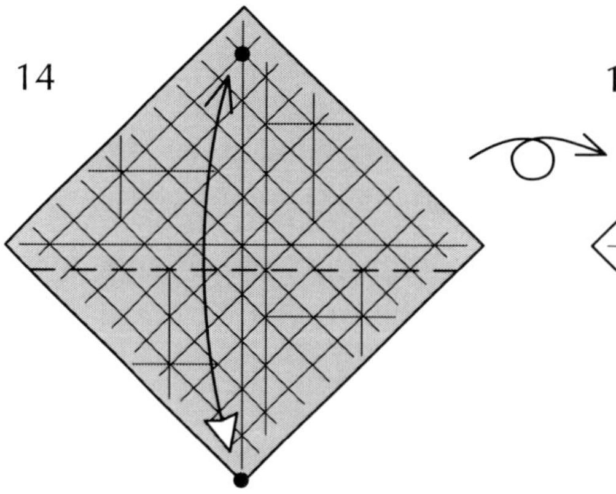

Repeat step 13 three times.

15

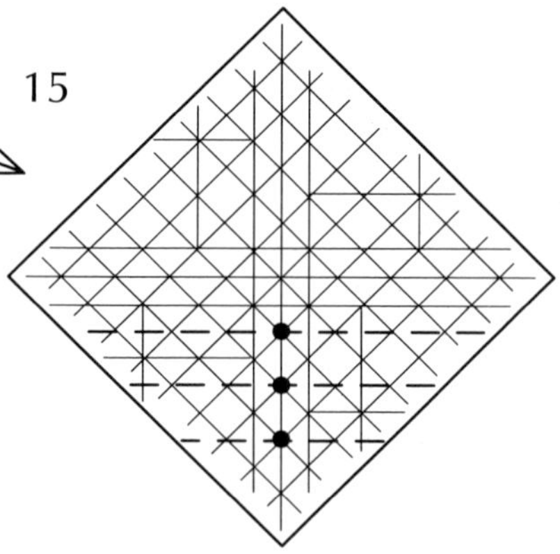

Fold and unfold.
Rotate 90°.

16

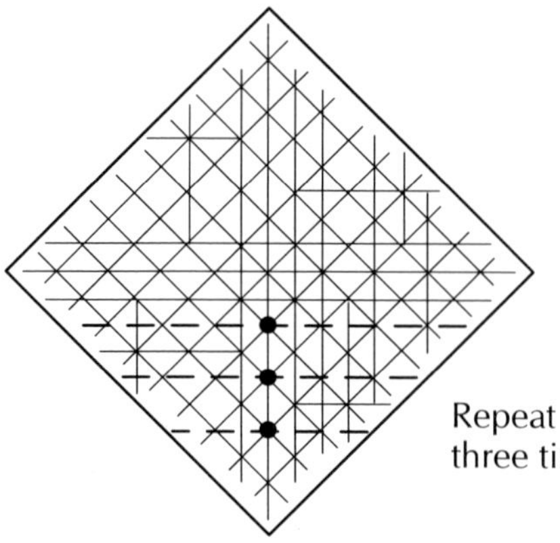

Repeat step 15 three times.

17

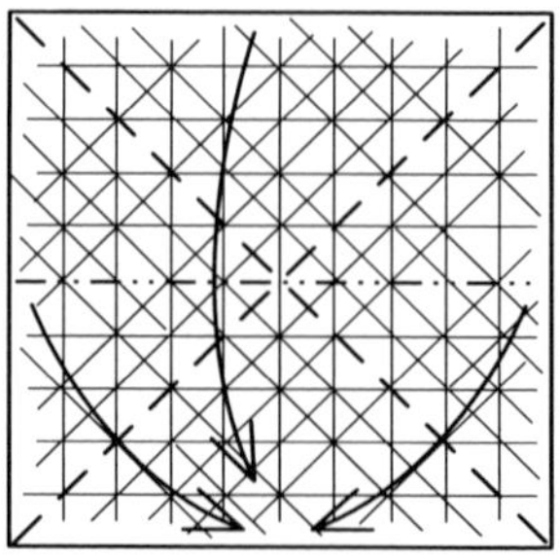

Fold along the creases to make a waterbomb base.

18

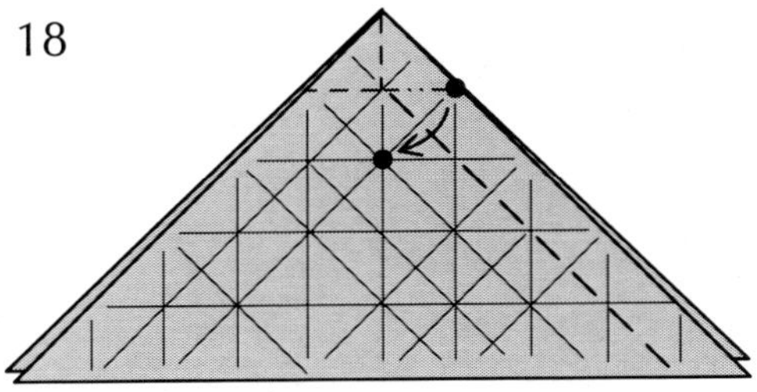

The dots will meet as the model opens. Repeat all around to form a 3D X in the center.

19

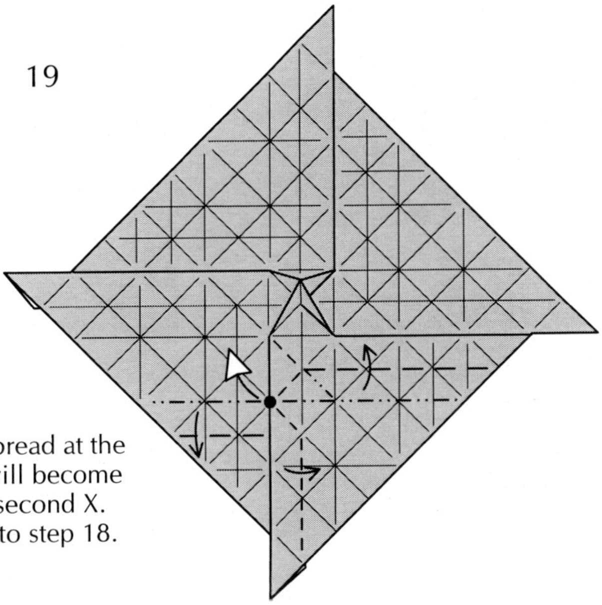

Puff out and spread at the dot. The dot will become the top of the second X. This is similar to step 18.

20

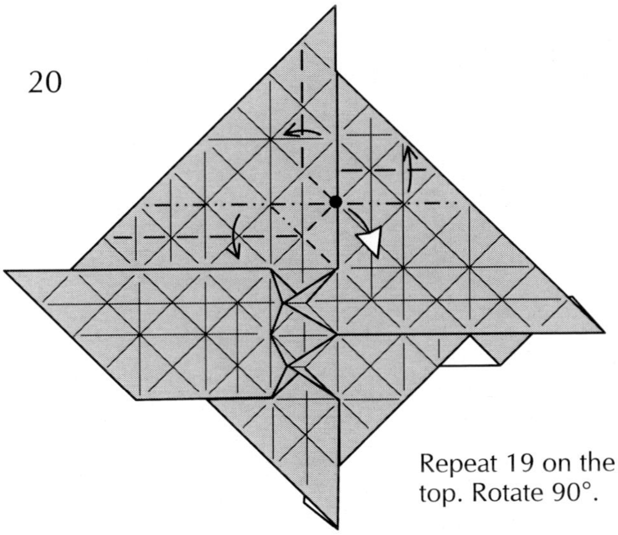

Repeat 19 on the top. Rotate 90°.

21

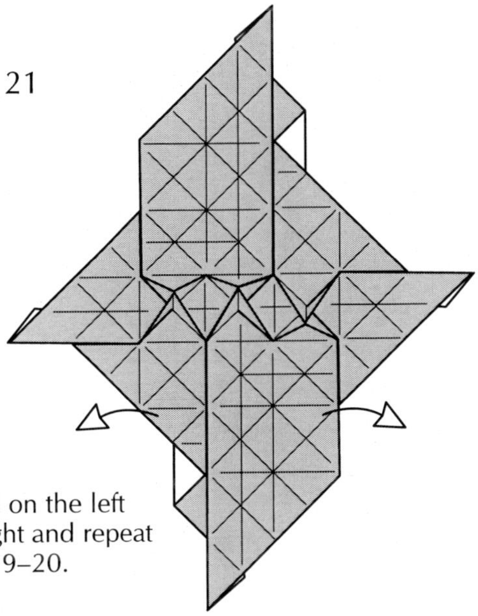

Spread on the left and right and repeat steps 19–20.

22

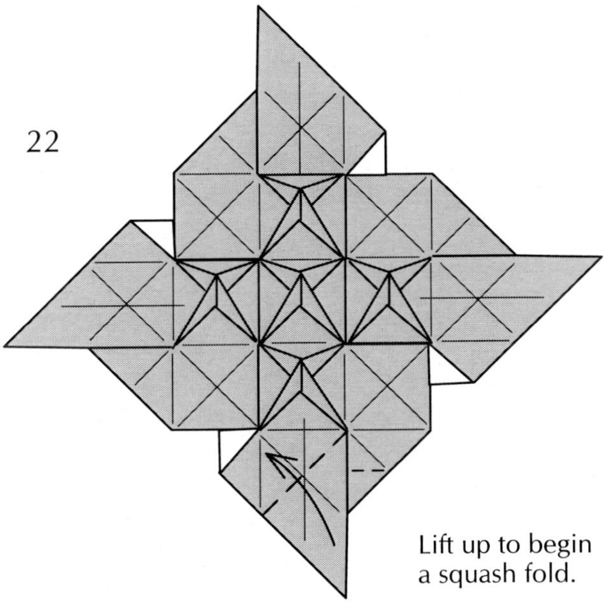

Lift up to begin a squash fold.

23

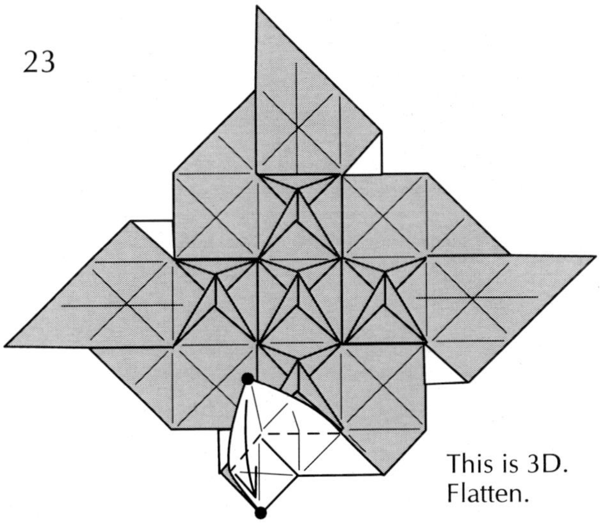

This is 3D. Flatten.

24

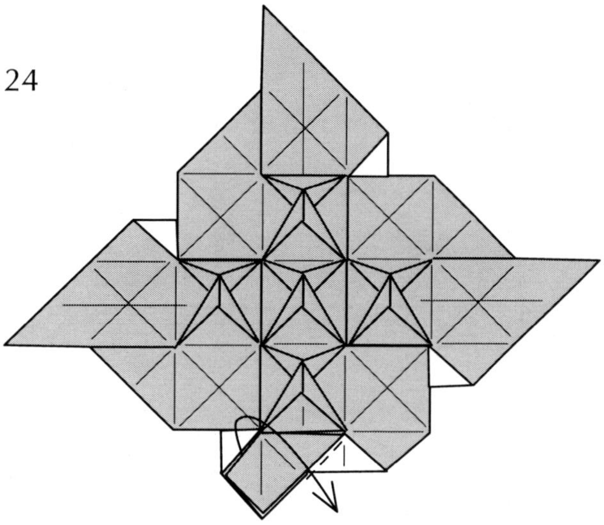

25

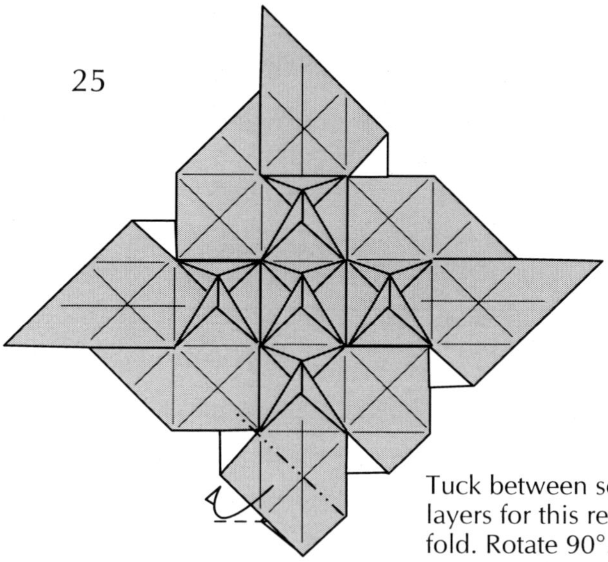

Tuck between some layers for this reverse fold. Rotate 90°.

26

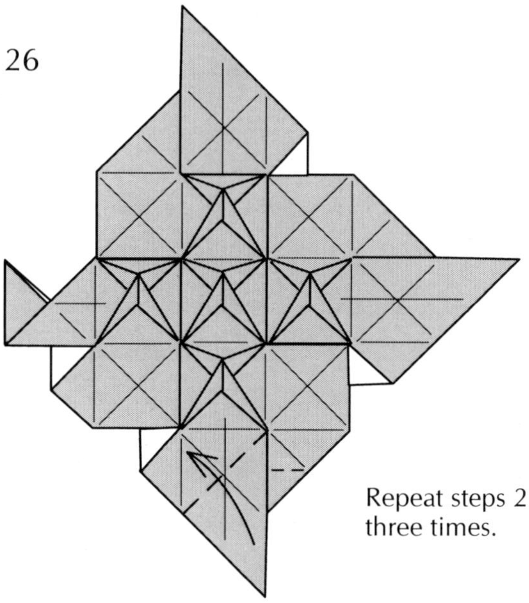

Repeat steps 22–25 three times.

27

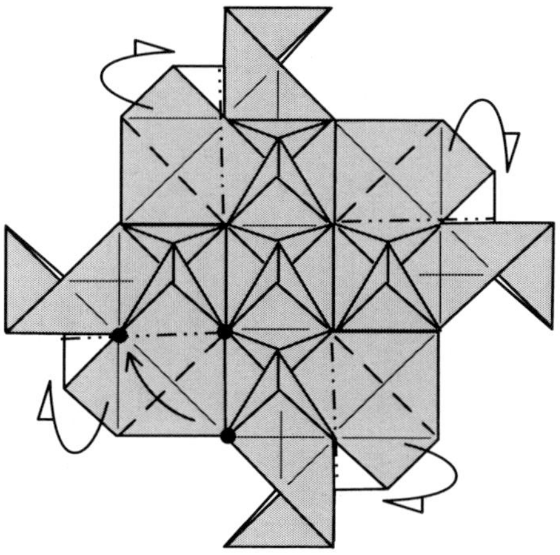

Two of the dots will meet. Puff out at the other (upper right) dot. Repeat three times and rotate the top to the bottom.

28

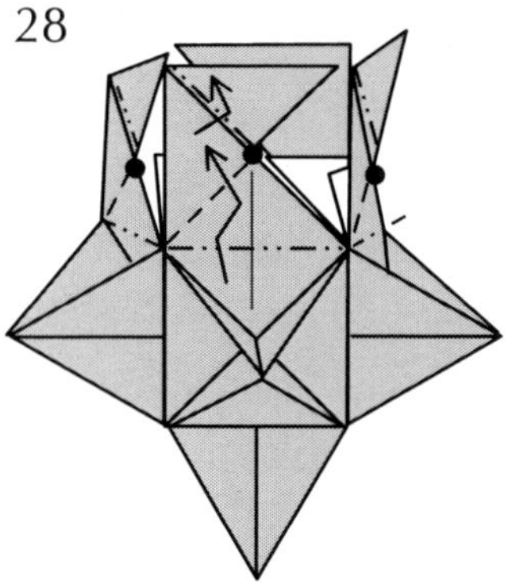

Fold the top flaps on four sides. Four dots will meet, three are shown.

29

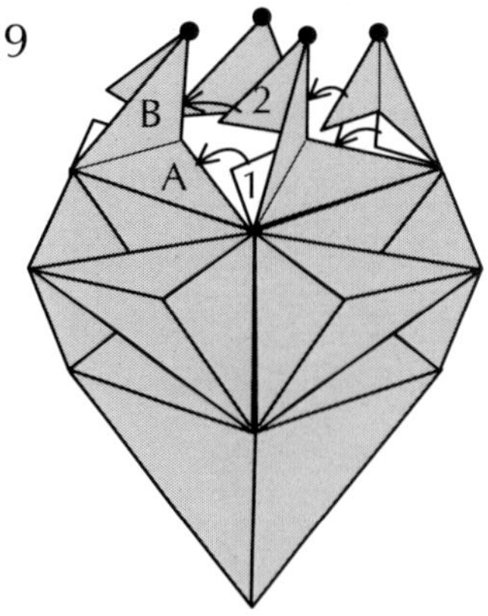

Tuck tabs 1 and 2 under A and into the pocket of B. Repeat all around so the four dots meet at the top. You can lift A and B up, place the tabs inside, then close A and B to lock it.

30

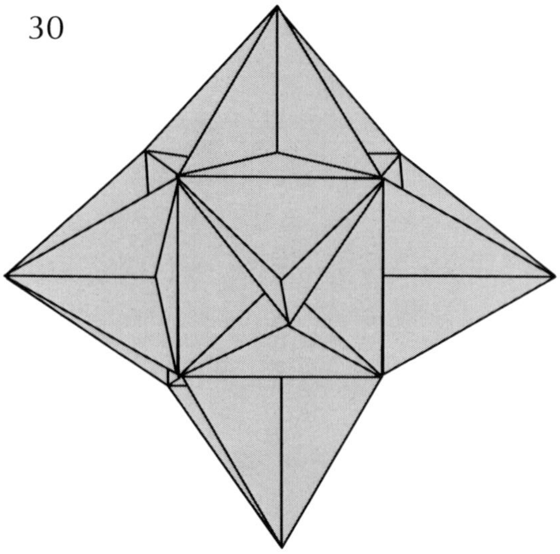

Sunken Stellated Cube

Pentagons in Polyhedra

The golden pentagonal antiprism has two pentagonal faces. For the pentagonal trapezohedron, if the top and bottom corners were cut at just the right place, the shape would be the dodecahedron. The dodecahedron has twelve pentagonal faces. These models are challenging.

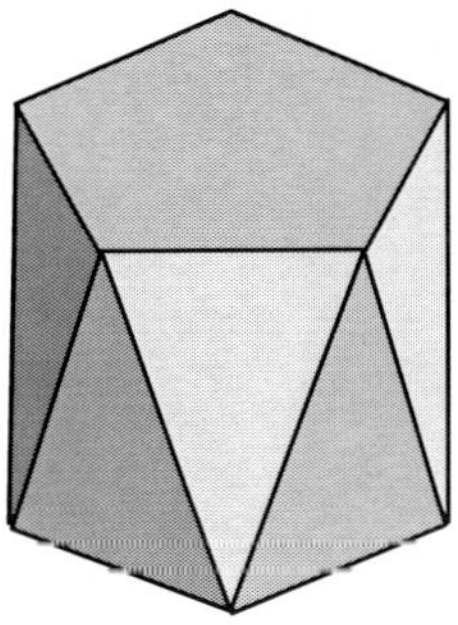

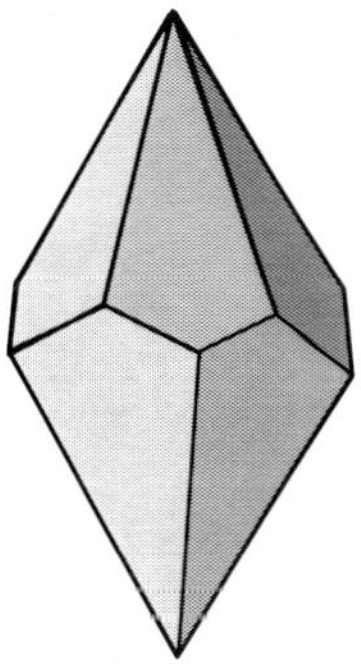

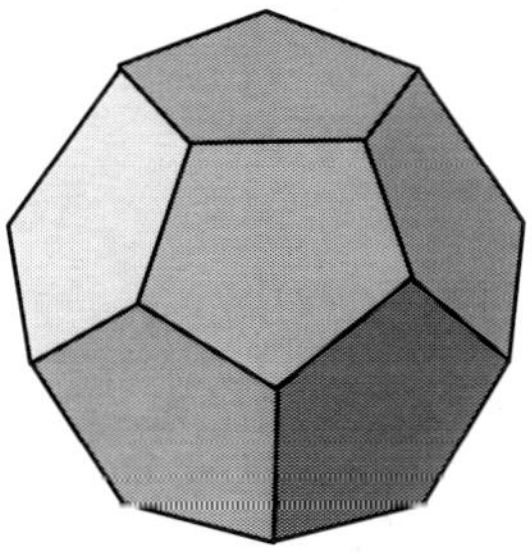

Golden Pentagonal Antiprism

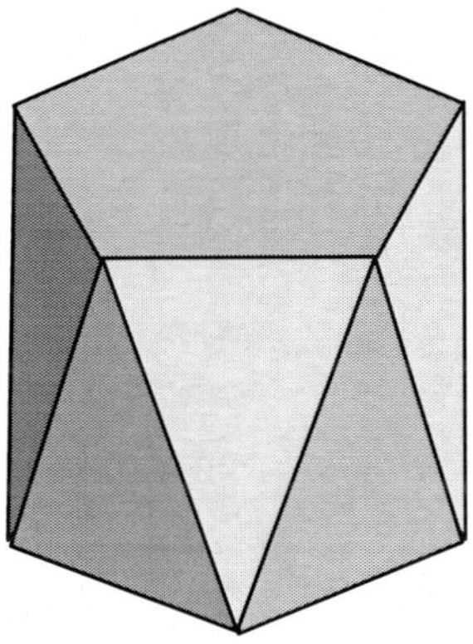

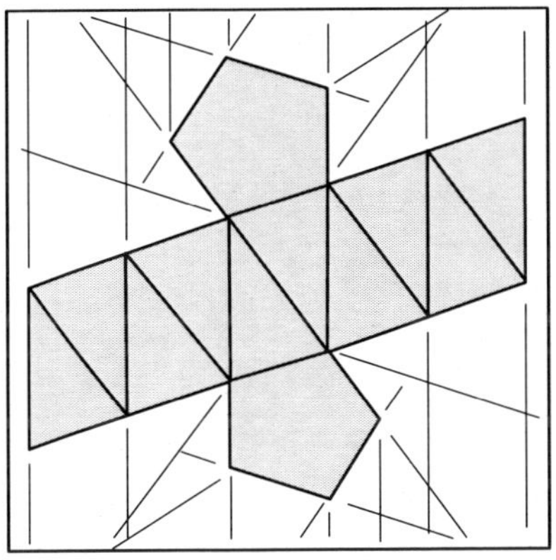

A pentagonal antiprism is composed of 10 triangles and two pentagons. In this one, the angles of each triangle are 36°, 72°, and 72°. The triangles form a band wrapping around.

1

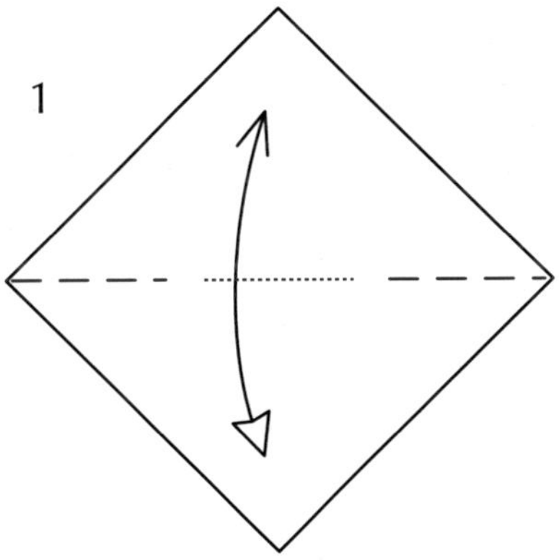

Fold and unfold.

2

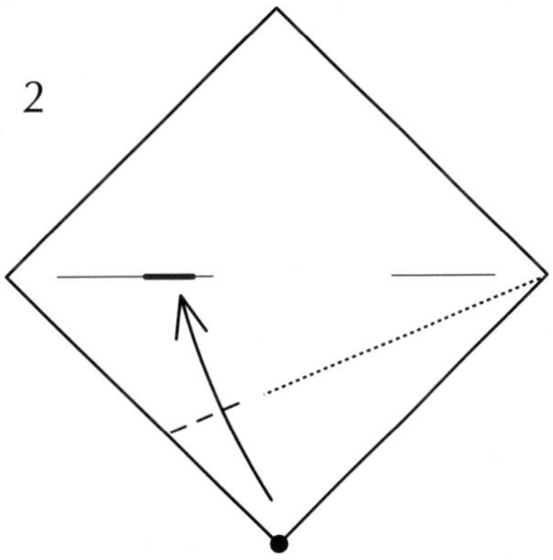

Crease on the left.

3

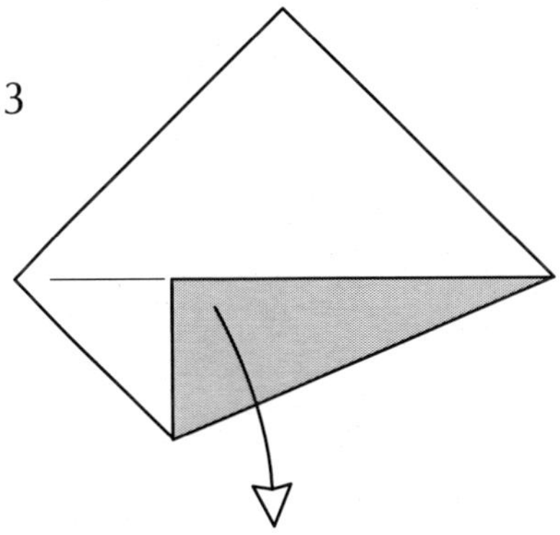

Unfold.

4

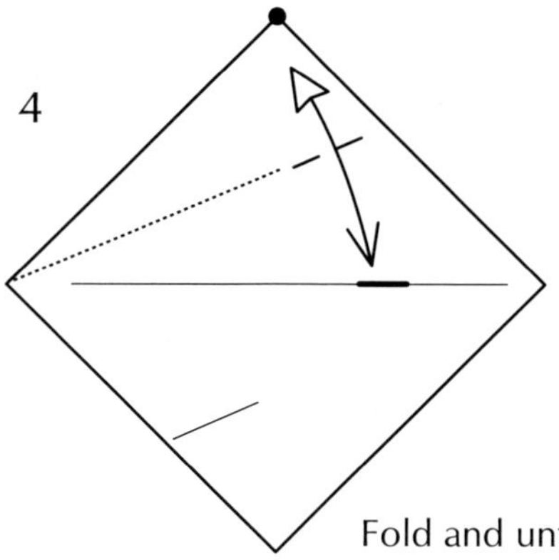

Fold and unfold.
Rotate.

5

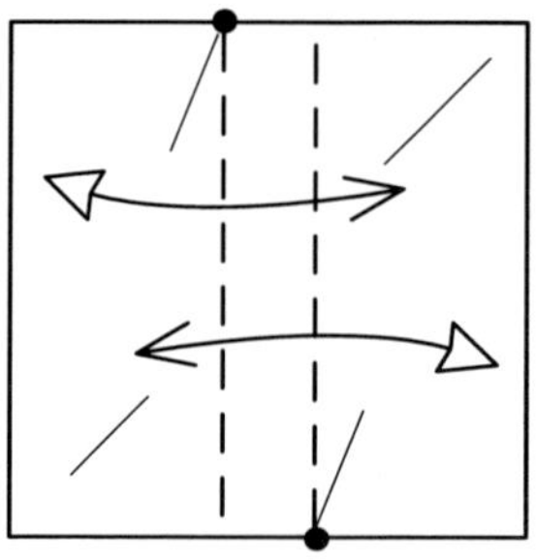

Fold and unfold.

6

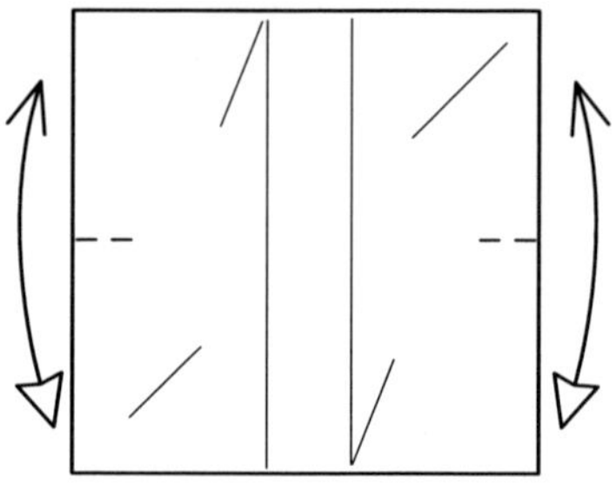

Fold and unfold in
half at the edges.

7

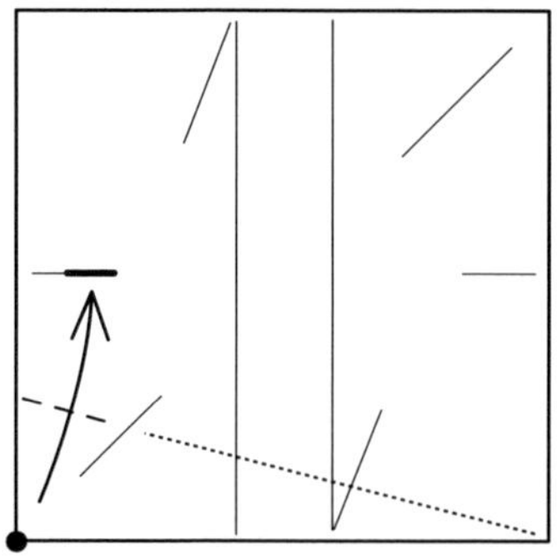

Crease on the left.

8

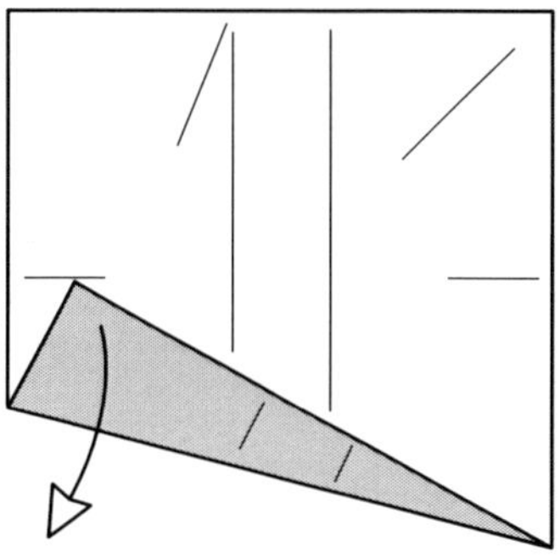

Unfold and rotate 180°.

9

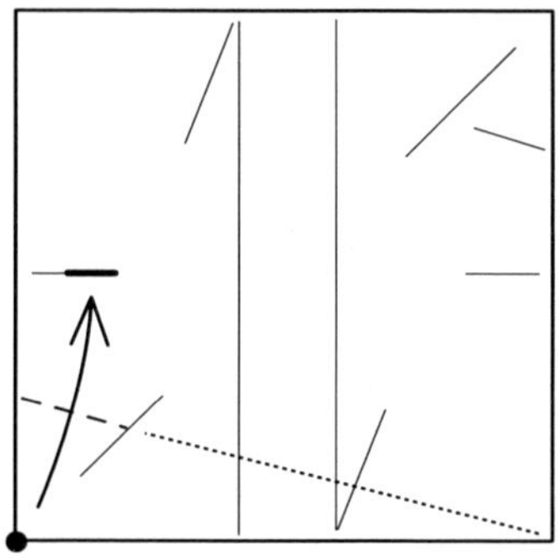

Repeat steps 7–8.

10

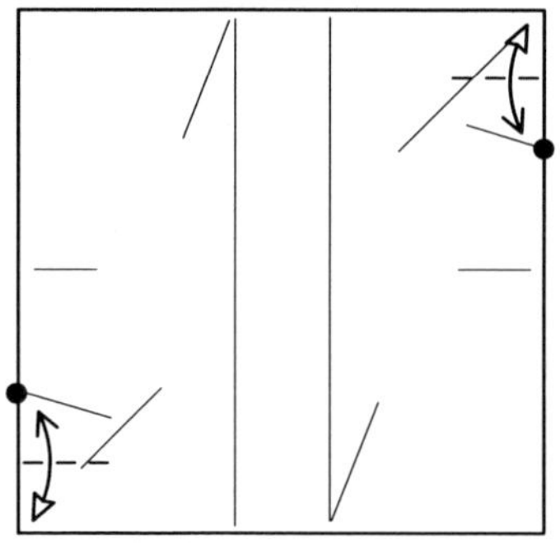

Fold and unfold.

11

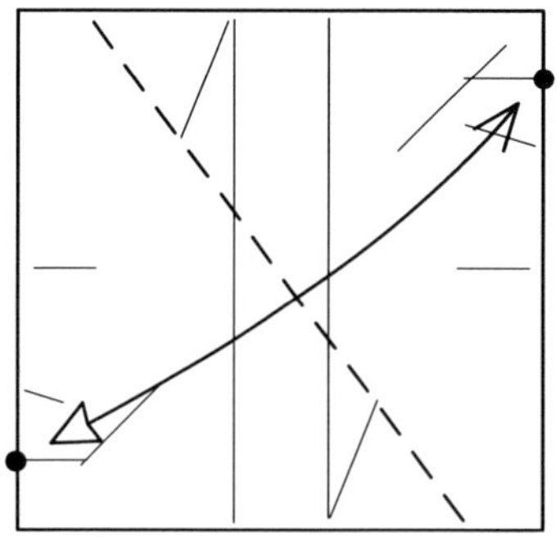

Fold and unfold.

12

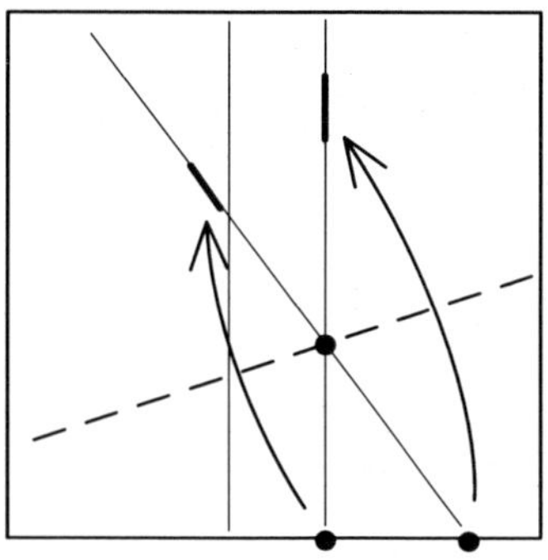

13

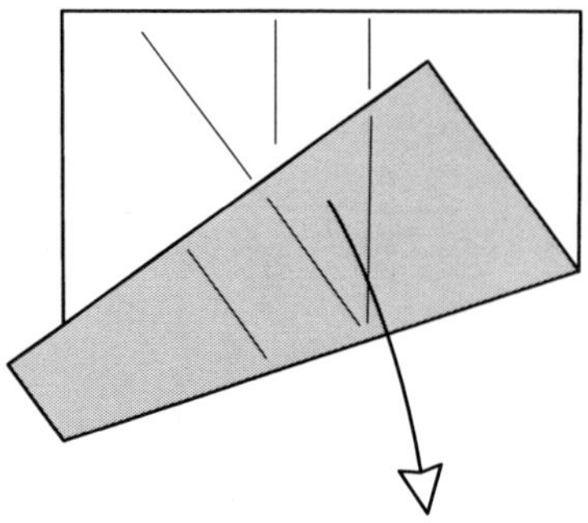

Unfold and rotate 180°.

14

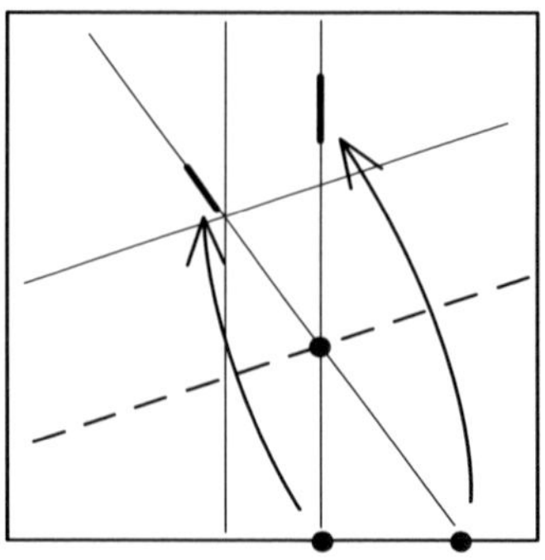

Repeat steps 12–13.

15

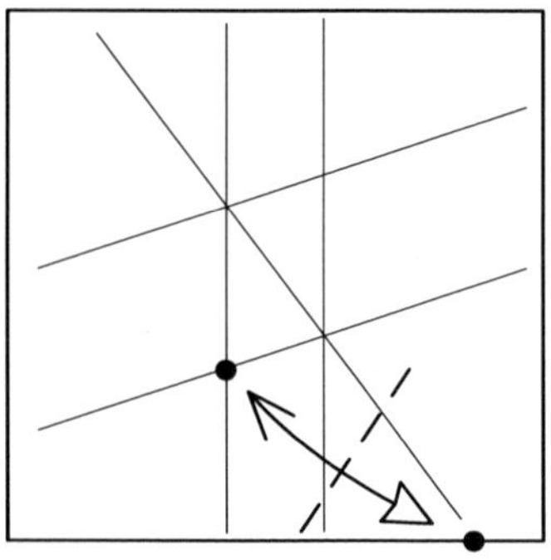

Fold and unfold.

16

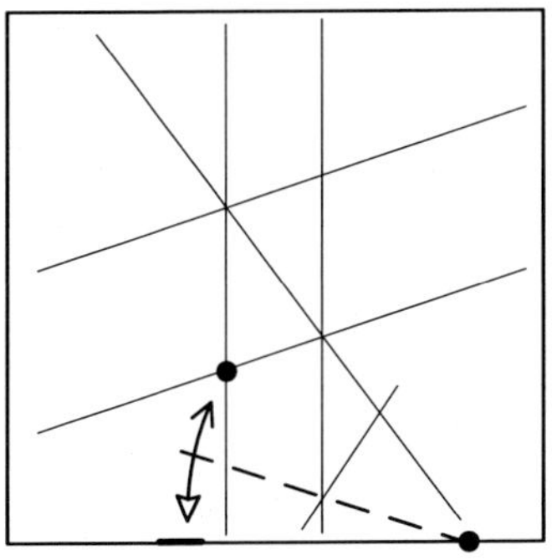

Fold and unfold.
Rotate 180°.

17

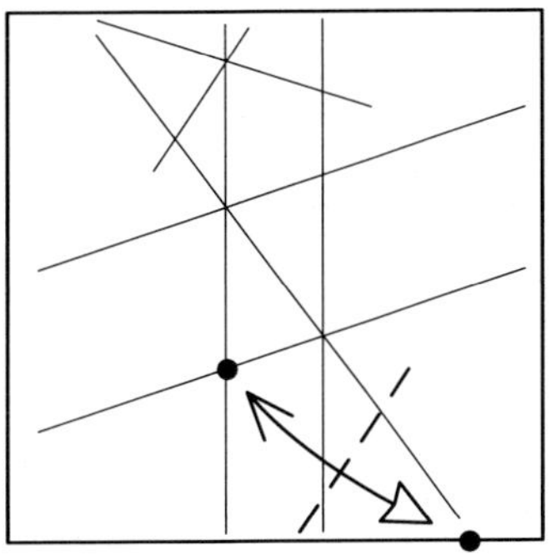

Repeat steps 15–16.

18

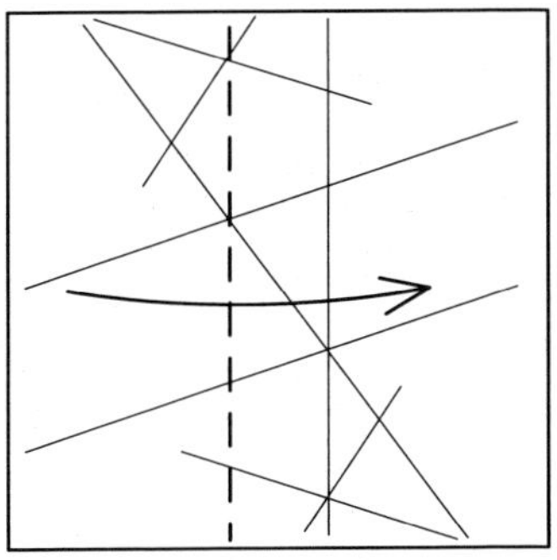

Fold along the crease.

19

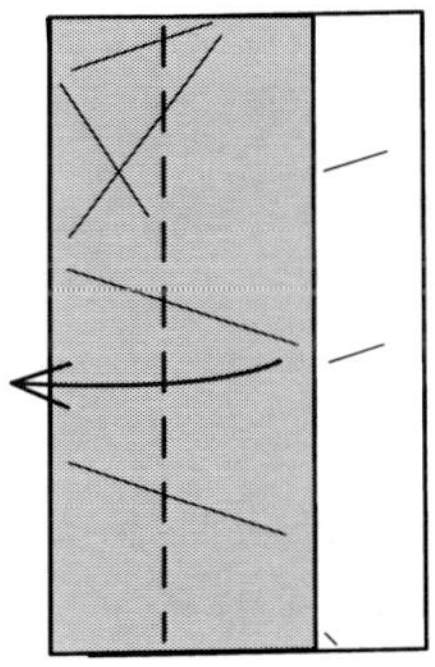

Fold along a
hidden crease.

20

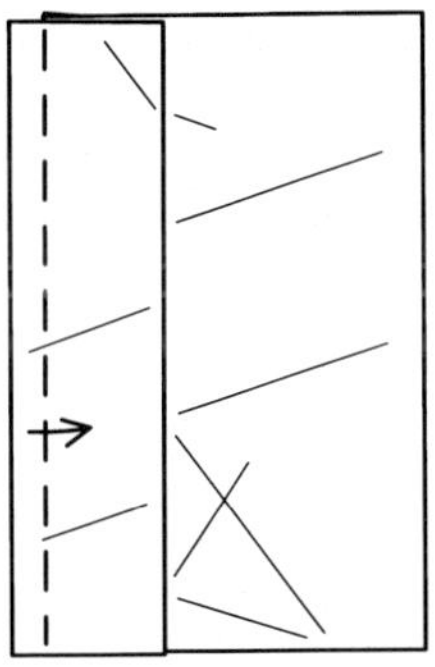

Fold along the
hidden edge.

21

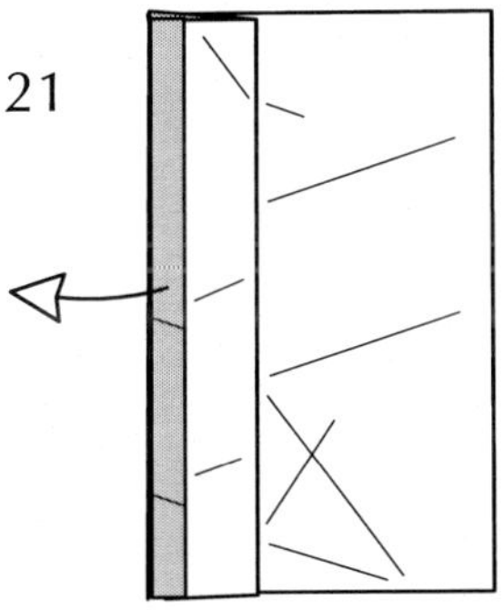

Unfold and rotate 180°.

22

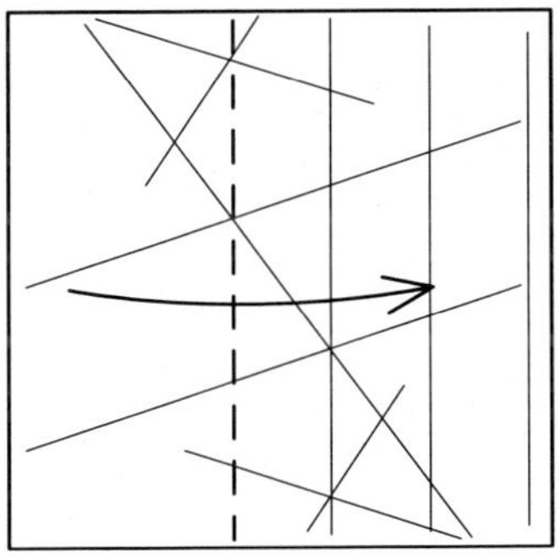

Repeat steps 18–21.

23

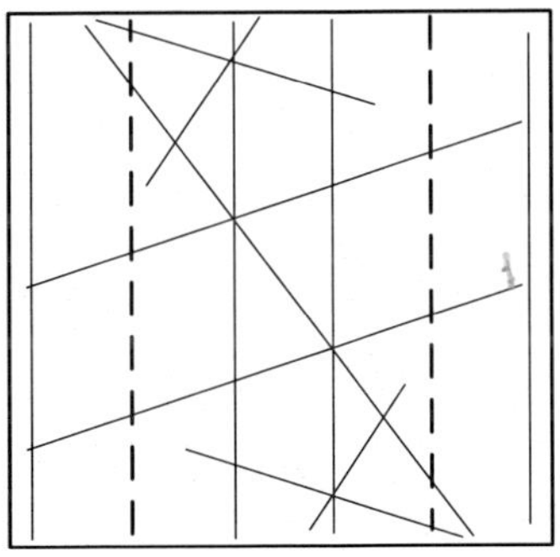

Fold and unfold
along the creases.

24

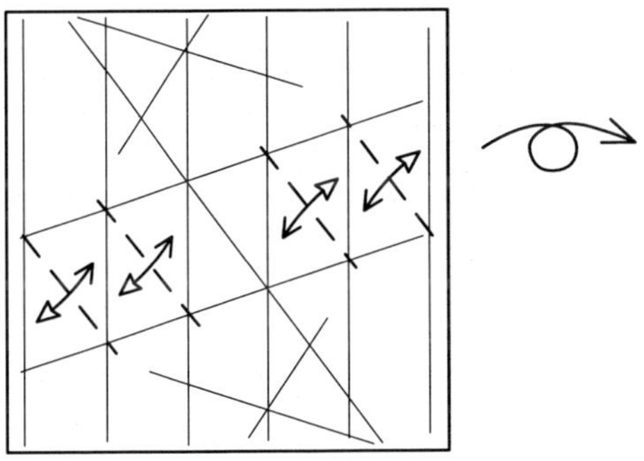

Fold and unfold.

25

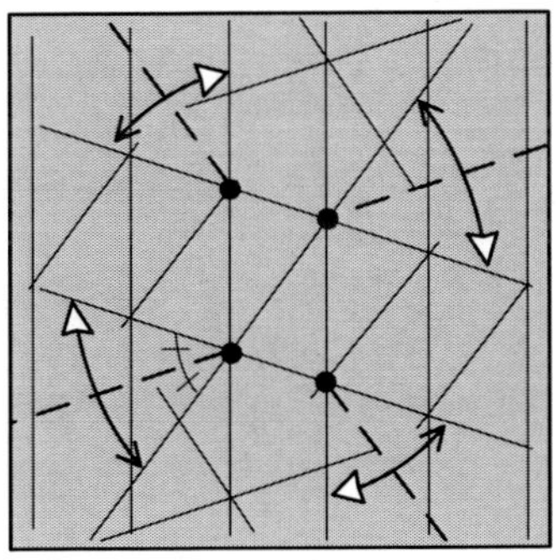

Fold and unfold.

26

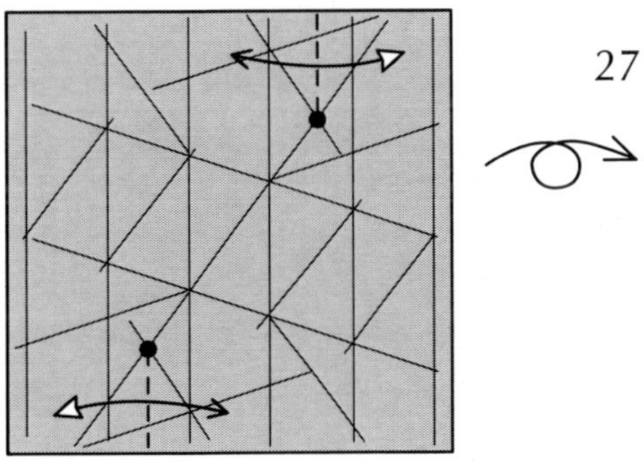

Fold and unfold.

27

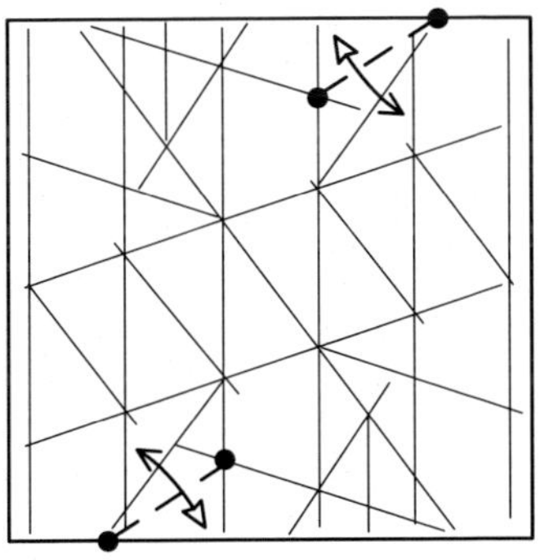

Fold and unfold.

28

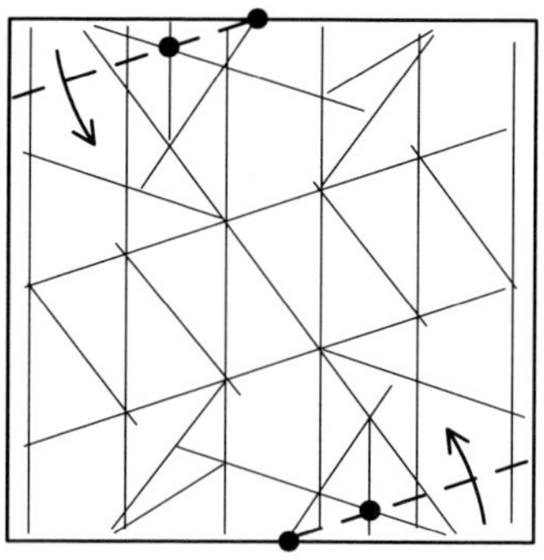

29

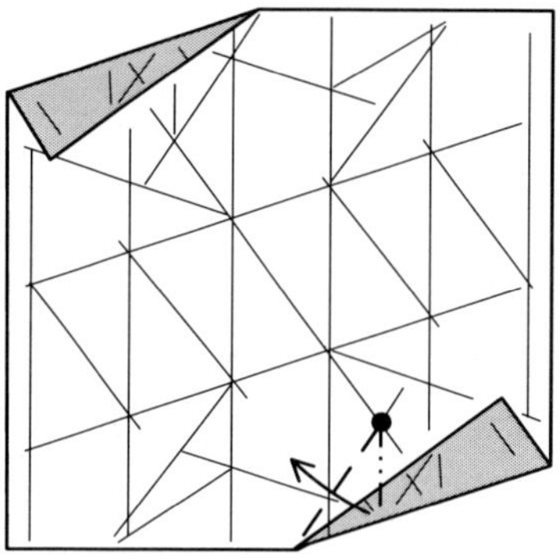

Push in at the dot.

30

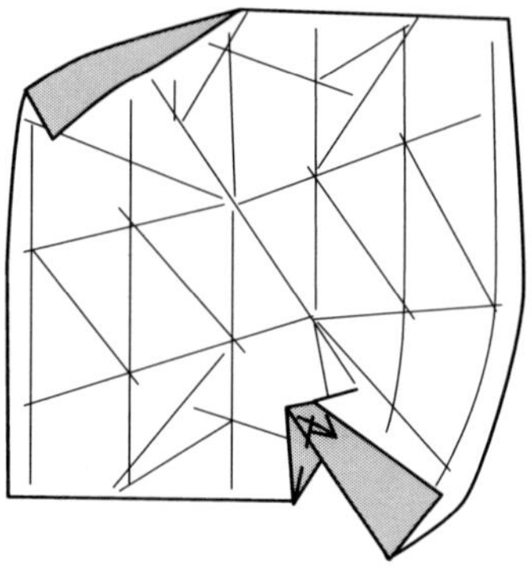

31

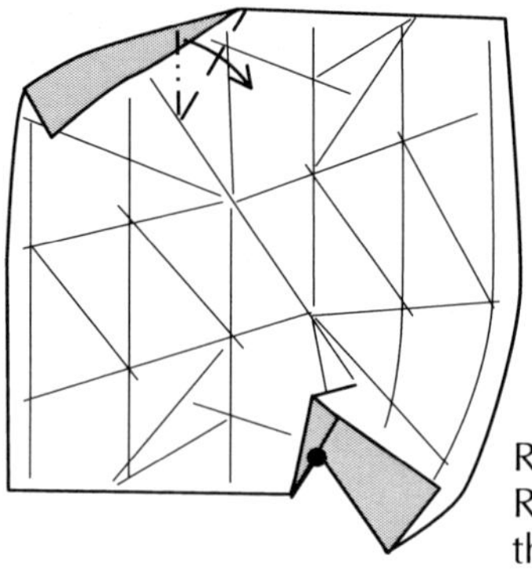

Repeat steps 29–30 above. Rotate to view the outside so the dot is front and center.

32

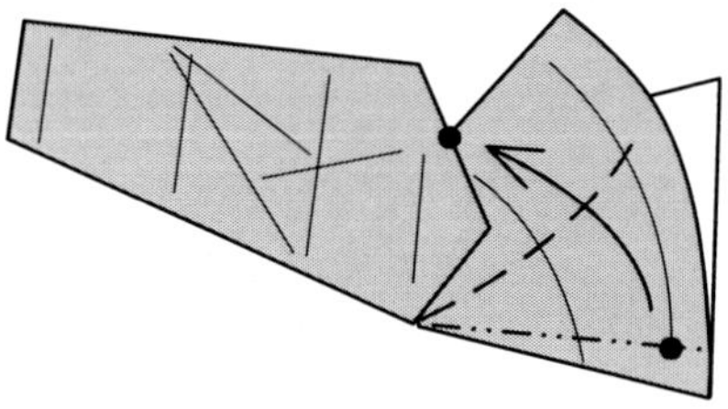

Push in on the valley line. Fold along the creases. Turn over and repeat.

33

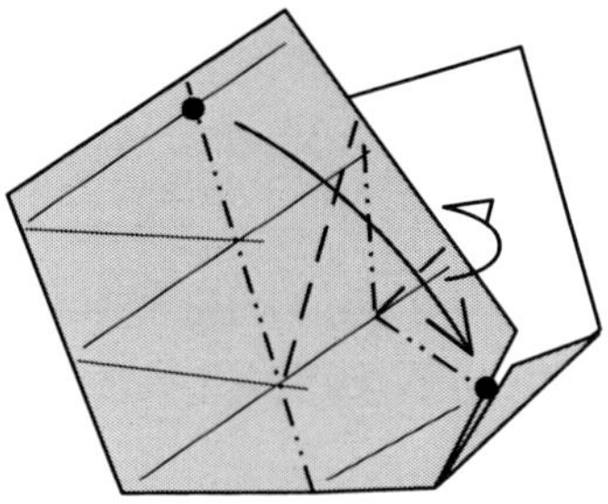

34

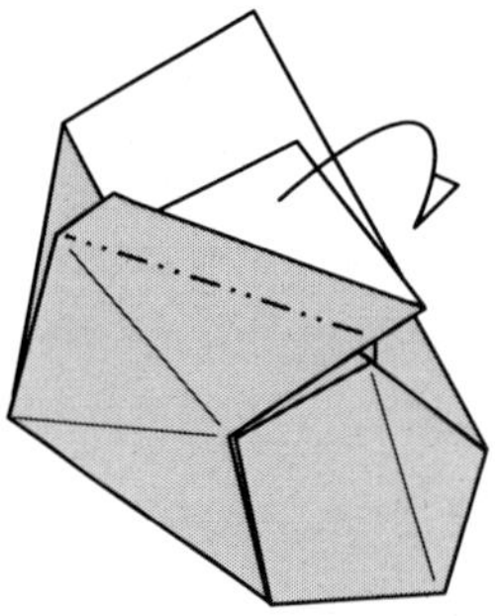

Tuck inside by wrapping around the layers.

35

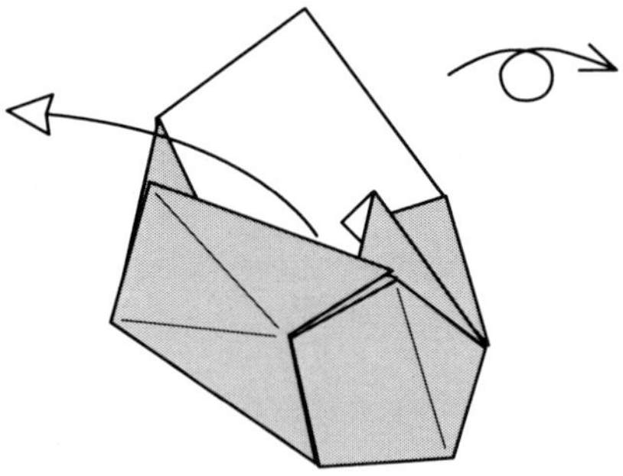

Unfold to step 33.

36

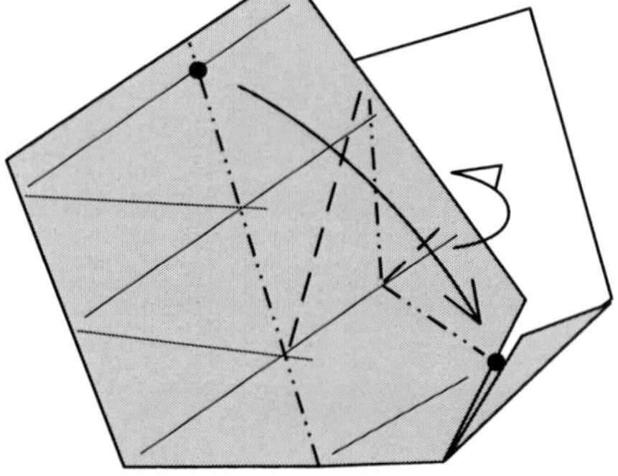

Repeat steps 33–34.

37

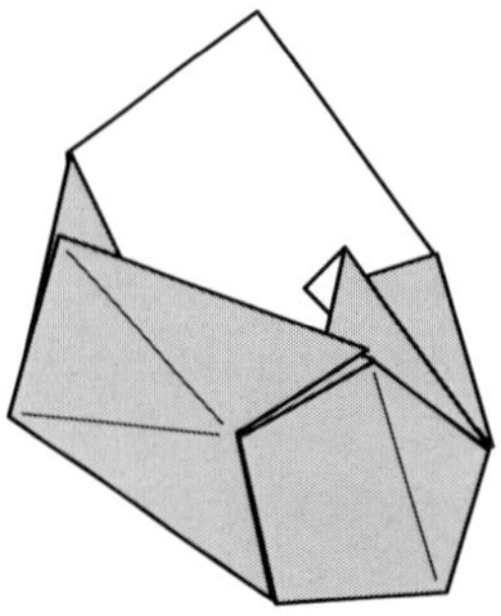

Refold steps 33–34 behind.

38

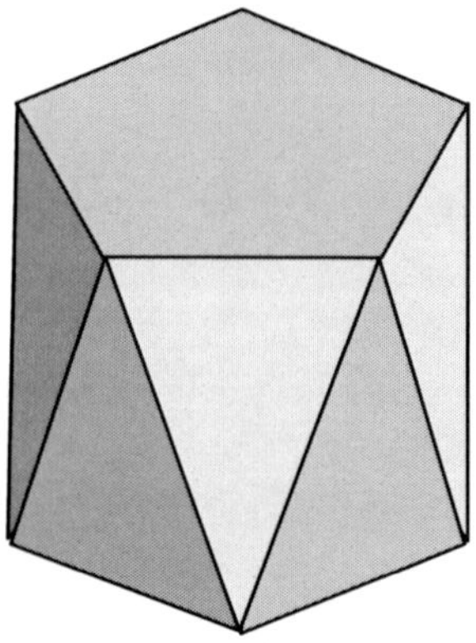

Golden Pentagonal Antiprism

Pentagonal Trapezohedron 36°

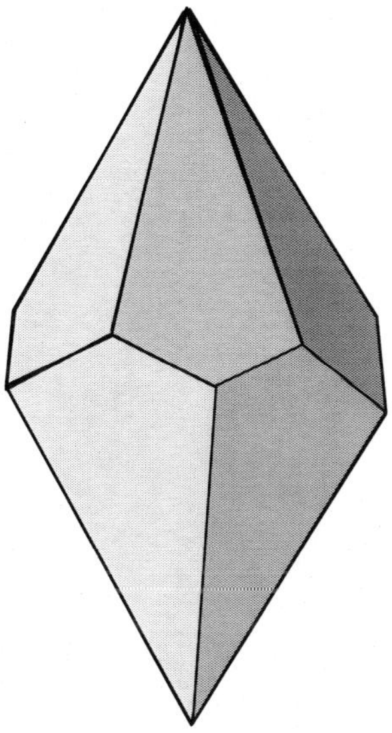

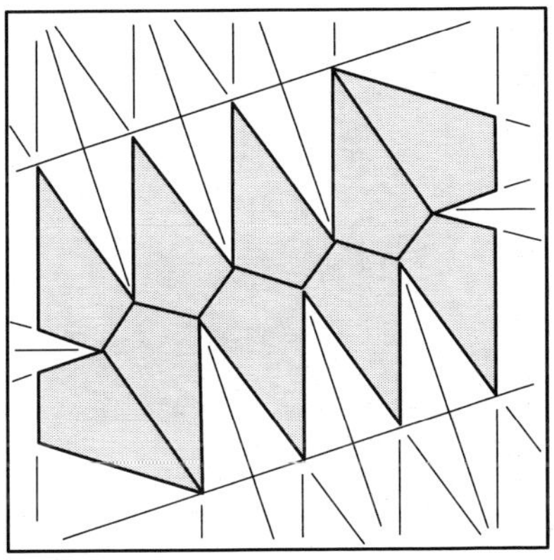

This trapezohedron, or antidiamond, is composed of ten quadrilaterals. In each quadrilateral, three of the angles are 108° and one is 36°. This is the dual of the pentagonal antiprism with equilateral triangles.

1

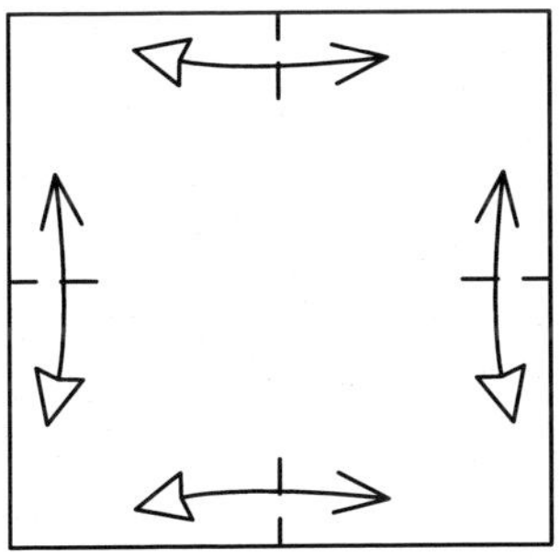

Fold and unfold at the edges.

2

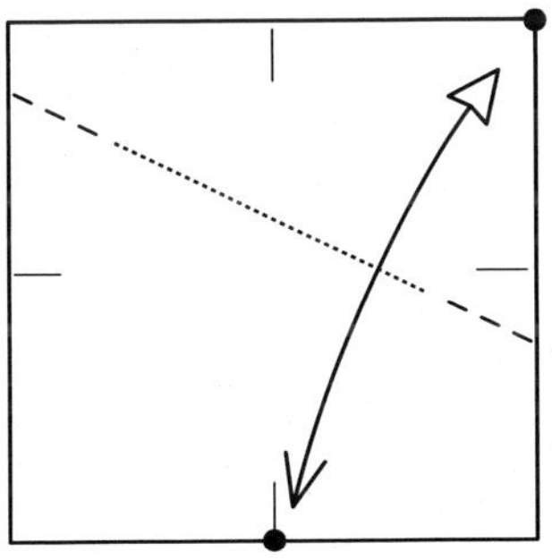

Fold and unfold on the edges.

3

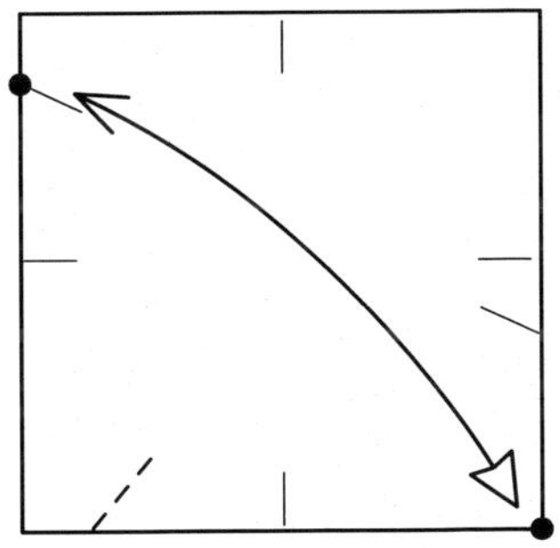

Fold and unfold at the bottom.

4

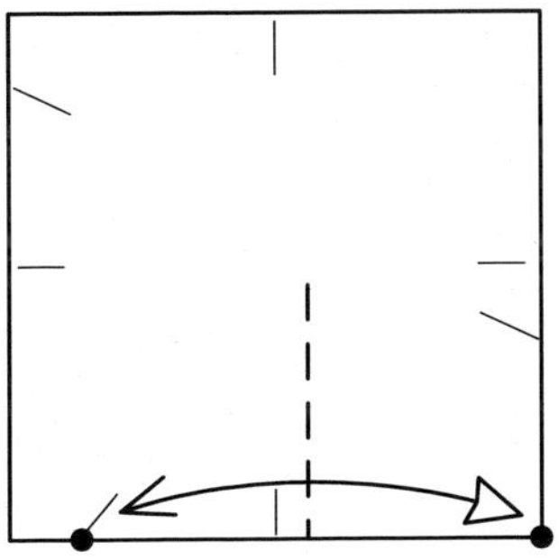

Fold and unfold. Rotate 180°.

5

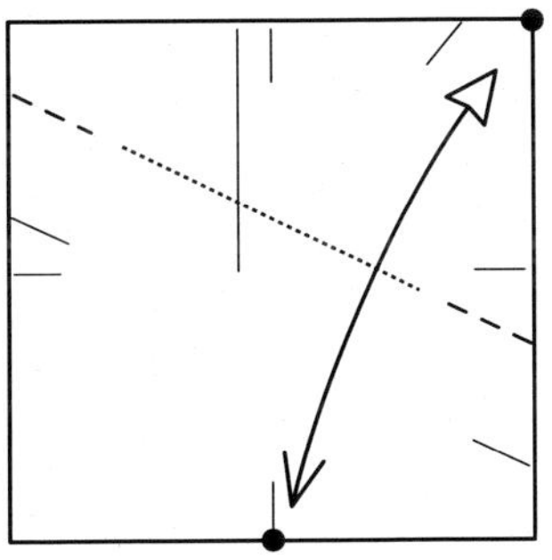

Repeat steps 2–4.

6

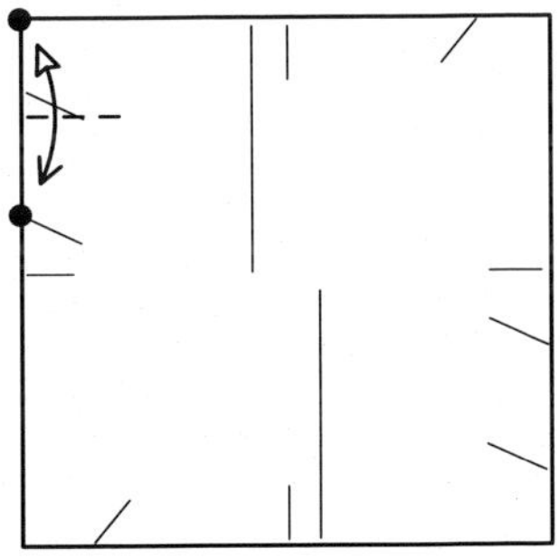

Fold and unfold on the left.

7

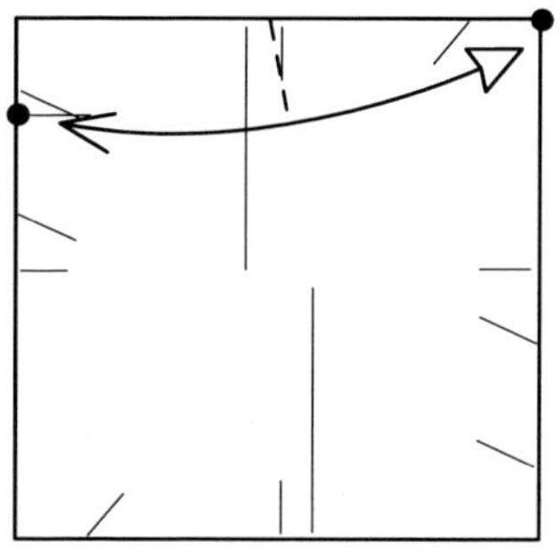

Fold and unfold
at the top.

8

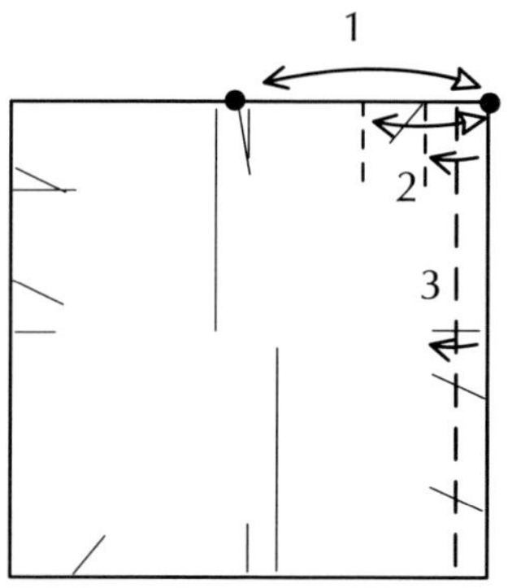

1 and 2. Fold and unfold
twice at the top.
3. Fold a thin strip.

9

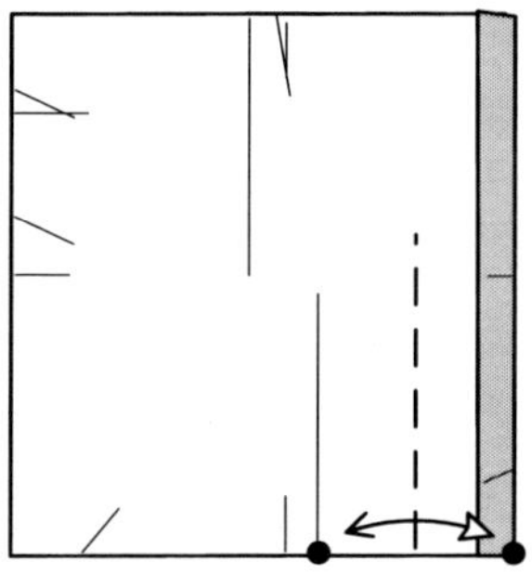

Fold and unfold
at the bottom.

10

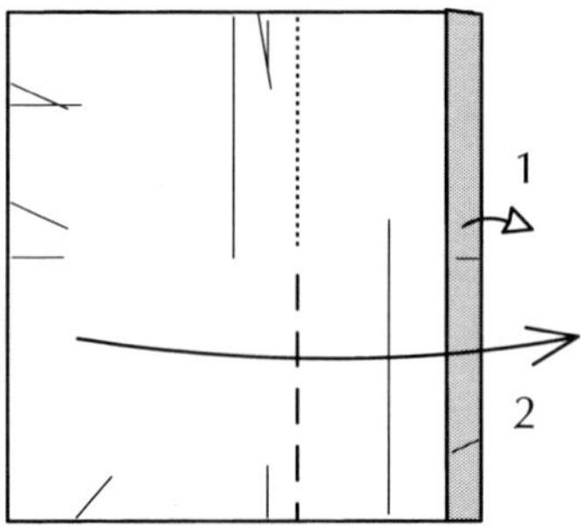

1. Unfold.
2. Fold along the crease.

11

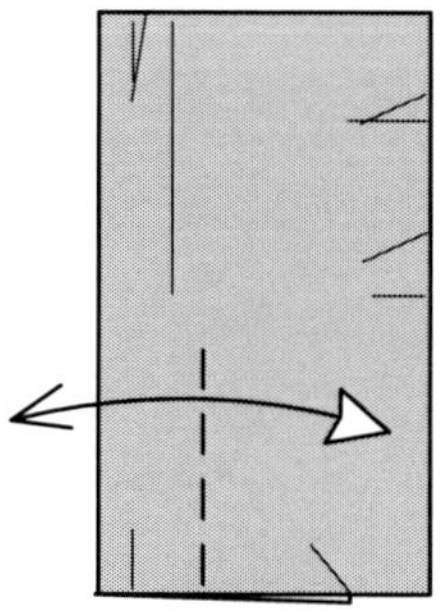

Fold and unfold along
a hidden crease.

12

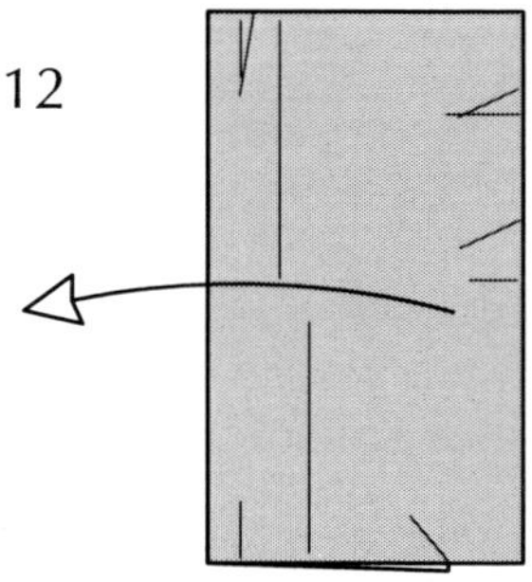

Unfold and rotate 180°.

13

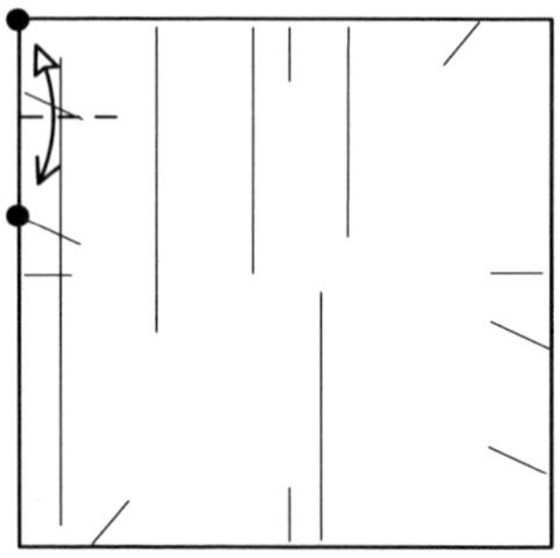

Repeat steps 6–12.

14

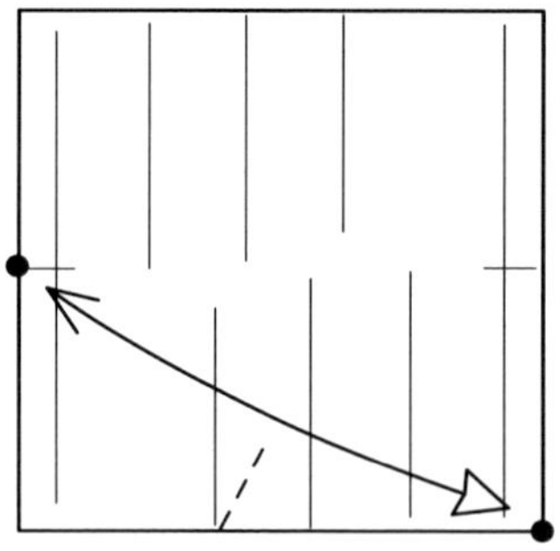

Fold and unfold
at the bottom.

15

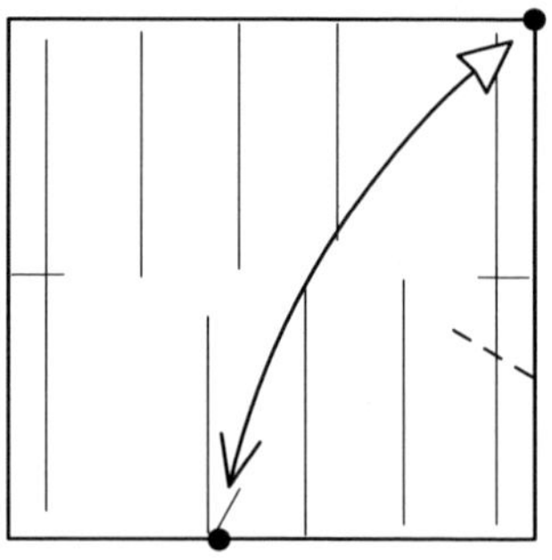

Fold and unfold
on the right.

16

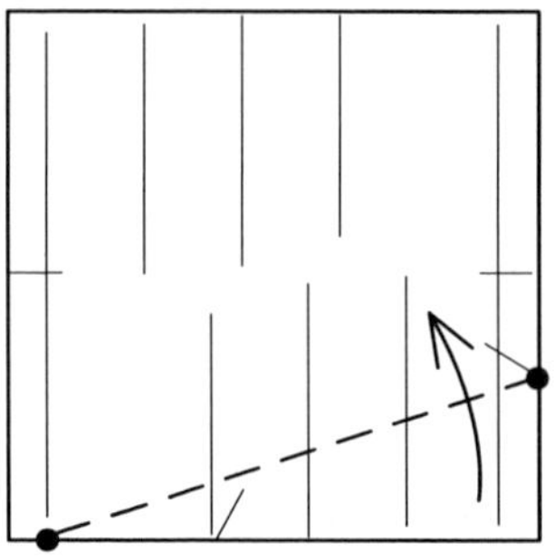

Rotate 180°.

17

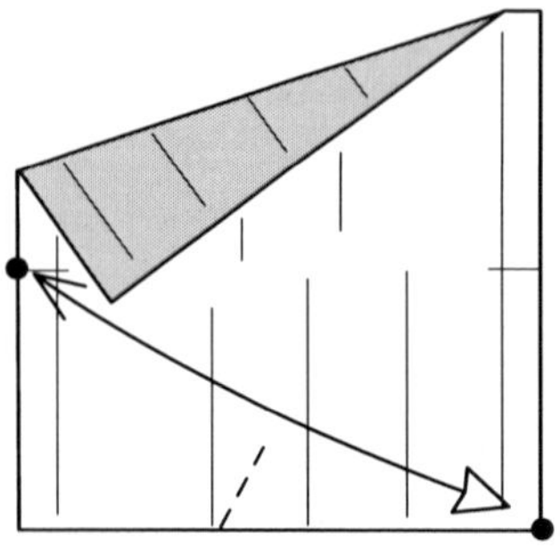

Repeat steps 14–16.

18

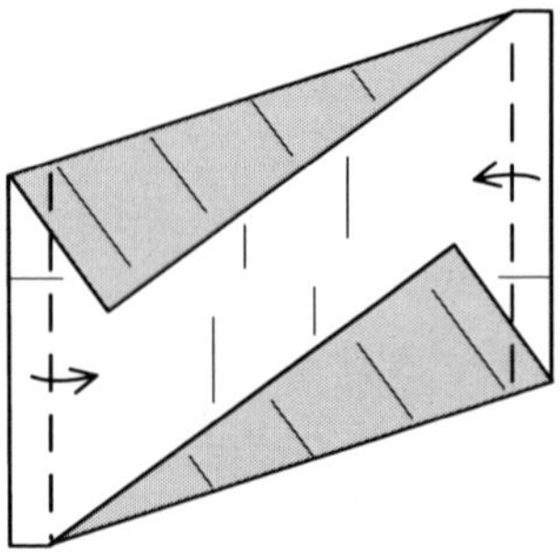

Fold along the creases.

19

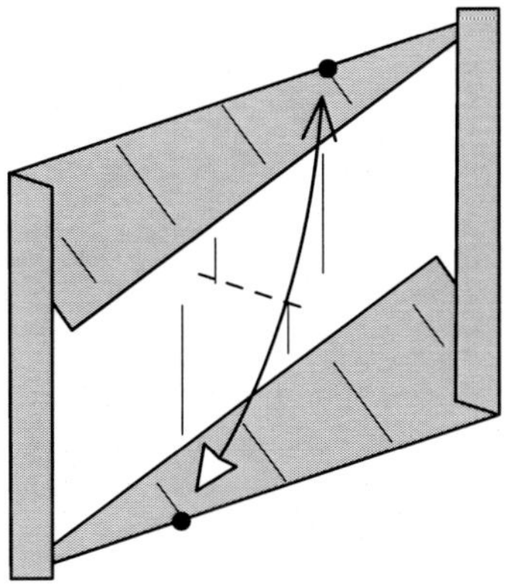

Fold and unfold in the center.

20

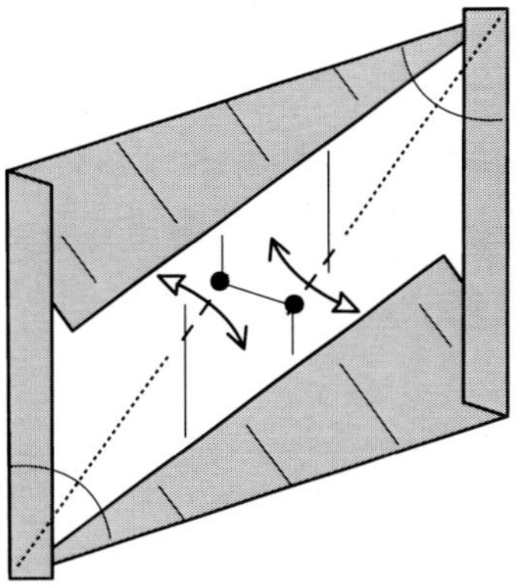

Fold and unfold in the center. Bisect the angles.

21

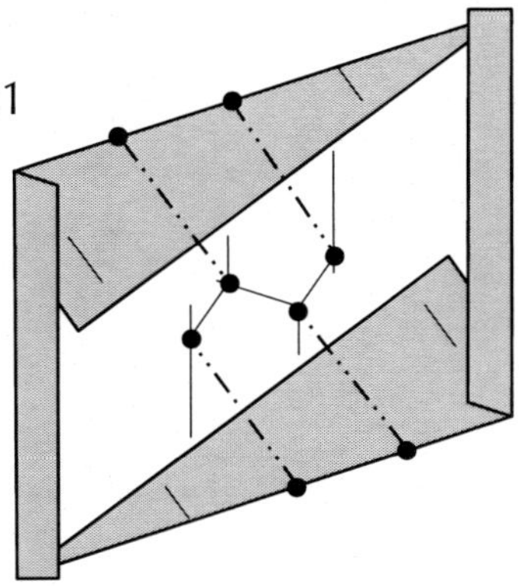

Fold and unfold along creases which start at the lower and upper dots.

22

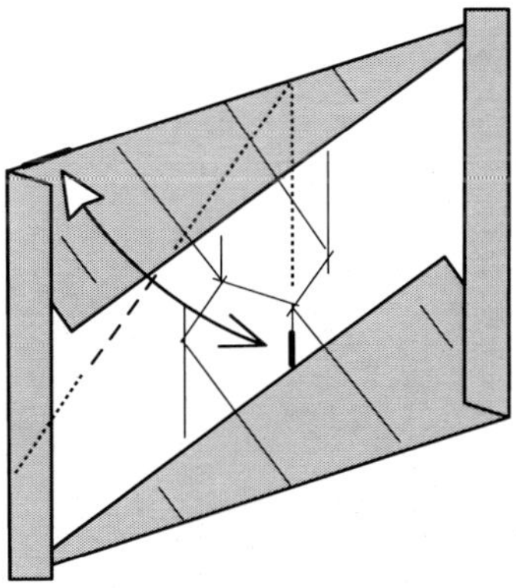

Bring the bold edge to the vertical line. Fold and unfold in the center.

23

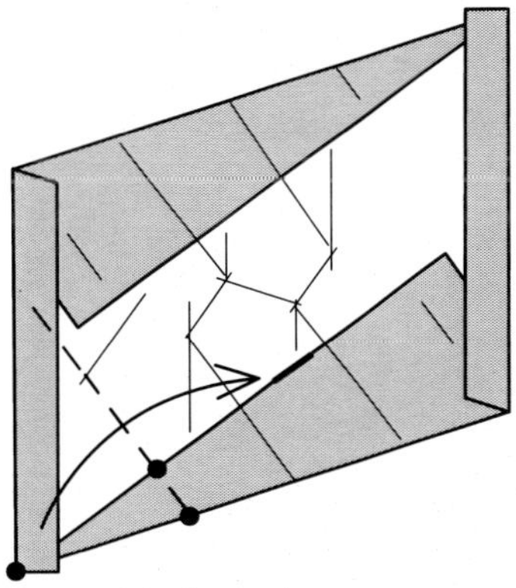

Fold along the crease by the right dots.

24

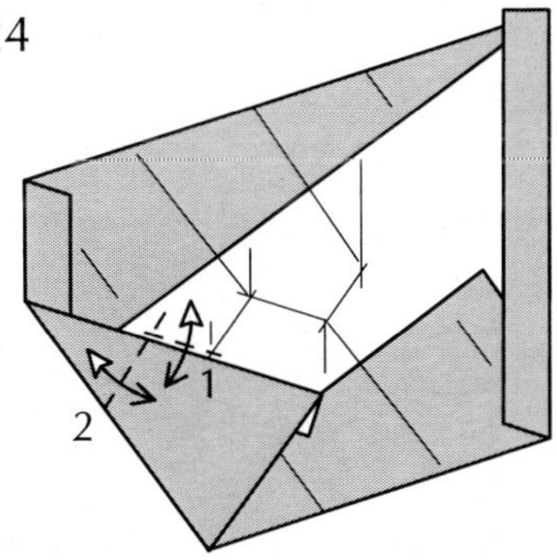

1. Fold and unfold in the center.
2. Fold and unfold all the layers along a hidden crease.

25

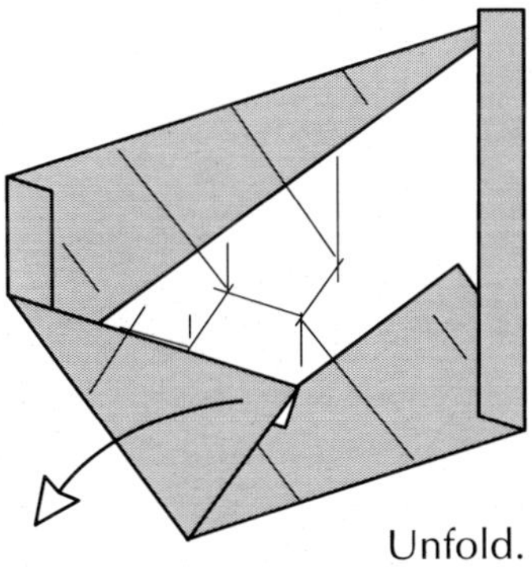

Unfold.

26

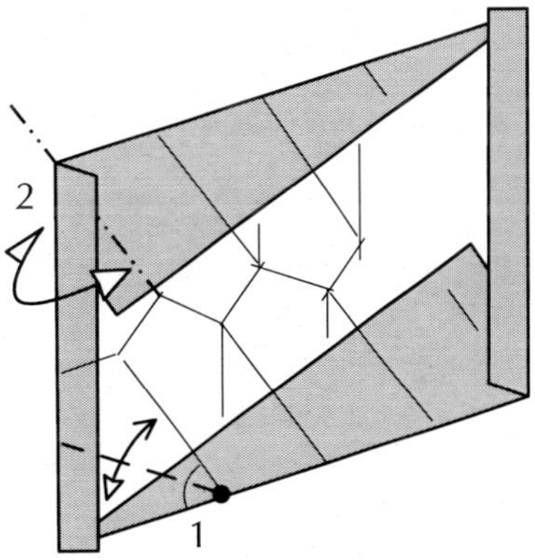

1. Fold and unfold.
2. Fold and unfold along the crease.

Rotate 180°.

27

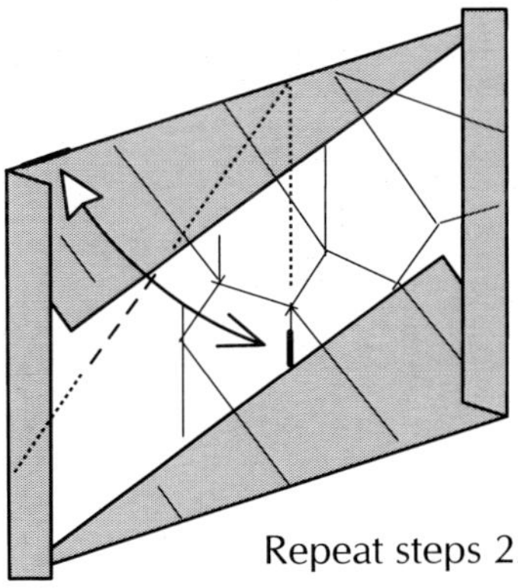

Repeat steps 22–26.

28

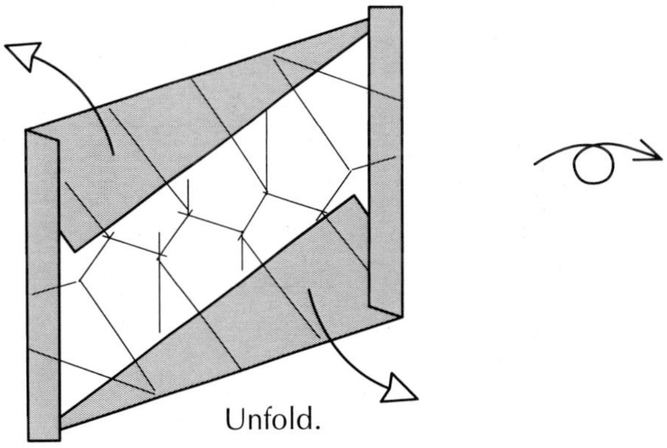

Unfold.

29

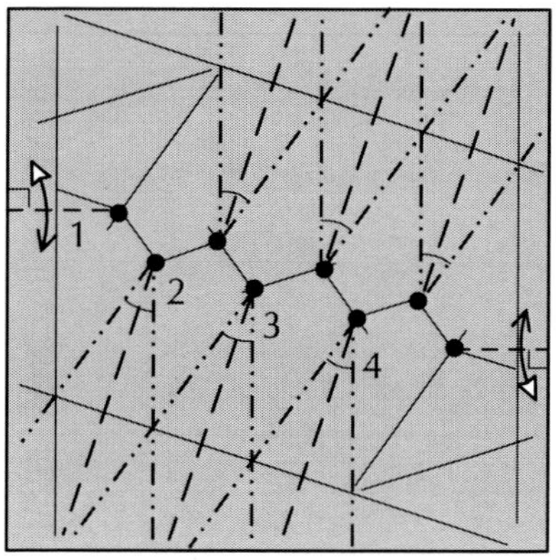

1. Fold and unfold.
2, 3, and 4. Fold and unfold to bisect the angles. Mountain-fold along the creases and extend some of them.
Rotate 90°.

30

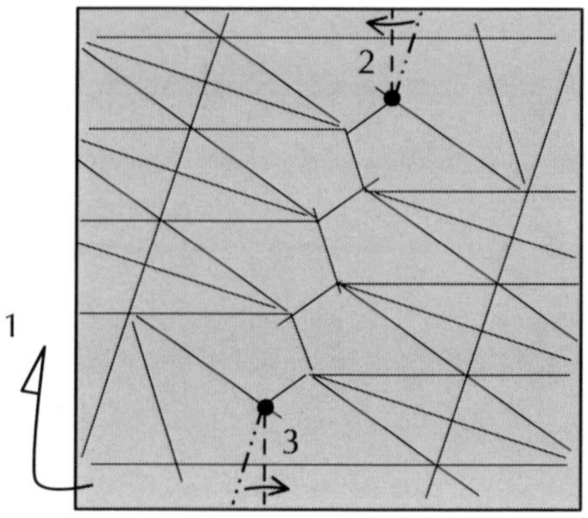

1. Begin to wrap the paper behind.
2. Puff out at the dot.
3. Turn over and repeat.

31

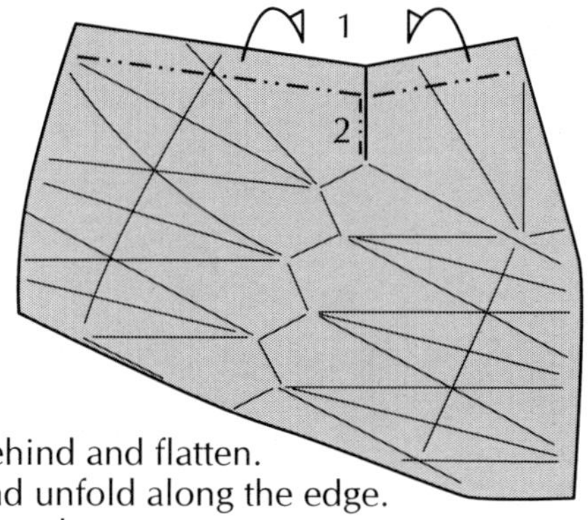

1. Fold behind and flatten.
2. Fold and unfold along the edge.
Turn over and repeat.

32

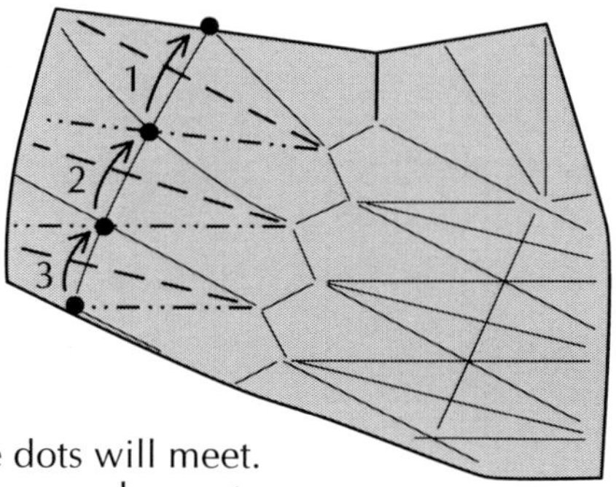

All the dots will meet.
Turn over and repeat.

33

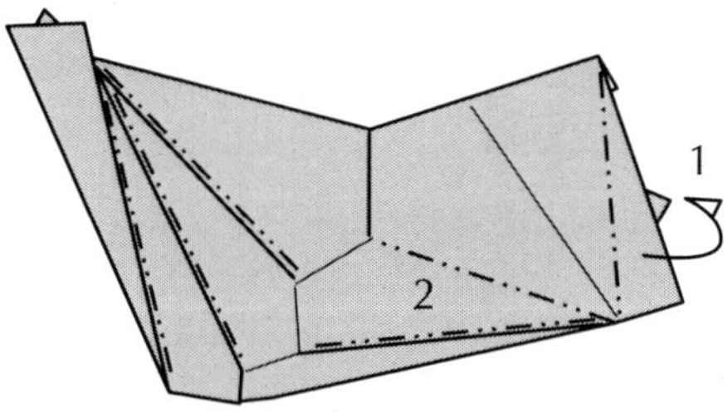

1. Fold behind.
2. Fold and unfold along the long edges.
Turn over and repeat.

34

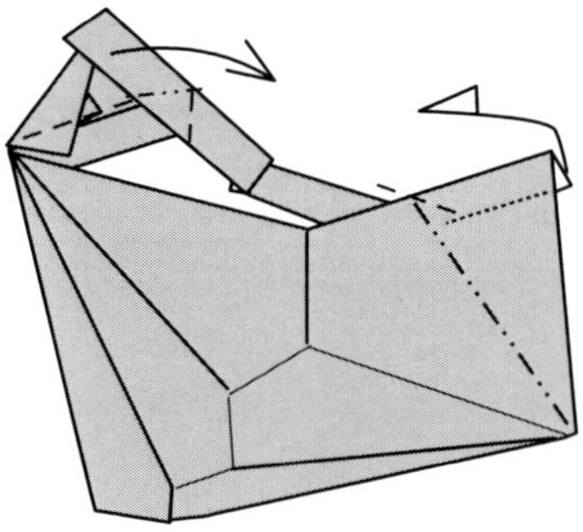

Squash-fold on both sides.

35

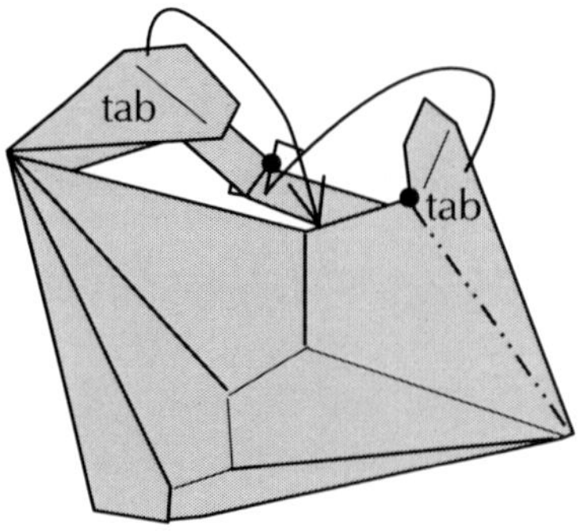

Slide the tabs through the pockets. The dots will meet.

36

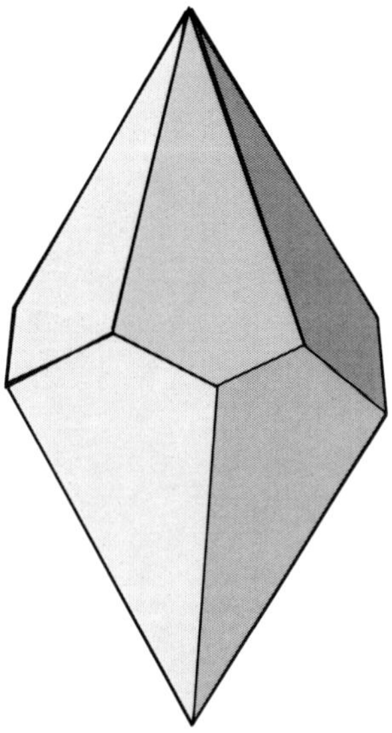

Pentagonal
Trapezohedron 36°

Dodecahedron

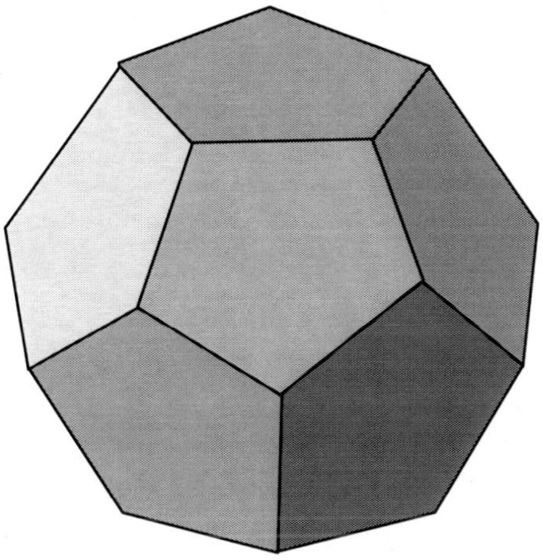

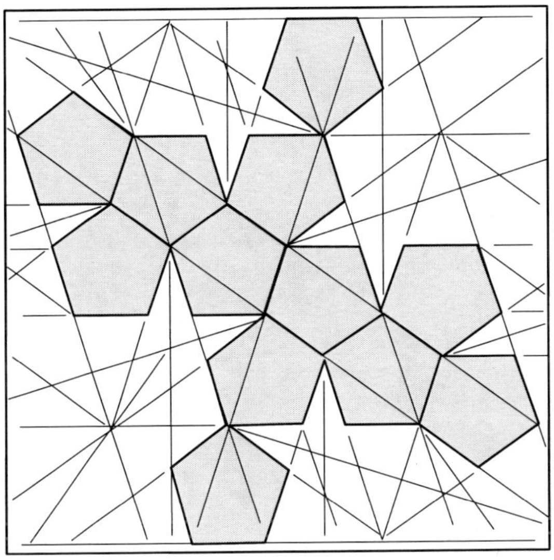

The dodecahedron has twelve pentagonal faces. To Plato, the dodecahedron, the quintessence (the "fifth being"), represented the whole universe.

This was a very difficult model to design. The crease pattern shows odd symmetry.

1

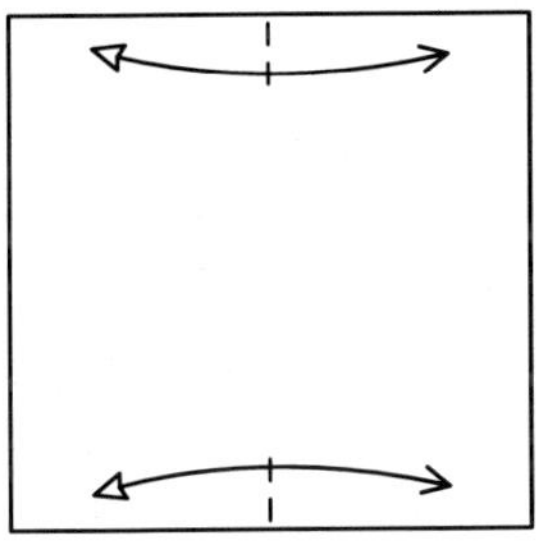

Fold and unfold at the top and bottom.

2

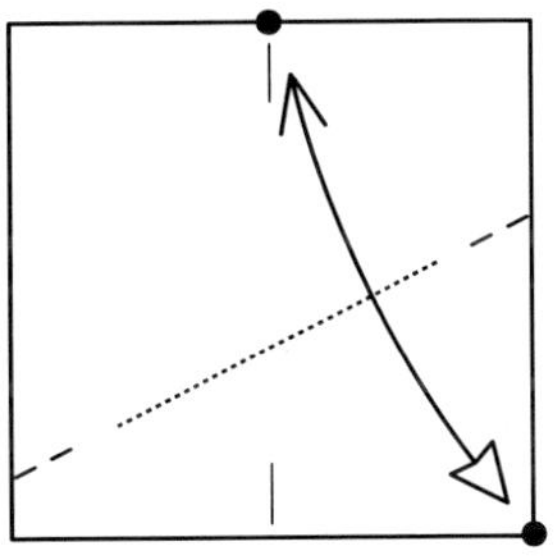

Fold and unfold on the left and right. Rotate 180°.

3

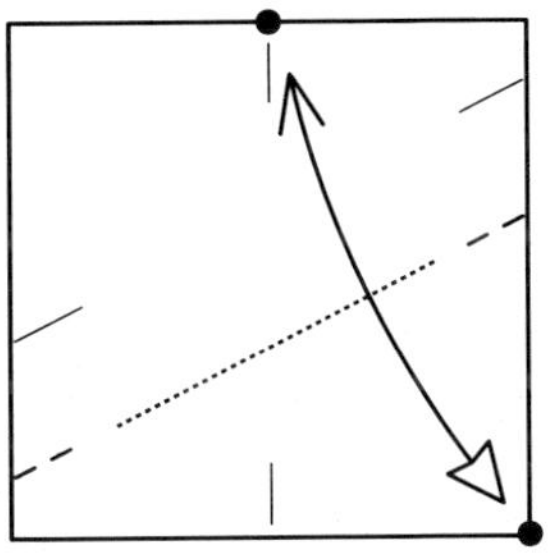

Repeat step 2.

4

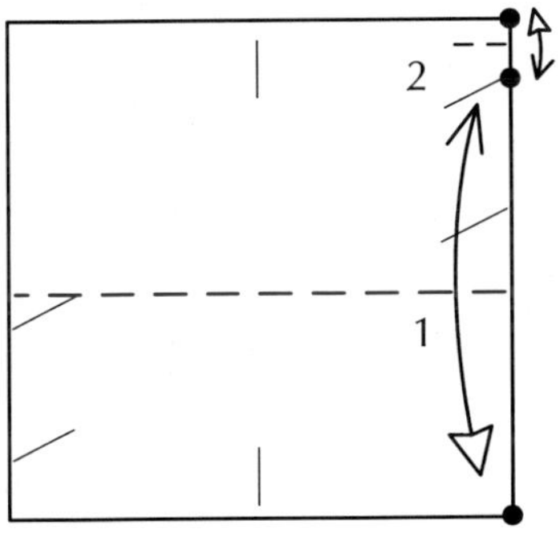

1. Fold and unfold.
2. Fold and unfold part way.
Rotate 180°.

5

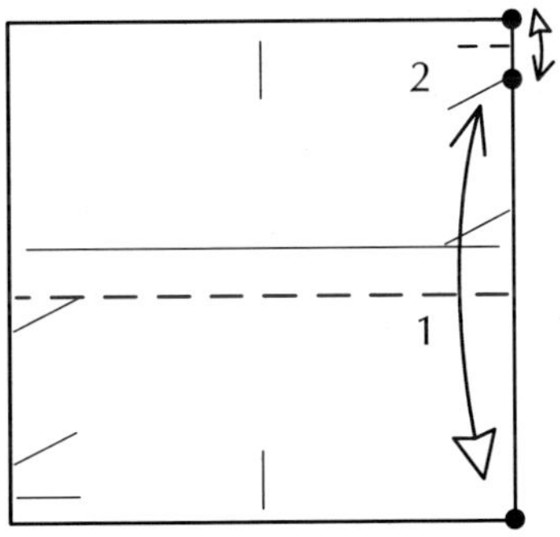

Repeat step 4.

6

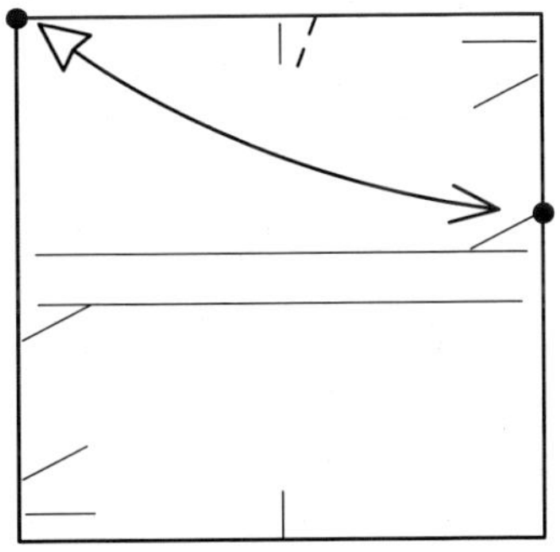

Fold and unfold at the top.

7

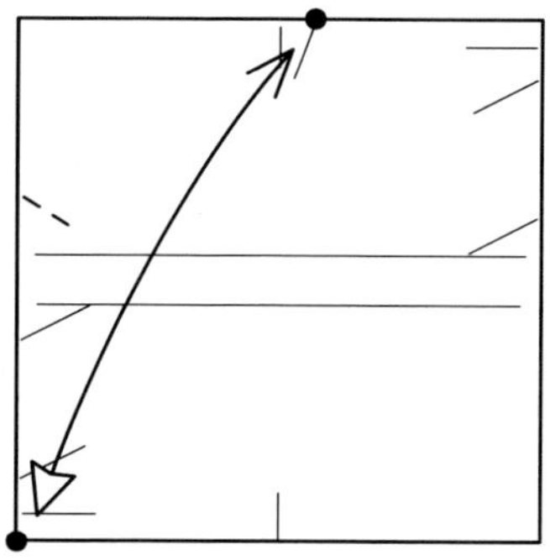

Fold and unfold on the left. Rotate 180°.

8

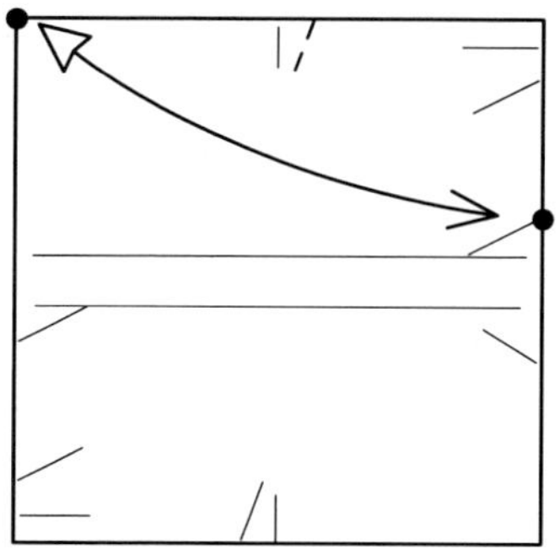

Repeat steps 6–7.

9

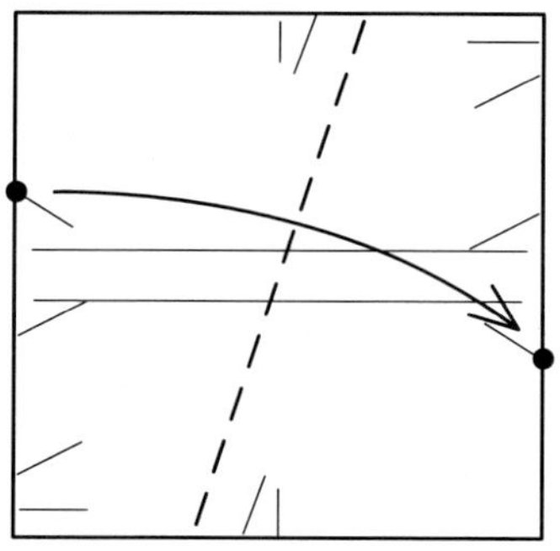

The dots will meet.

10

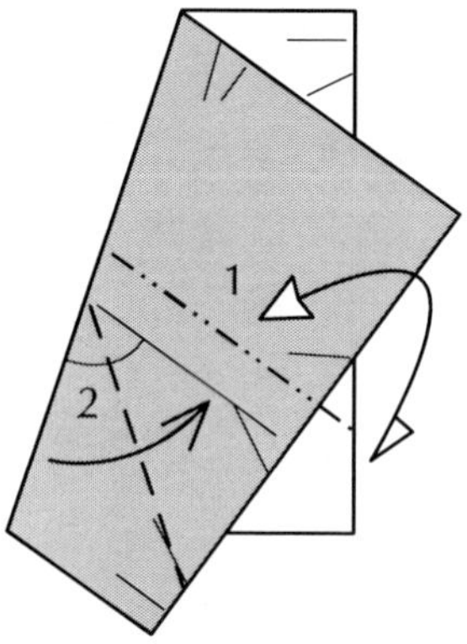

1. Fold and unfold along the crease.
2. Fold to the crease.

11

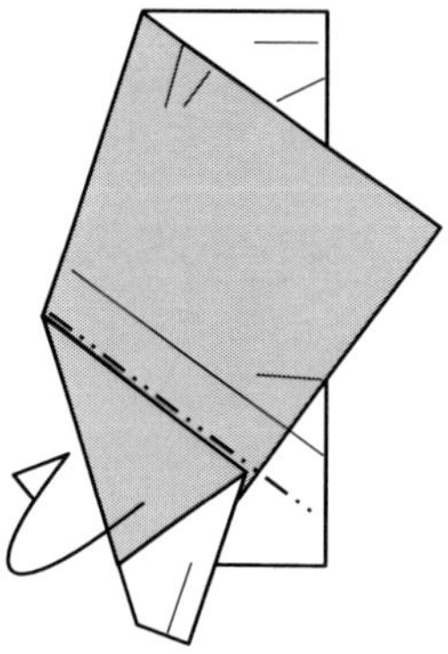

Fold along the crease.

12

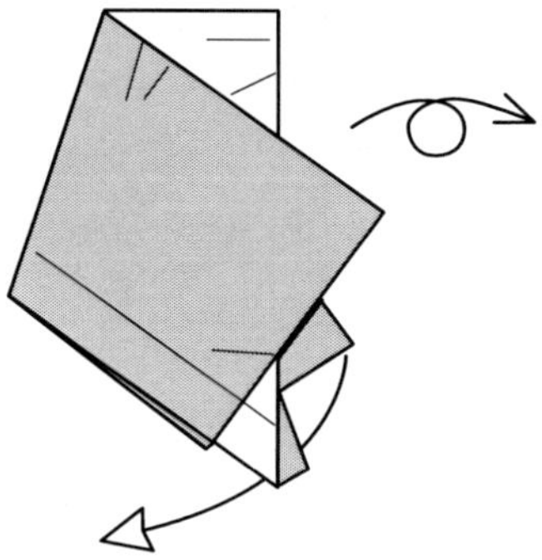

Unfold to step 10.

13

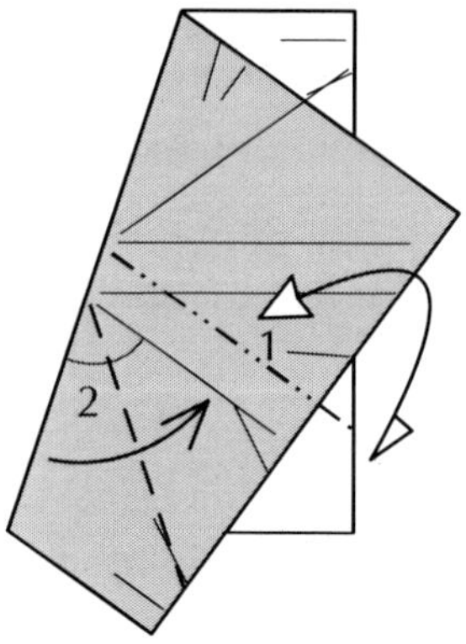

Repeat steps 10–12.

14

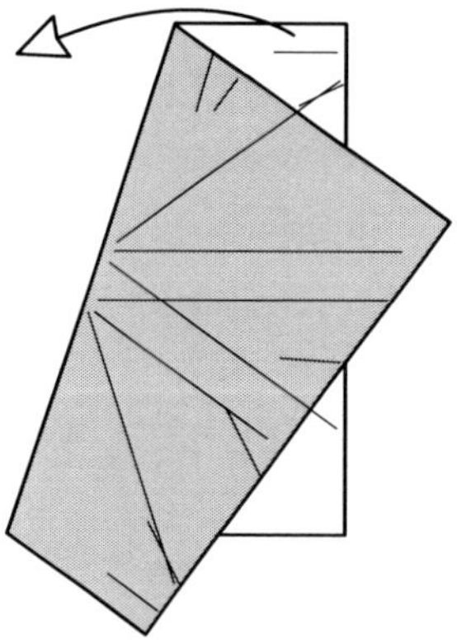

Unfold and rotate.

15

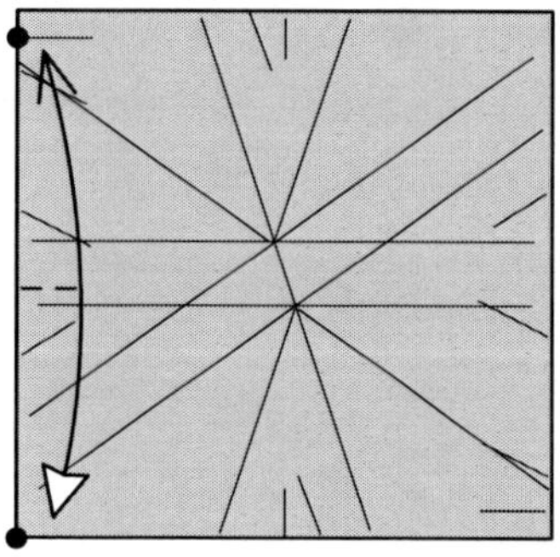

Fold and unfold on the left.

16

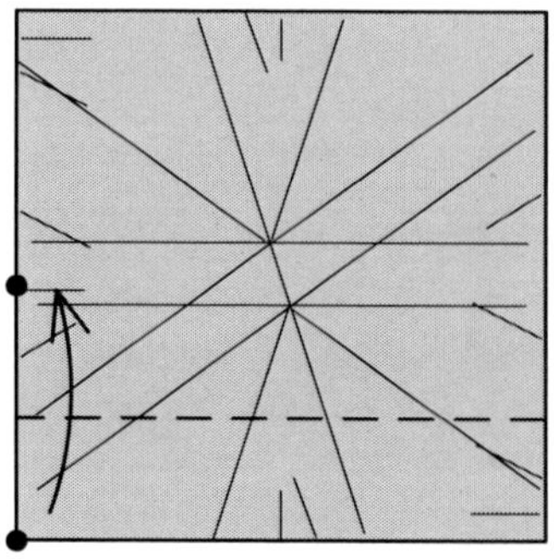

Fold up to the crease.

17

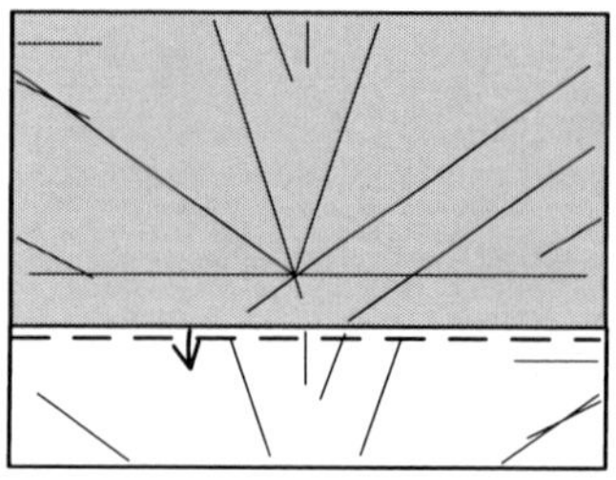

Fold a thin strip along the hidden crease.

18

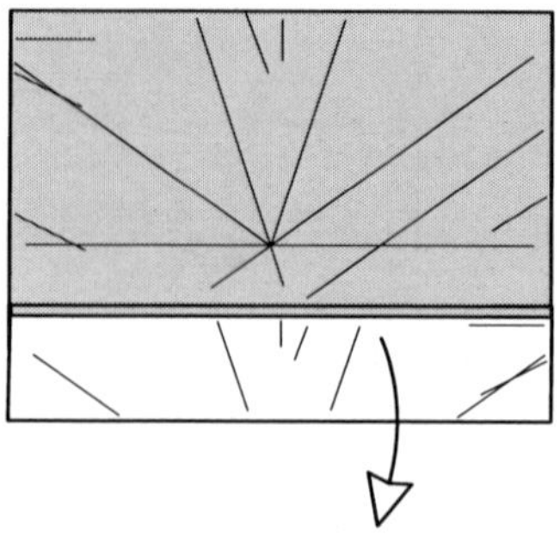

Unfold and rotate 180°.

19

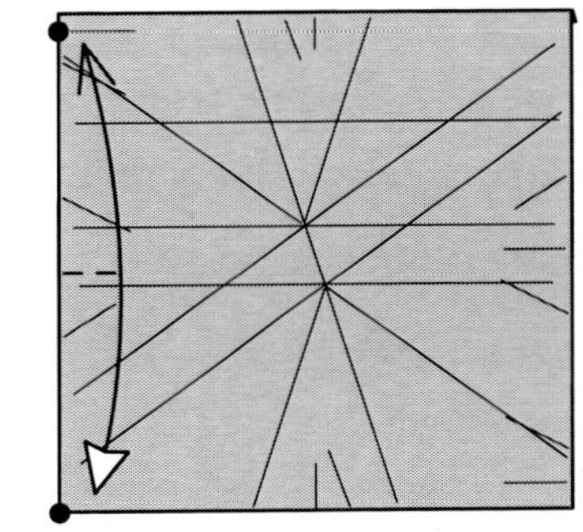

Repeat steps 15–18.

20

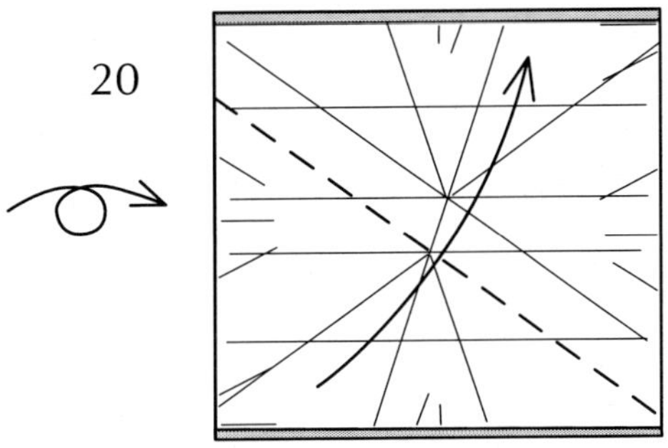

Fold along the crease.

21

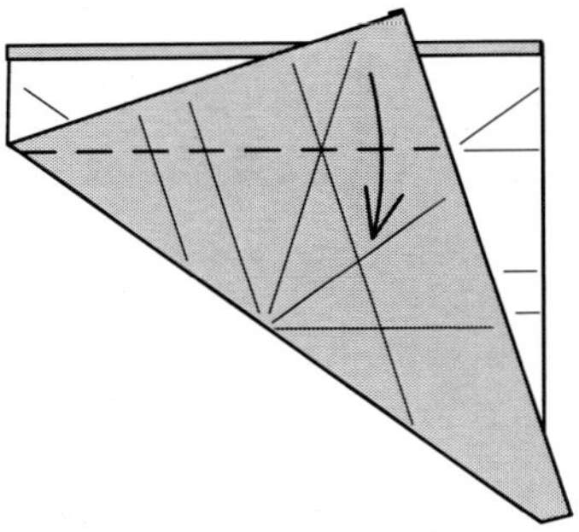

Fold along a hidden crease.

22

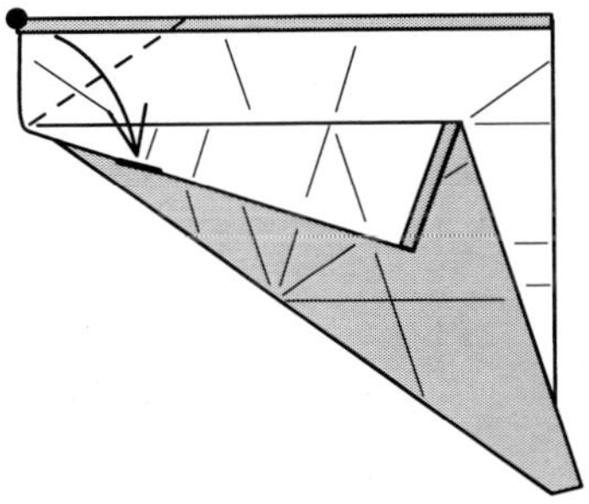

This is 3D. Flatten.

23

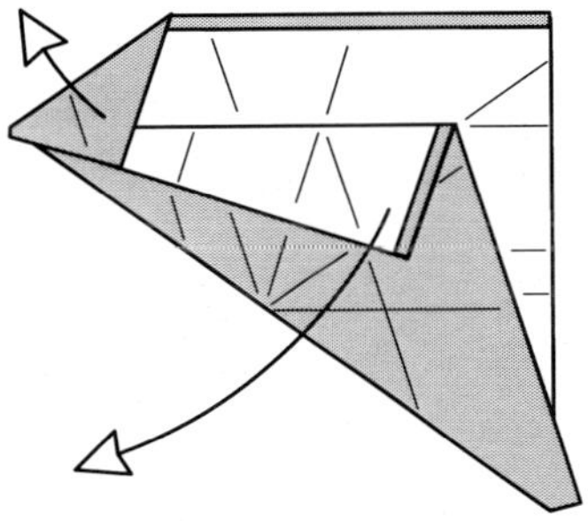

Unfold and rotate 180°.

24

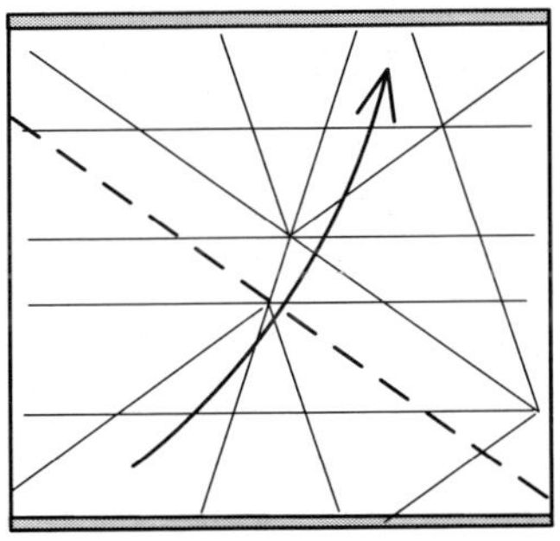

Repeat steps 20–23.

25

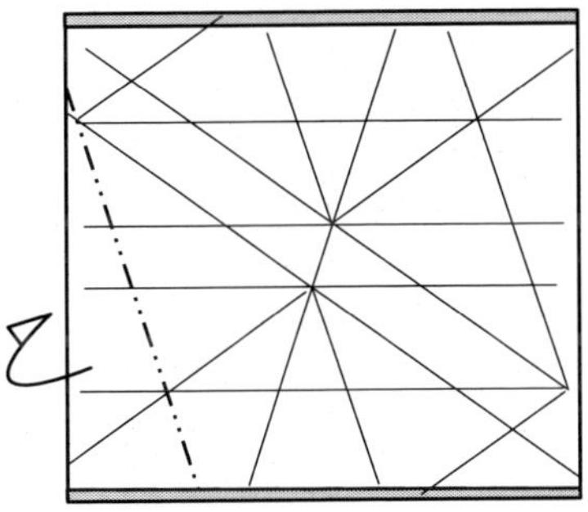

Fold behind
along the crease.

26

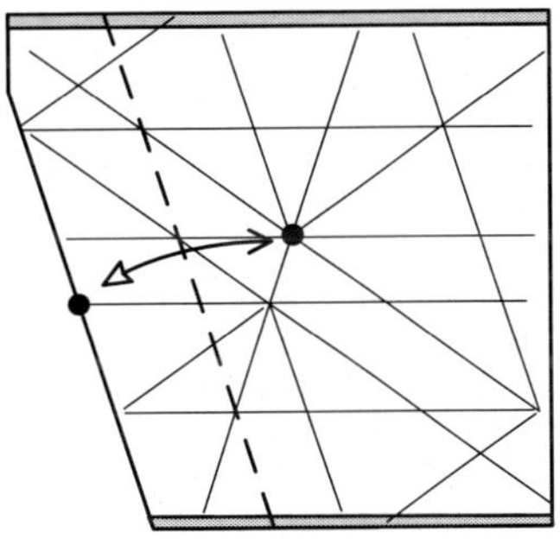

Fold and unfold.

27

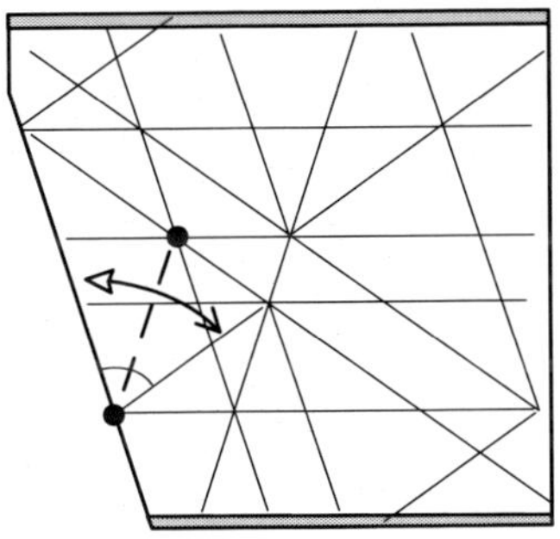

Fold and unfold.

28

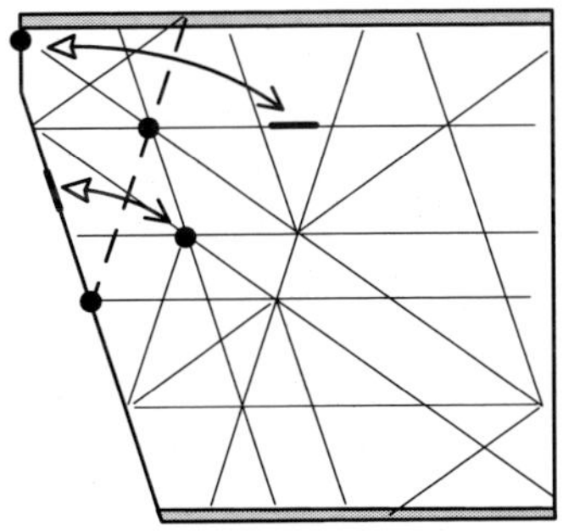

Fold and unfold.

29

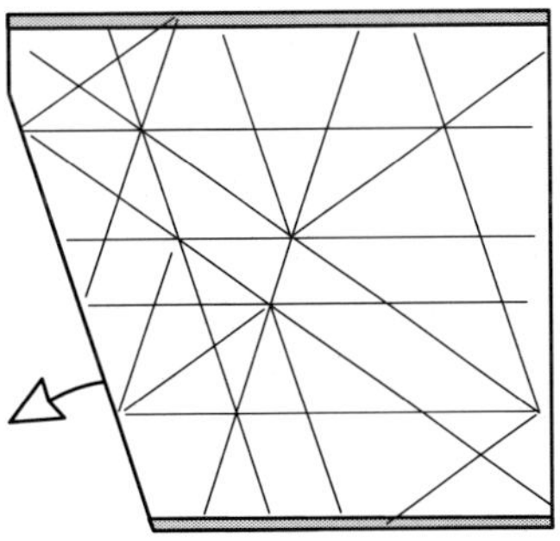

Unfold and rotate 180°.

30

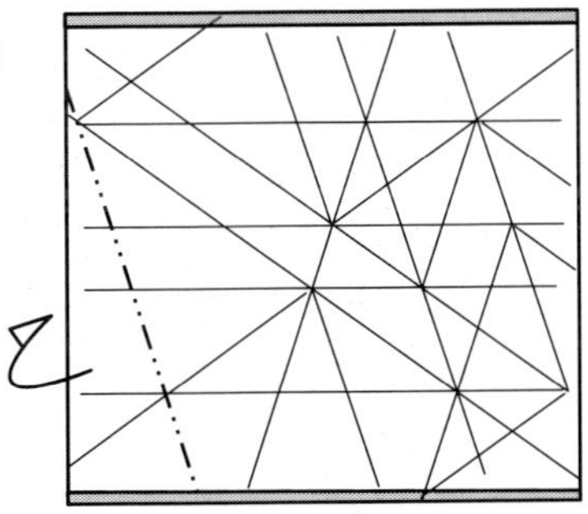

Repeat steps 25–29.

31

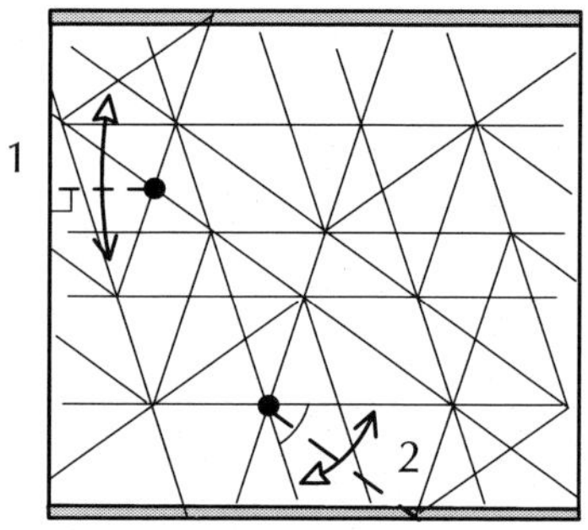

Fold and unfold.

32

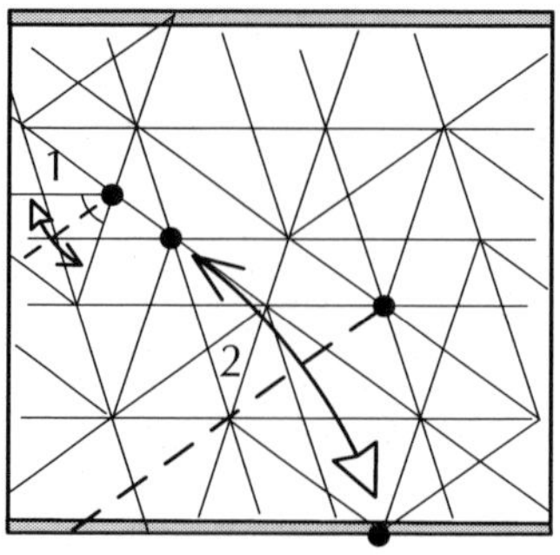

Fold and unfold.

33

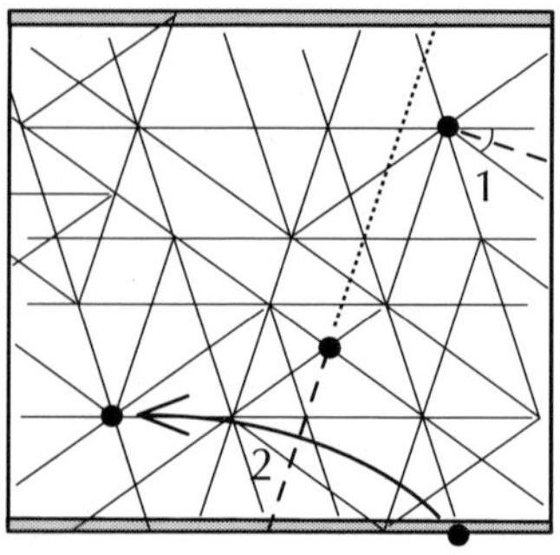

1. Fold and unfold.
2. The dots will meet.

34

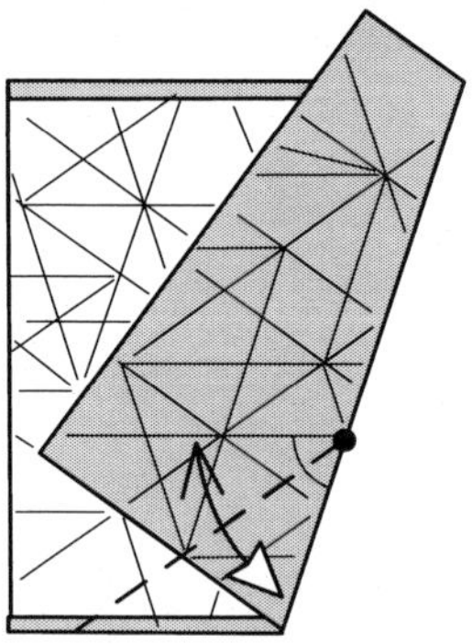

Fold and unfold along a partially hidden crease.

35

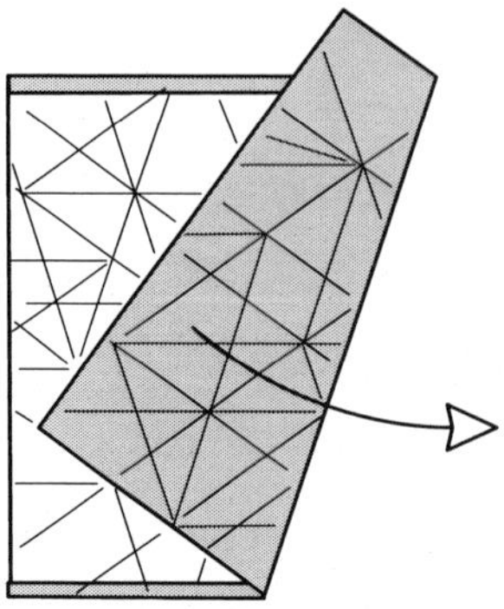

Unfold.

36

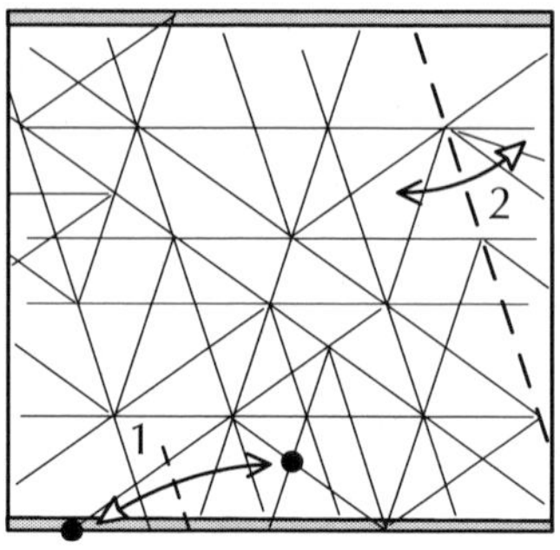

1. Fold and unfold.
2. Fold and unfold along the crease.

Rotate 180°.

37

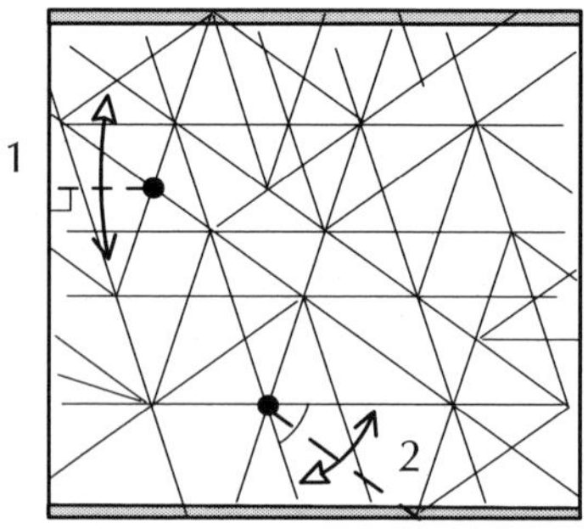

Repeat steps 31–36.

38

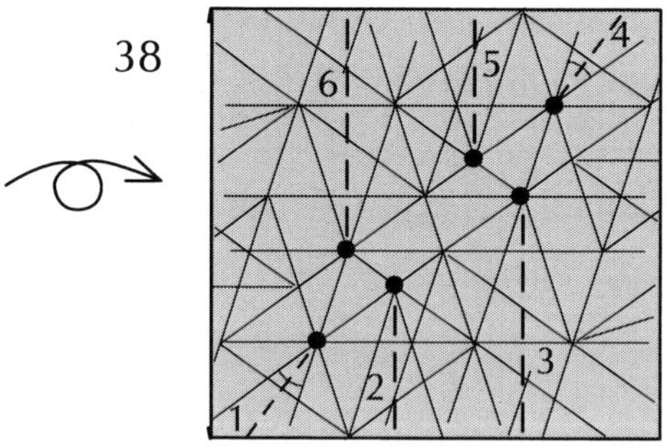

Fold and unfold six times.

39

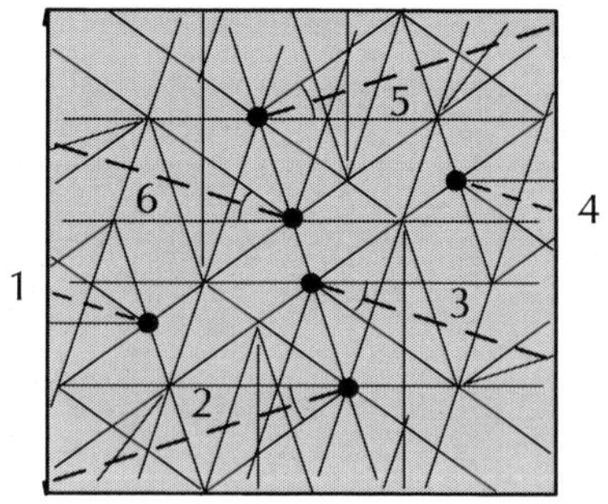

Fold and unfold.
Rotate 90°.

40

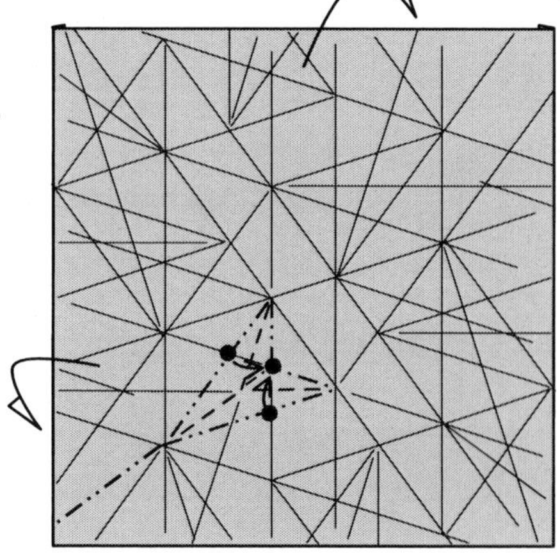

The dots will meet as the model becomes 3D.

41

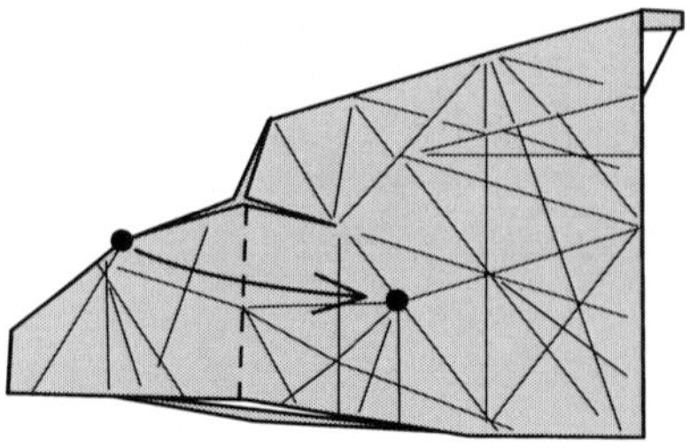

42

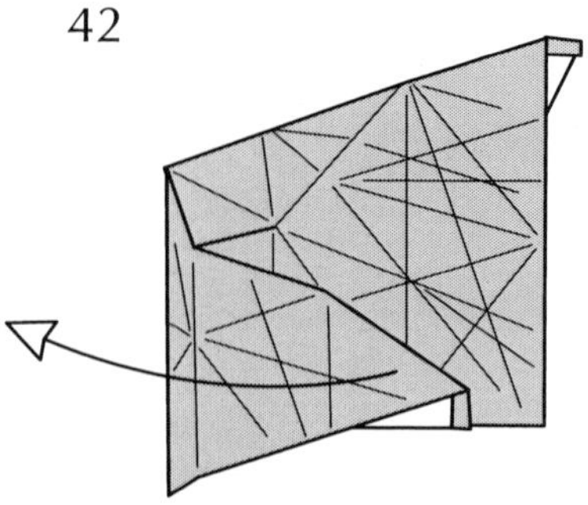

Unfold to step 40.

43

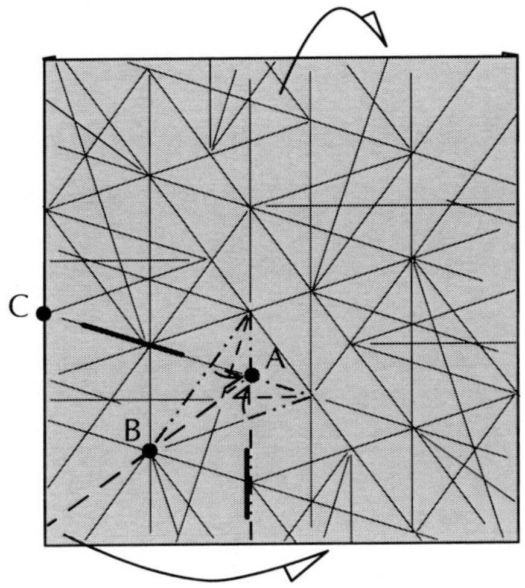

Refold along the creases. Push in at dot B, puff out at dot A. The bold lines will meet and the model will become 3D. Rotate dot C to the bottom and center.

44

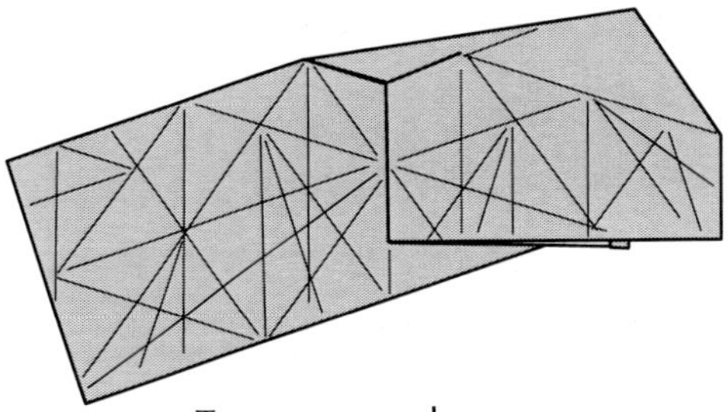

Turn over and repeat steps 40–43.

45

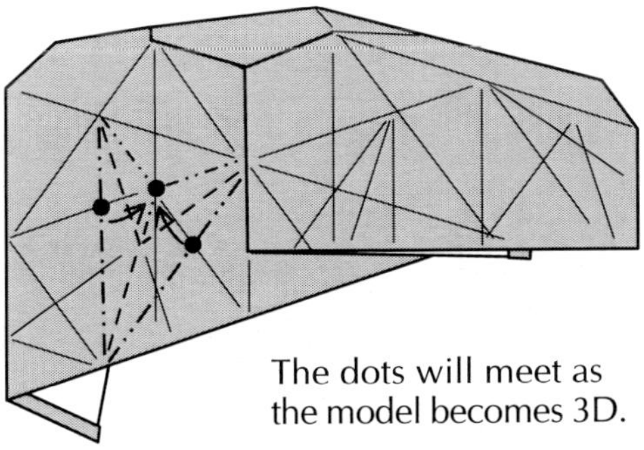

The dots will meet as the model becomes 3D.

46

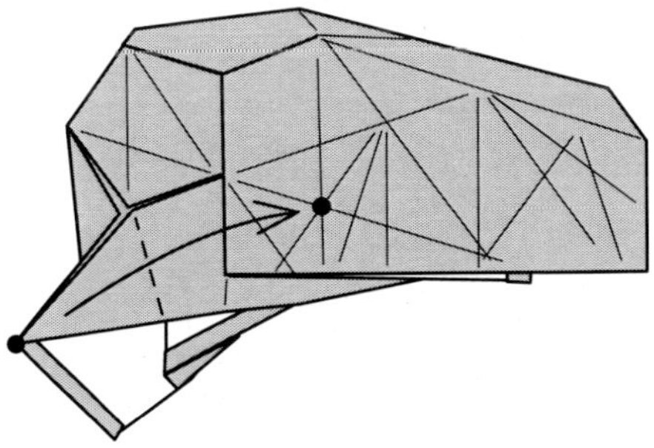

47

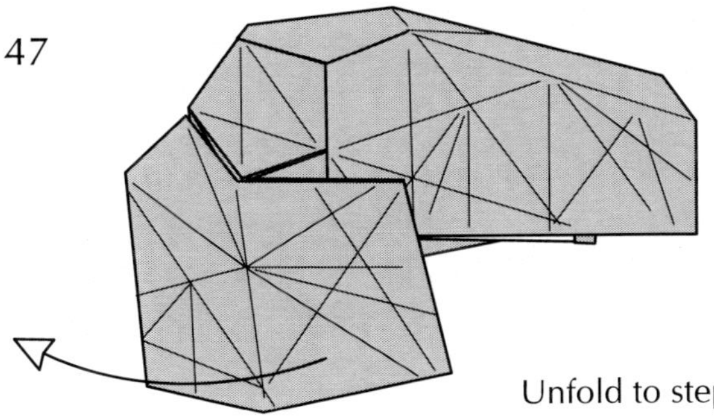

Unfold to step 45.

48

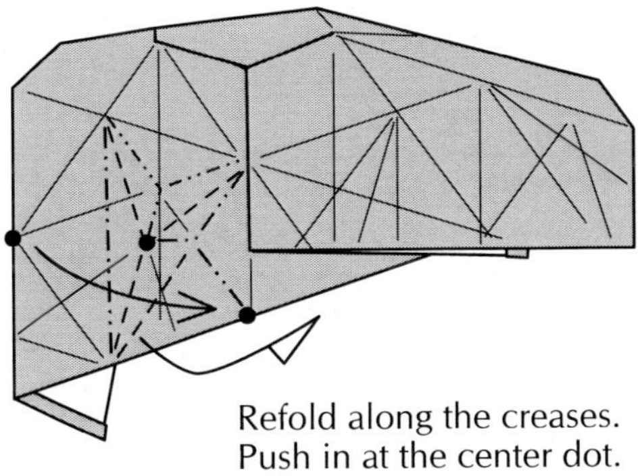

Refold along the creases. Push in at the center dot. The other two dots will meet.

49

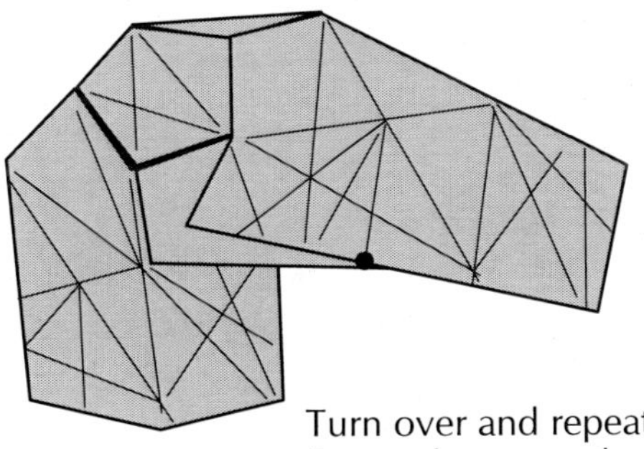

Turn over and repeat steps 45–48. Rotate the top to the bottom so the dot will be at the top and center.

50

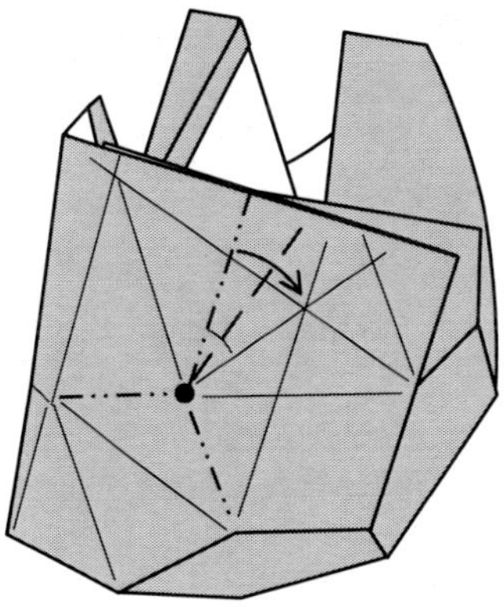

Fold two layers together. Puff out at the dot.

51

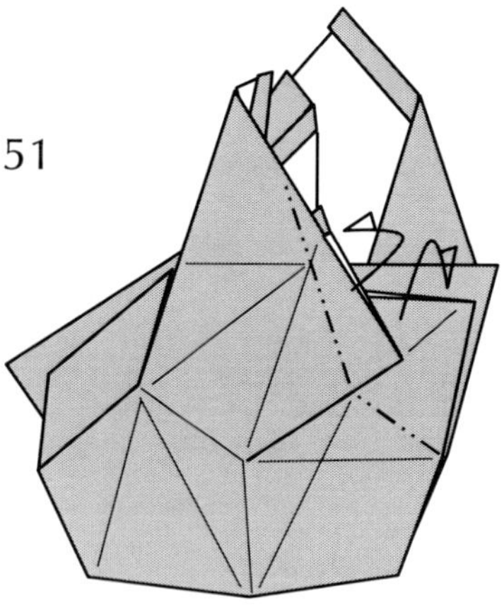

Fold along the creases.
Fold behind and flatten.

52

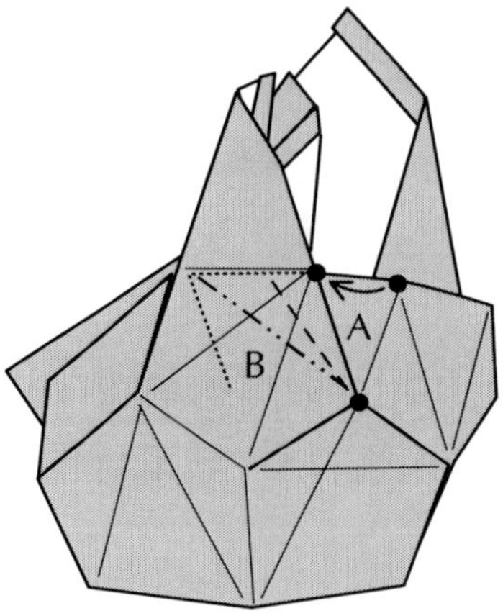

Puff out at the lower dot. The upper dots will meet. The folds are done on panel A, which is covered by region B. Mountain-fold along the crease.

53

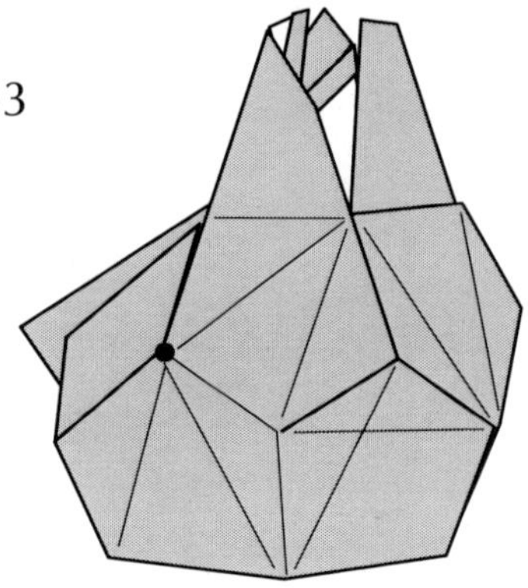

Turn over and repeat steps 50–52. Rotate the dot to the center.

54

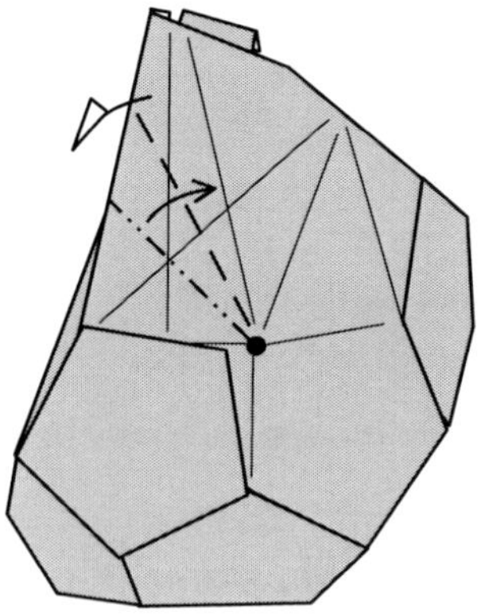

Fold along the creases and puff out at the dot. Turn over and repeat.

55

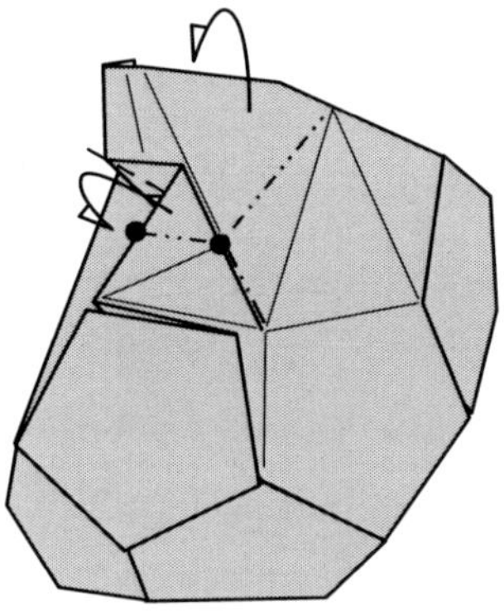

Fold inside but do not completely flatten. There is no crease for the mountain-fold between the dots. Turn over and repeat.

56

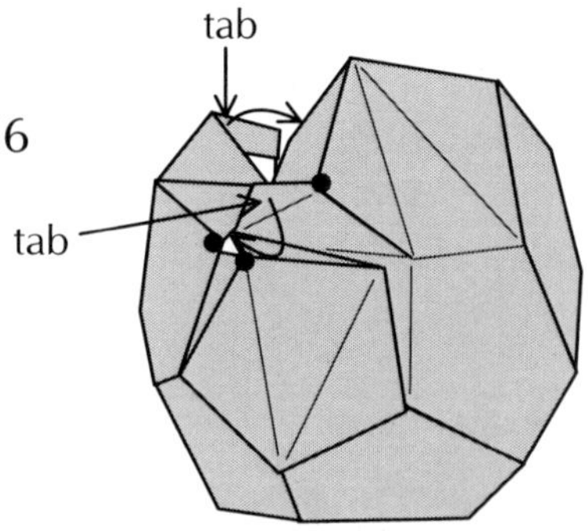

Tuck the two tabs into the pockets. The three dots will meet. You can also inflate into it to round it out.

57

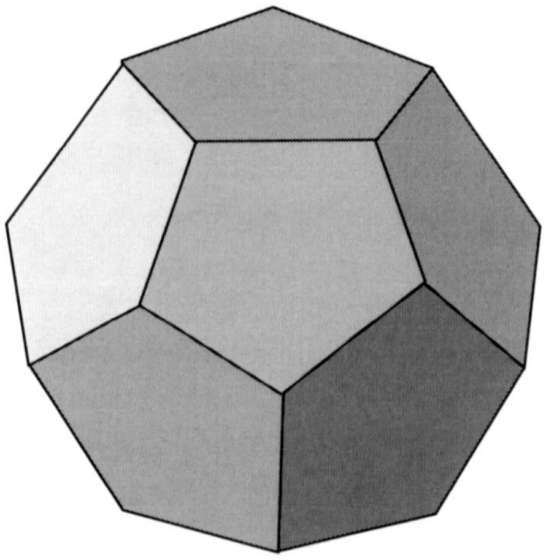

Dodecahedron

Made in United States
North Haven, CT
05 November 2026

21771710R00067